Advances in Mapping from Remote Sensor Imagery

Techniques and Applications

Advances in Mapping from Remote Sensor Imagery

Techniques and Applications

EDITED BY

Xiaojun Yang ▪ Jonathan Li

CRC Press
Taylor & Francis Group
Boca Raton London New York

CRC Press is an imprint of the
Taylor & Francis Group, an **informa** business

MATLAB® and Simulink® are trademarks of The MathWorks, Inc. and are used with permission. The MathWorks does not warrant the accuracy of the text or exercises in this book. This book's use or discussion of MATLAB® and Simulink® software or related products does not constitute endorsement or sponsorship by The MathWorks of a particular pedagogical approach or particular use of the MAT-LAB® and Simulink® software.

CRC Press
Taylor & Francis Group
6000 Broken Sound Parkway NW, Suite 300
Boca Raton, FL 33487-2742

First issued in paperback 2017

© 2013 by Taylor & Francis Group, LLC
CRC Press is an imprint of Taylor & Francis Group, an Informa business

No claim to original U.S. Government works

ISBN-13: 978-1-4398-7458-5 (hbk)
ISBN-13: 978-1-138-07294-7 (pbk)

Visit the Taylor & Francis Web site at
http://www.taylorandfrancis.com

and the CRC Press Web site at
http://www.crcpress.com

Contents

Preface

Remote sensing is the science and technology of acquiring the information about physical objects and the environment through recording, measuring, and interpreting imagery and digital representations of electromagnetic radiations derived from noncontact sensors. With the recent advancement in data, technologies, and theories in the broad arena of remote sensing, the use of different aerospace remote sensors and related information extraction techniques to support various mapping applications has received more attention than ever.

This book reviews some of the latest developments in remote sensing and information extraction techniques when applying for topographic and thematic mapping. This is an area in which many exciting advancements have been made over the past decade. This book includes a total of 16 chapters, falling into three major parts. The first part (Chapters 1–3) provides an overview on modern photogrammetry, light detection and ranging (LiDAR) remote sensing, and advanced image classification methods. The second part (Chapters 4–7) reviews the utilities of remote sensing and related image-processing techniques for extracting several types of geographic features that are essential for topographic map production, including elevation, shorelines, and human settlements. The third and final part (Chapters 8–16) showcases some of the latest developments in the synergistic use of remote sensing and relevant data-processing techniques for thematic mapping in environmental and ecologic domains.

This book is the result of extensive research by interdisciplinary experts and will appeal to students, researchers, and professionals dealing with remote sensing, photogrammetry, cartography, geographic information systems, geography, and environmental science. The editors are grateful to all of those who contributed papers and revised their papers one or more times as well as those who reviewed papers according to our requests and timelines. The book project would not have been completed without the help and assistance from several staff members at CRC/Taylor & Francis Group, especially LiMing Leong, Simon Bates, Jennifer Stair, Irma Shagla-Britton, and Saprik Khairunnisa. Acknowledgment also is due to

Haowen Han and Marc Johnston of Cenveo Publisher Services for their help and assistance in this book project.

Xiaojun Yang
Tallahassee, Florida

Jonathan Li
Waterloo, Canada

Editors

Xiaojun Yang is with the Department of Geography at Florida State University. His research focuses on the development of geospatial science and technology to support geographic inquiries in urban and environmental domains. He has authored or coauthored more than 80 English publications, including five books with John Wiley, Springer, and Taylor & Francis. He currently serves the second term as chair of the ICA Commission on Mapping from Remote Sensor Imagery (2011–2015) and director of Cartography and Geographic Information Society (2012–2016).

Jonathan Li is professor of geomatics at the University of Waterloo in Canada. His research interests include remote sensing, mobile mapping, and geographic information systems. He has authored more than 180 publications. He is vice chair of the International Cartographic Association (ICA) Commission on Mapping from Remote Sensor Imagery and chair of the ISPRS Intercommission Working Group V/I on Land-Based Mobile Mapping Systems.

Contributors

Piero Boccardo is an associate professor in surveying and mapping at Politecnico di Torino, Italy. He is director of information technology for humanitarian assistance, cooperation and action. He has served as president of the Italian Association for Remote Sensing since 2011 and chair of Working Group VIII/1 Disaster Management of the International Society for Photogrammetry and Remote Sensing (ISPRS). He has authored more than 110 publications.

Guolin Cai is a lecturer of remote sensing at the Southwest Jiaotong University in China. His research interest centers on radar interferometry for mapping regional topography and deformation.

Glenn Campbell currently lectures in surveying and spatial science at the University of Southern Queensland, Australia. He obtained a doctorate from the University of Queensland in remote sensing of water quality. His research focuses on developing remote-sensing applications for monitoring environmental conditions in a range of aquatic and terrestrial environments, and he is a member of the Spatial Analysis and Modelling research group of the Australian Centre for Sustainable Catchments.

Qiang Chen is an associate professor of photogrammetry and remote sensing at the Southwest Jiaotong University, China. His research interest centers on topographic mapping and deformation monitoring with digital photogrammetry and radar interferometry. He is the author of 2 books and 30 papers.

Zheng Cui is a research associate and IT administrator of the International Hurricane Research Center. He is a doctoral candidate of the School of Computing and Information Sciences at Florida International University in Miami, Florida. His research focuses on airborne LiDAR and ranging data-processing and algorithm development.

Nate Currit is an associate professor of geography at Texas State University-San Marcos, in San Marcos, Texas. His main research interest is in applying advanced GIS methodologies to understanding human–environment interaction across spatiotemporal scales of analysis.

Xiaoli Ding is chair professor of geomatics at the Hong Kong Polytechnic University. His main research interests are in satellite positioning, radar interferometry, and geohazards studies. He is the author of one book and more than 300 papers. He is a fellow of the International Association of Geodesy.

Daniel Dzurisin is a geologist at the U.S. Geological Survey (USGS) Cascades Volcano Observatory (CVO) in Vancouver, Washington, specializing in volcano deformation. He served as the CVO scientist-in-charge from 1994 to 1997 and currently is chief of the Interferometric Synthetic Aperture Radar (InSAR) Applied to Volcano Studies project. His research is directed toward understanding volcanic unrest using various geodetic techniques, including leveling, global navigation satellite systems, and satellite InSAR. His study areas include Mount St. Helens, the Three Sisters volcanic center, Yellowstone caldera, and volcanoes in the Aleutian arc. His book, *Volcano Deformation: Geodetic Monitoring Techniques*, was published by Springer-Praxis in 2007.

Amy E. Frazier is a doctoral candidate in the Department of Geography at the State University of New York at Buffalo. Her research interests focus on modeling the spatial patterns of invasive species and integrating subpixel remote-sensing classifications with landscape ecology.

Dean B. Gesch is a research physical scientist with the USGS at the Earth Resources Observation and Science Center in Sioux Falls, South Dakota. His research interests include topographic change detection and monitoring, sea-level rise impact assessment, elevation data accuracy assessment, and development of topographic data and derivative products. He was the lead scientist for the development of several national and global topographic data sets, including the U.S. National Elevation Dataset, and the global products GTOPO30 and GMTED2010. He served as guest editor for a special issue of *Photogrammetric Engineering & Remote Sensing* on the Shuttle Radar Topography Mission.

Ayman F. Habib is a professor at the Department of Geomatics Engineering at the University of Calgary, Canada. His research interests span the fields of terrestrial and aerial mobile mapping systems, modeling the perspective geometry of nontraditional imaging scanners, automatic matching and change detection, automatic calibration of low-cost digital cameras, object recognition in imagery, LiDAR mapping, and integrating photogrammetric data with other sensor data sets. He has authored more than 250 publications. He has received several awards from the American Society for Photogrammetry and Remote Sensing (ASPRS) and ISPRS.

Collin Homer is the chief for the Land Characterization and Trends Team for the USGS at Earth Resources Observation and Science in Sioux Falls, South Dakota. His has focused the past 20 years on multiscale remote-sensing applications for land cover characterization and change analysis throughout the United States. His research efforts have focused on developing landscape-scale remote-sensing models, methods, and databases such as the National Land Cover Database. He is the coordinator of the federal agency Multi-Resolution Land Characteristics Consortium that coordinates national land cover production for 10 agencies.

Patricia A. Houle is an instructor in the Department of Earth and Environment at Florida International University in Miami, Florida. She currently teaches courses in environmental science and social science and is engaged in postgraduate studies in curriculum and instruction with a concentration in science education.

Hyung-Sup Jung is an assistant professor in the Department of Geoinformatics at the University of Seoul. He is a principal investigator or coinvestigator of several projects on the development of SAR, InSAR, and InSAR altimeter techniques for the measurement of surface deformation and moving object's velocity, precise 3D geopositioning, and forest height estimation.

Tae-Jung Kwon is a doctoral student in the Department of Civil and Environmental Engineering at the University of Waterloo in Canada. His research interests include remote sensing, geographic information systems, and system modeling and optimization.

Chang-Wook Lee is a research professor in the Department of Geoinfomatics at the University of Seoul. His research interests include the development of small baseline subsets (SBAS) and persistent scatterer (PS) InSAR techniques and their applications to the study of earthquake, volcanic activities, and land subsidence.

Wonjin Lee is a doctoral candidate in the Department of Geoinformatics at the University of Seoul. His research interests include developing InSAR and GPS technologies for mapping ground deformation.

Jonathan Li is professor of geomatics at the University of Waterloo in Canada. His research interests include remote sensing, mobile mapping, and geographic information systems. He has authored more than 180 publications. He is vice chair of the International Cartographic Association (ICA) Commission on Mapping from Remote Sensor Imagery and chair of the ISPRS Intercommission Working Group V/I on Land-Based Mobile Mapping Systems.

Chunhua Liao is pursuing her Master of engineering degree in photogrammetry and remote sensing at Peking University in Beijing. Her research interests include hyperspectral data processing and analysis and vegetation dynamics mapping. She is also the author of more than 10 papers.

Guoxiang Liu is professor of remote sensing at the Southwest Jiaotong University in China. His research interests are radar interferometry, photogrammetry, and radargrammetry for mapping regional topography and deformation. He is the author of two books and more than 100 papers. He is a member of the ICA Commission on Mapping from Remote Sensor Imagery and a member of the ISPRS Inter-Commission Working Group V/I.

Zhong Lu is a physical scientist with the USGS in Vancouver, Washington. He is a principal investigator of several projects funded by the USGS, NASA, and European, Japan and German Space Agencies on the study of land surface deformation using satellite InSAR. His research interests include technique developments of SAR, InSAR, and persistent-scatterer InSAR processing, and applications of InSAR on natural hazards monitoring and natural resources management. He has authored or coauthored more than 100 publications.

Xiaojun Luo is a lecturer of remote sensing at the Southwest Jiaotong University, China. His research interest centers on topographic mapping and deformation monitoring with radar interferometry.

Xuelian Meng is an assistant professor in the Department of Geography and Anthropology at Louisiana State University. Her research interests include land use and land cover mapping, LiDAR-based urban structure analysis, image processing and feature extraction, and 3D topographic modeling.

Soe W. Myint is associate professor of geographical sciences at Arizona State University. He is a senior sustainability scientist at the Global Institute of Sustainability and is affiliated with the GeoDa Center for Geospatial Analysis and Computation, Decision Center for a Desert City, and Central Arizona Phoenix–Long Term Ecological Research at ASU. His research interest centers on geographic information systems, geospatial statistics, spatial modeling, and remote sensing. He is currently serving as a lead scholar of the International Research Training Group of the University of

Kaiserslautern, Germany, and chair of the Remote Sensing Specialty Group of the Association of American Geographers (AAG).

Mahesh Pal received his doctoral degree from the University of Nottingham. He is an associate professor with the Department of Civil Engineering, National Institute of Technology, Kurukshetra, India. His major research areas include land–cover classification, feature selection, and application of artificial intelligence techniques in various civil engineering applications. He has published more than 80 research papers and is in the editorial board of Remote Sensing Letters. He is a fellow of Institution of Engineers (India).

Christopher E. Parrish is the lead physical scientist in the Remote Sensing Division of NOAA's National Geodetic Survey (NGS) and NGS project manager for Integrated Ocean and Coastal Mapping (IOCM). He holds an appointment as affiliate professor of Earth sciences and ocean engineering at the University of New Hampshire (UNH) and is based at the NOAA-UNH Joint Hydrographic Center–Center for Coastal and Ocean Mapping (JHC-CCOM). His primary research interests include full-waveform light detection and ranging, sensor modeling and calibration, uncertainty analysis, and coastal mapping applications. He is serving as assistant director of the ASPRS LiDAR Division and past president of ASPRS Potomac Region.

Ruiliang Pu is an associate professor of geography, environment, and planning at the University of South Florida, Tampa Bay. His research interest centers on applications of remote sensing, geographic information systems, and spatial statistics to natural hazard monitoring, land use and cover change detection, biophysical and biochemical parameters extraction, and coastal and terrestrial ecosystems modeling. He is the author and coauthor of two books and more than 80 papers.

Fabio Giulio Tonolo is a senior researcher at Information Technology for Humanitarian Assistance, Cooperation and Action, Italy. His research interest is mainly on the use of satellite imagery and geospatial data to support emergency response activities. He is secretary of the ISPRS Working Group VIII/1 Disaster Management.

Le Wang is an associate professor of geography at the State University of New York, Buffalo. His research interests include the development of advanced remote-sensing techniques for estimating small-area urban populations, mapping and monitoring coastal mangrove forests, and mapping and modeling the spread of invasive species. He has authored more than 40 refereed journal articles and four book chapters. He was the recipient of the 2008 Early Career Award from the Remote Sensing Specialty Group of the AAG.

George Xian is a research scientist at the USGS Earth Resources Observation and Science Center. His research interests include land remote-sensing data application for terrestrial ecosystems change analysis, climate change, and environmental modeling.

Xiaojun Yang is with the Department of Geography at Florida State University. His research focuses on the development of geospatial science and technology to support geographic inquiries in urban and environmental domains. He has authored or coauthored more than 80 English publications, including five books with John Wiley, Springer, and Taylor & Francis. He currently serves the second term as chair of the ICA Commission on Mapping from Remote Sensor Imagery (2011–2015) and director of Cartography and Geographic Information Society (2012–2016).

Keqi Zhang is the interim director of the International Hurricane Research Center and an associate professor in the Department of Earth and Environment at Florida International University in Miami, Florida. His research focuses on airborne light detection and ranging mapping, beach erosion, storm surge modeling, coastal responses to climate changes and human activity, and 3D visualization animation. He has authored more than 70 publications.

Lei Zhang is a research associate at the Hong Kong Polytechnic University. His research interests include developments of advanced multitemporal InSAR processing techniques and their applications on natural hazard monitoring and natural resources management.

Xianfeng Zhang is an associate professor at the Institute of Remote Sensing and Geographic Information Systems, Peking University, China. His research interest is focused on remote sensing of ecology, hyperspectral data processing, and data assimilation, and he is the author of 4 books and more than 60 papers. He is a member of the Scientific Committee of the ISRS GI4DM and a corresponding member of the ICA Commission on Mapping from Remote Sensor Imagery.

Chapter I

Modern photogrammetric mapping

Ayman F. Habib

CONTENTS

Photogrammetric mapping is traditionally defined as the art and science of generating three-dimensional (3D) spatial and descriptive information from imagery. Photogrammetric mapping is a well-established discipline. In the past few years, however, several technical advances have had an impact on the photogrammetric mapping process—for example, integrated Global Navigation Satellite System and Inertial Navigation System (GNSS/INS) units, medium-format digital cameras, light detection and ranging (LiDAR) systems, and advanced image-processing techniques. This chapter will provide insight into the impacts of recent technical advances in the data acquisition and processing systems on photogrammetric mapping operations. The chapter begins with a brief introduction of the photogrammetric mapping principles. Then, it proceeds with a discussion of the impact of these advances on the photogrammetric mapping practices.

1.1 INTRODUCTION

Photogrammetric mapping is traditionally defined as the art and science of generating three-dimensional (3D) spatial and descriptive information from imagery. Photogrammetric mapping is a well-established discipline. In the past few years, however, several technical advances have had an impact on the photogrammetric mapping process—for example, integrated Global Navigation Satellite System and Inertial Navigation System (GNSS/INS) units, medium-format digital cameras, light detection and ranging (LiDAR) systems, and advanced image-processing techniques. This chapter will provide some insight into the impacts of recent technical advances in the data acquisition and processing systems on photogrammetric mapping operations. The chapter begins with a brief introduction of the photogrammetric mapping principles. Then, it proceeds with some discussion related to the following issues:

1. Calibration and stability analysis of medium-format digital cameras for photogrammetric mapping
2. The incorporation of GNSS/INS-based position and orientation information in object space reconstruction
3. System calibration of GNSS/INS-assisted photogrammetric mapping units
4. Line cameras for photogrammetric mapping
5. Feature-based photogrammetric triangulation
6. Integration of photogrammetric and LiDAR data for orthorectification and 3D visualization

1.2 PHOTOGRAMMETRIC PRINCIPLES

The main objective of photogrammetric mapping is the derivation of 3D coordinates of features of interest by observing their image coordinates in overlapping imagery. The mathematical model for the photogrammetric point positioning is based on the collinearity of the camera's perspective/projection center, the object point, and the corresponding image point. This model is embodied in the collinearity equations (Equation 1.1). The collinearity equations provide the mathematical expression relating the image coordinates of an image point (x_i, y_i), the ground coordinates of the corresponding object point (X_I, Y_I, Z_I), the ground coordinates of the perspective center (X_o, Y_o, Z_o), the attitude of the image coordinate system relative to the ground coordinate system as defined by the elements of the rotation matrix $(r_{11}, r_{12}, r_{13}, \ldots r_{33})$, and the internal characteristics of the camera (principal point coordinates – x_p, y_p, principal distance – c, and deviations from the assumed collinearity condition representation by $dist_x$ and $dist_y$) (Brown, 1966):

$$x_i = x_p - c \frac{r_{11}(X_I - X_o) + r_{21}(Y_I - Y_o) + r_{31}(Z_I - Z_o)}{r_{13}(X_I - X_o) + r_{23}(Y_I - Y_o) + r_{33}(Z_I - Z_o)} + dist_x, \qquad (1.1a)$$

$$y_i = y_p - c \frac{r_{12}(X_I - X_o) + r_{22}(Y_I - Y_o) + r_{32}(Z_I - Z_o)}{r_{13}(X_I - X_o) + r_{23}(Y_I - Y_o) + r_{33}(Z_I - Z_o)} + dist_y. \qquad (1.1b)$$

Alternatively, the collinearity equations can be derived through the vector summation principle illustrated in Figure 1.1. Before discussing the vectors that are summed up, one should define the utilized symbols for defining vectors and rotation matrices throughout this chapter. The vector notation (r_a^b) stands for the coordinates of point (a) relative to point (b)—this vector is defined relative to the coordinate system associated with point (b). The rotation matrix notation (R_a^b) stands for the rotation matrix that transforms a vector defined relative to the coordinate system (a) into a vector defined relative to the coordinate system (b). Using such a terminology, Equation 1.2a can be explained as follows: The coordinates of an object point (I) relative to the mapping frame (r_I^m) is the summation of the position of the camera's perspective center relative to the mapping frame (r_c^m) and the scaled vector that connects the perspective center and the image point (i)—r_i^c—after applying the rotation matrix (R_c^m), which is defined by the rotation angles $(\omega, \varphi, \kappa)$. The scale factor (S_i) represents the ratio between the magnitudes of the vector connecting the perspective center and the object point and the vector connecting the perspective center and the image point. This scale factor would change from one point to another within a given image and from one image to another for a given point. Equation 1.2b

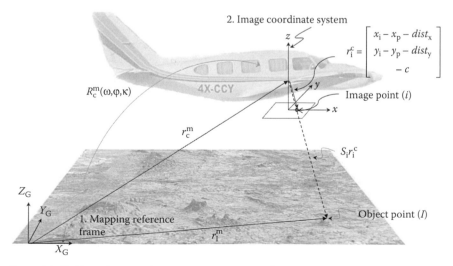

$$r_i^c = \begin{bmatrix} x_i - x_p - dist_x \\ y_i - y_p - dist_y \\ -c \end{bmatrix}$$

Figure 1.1 Collinearity equations as a vector summation process.

is another version of Equation 1.2a, which shows the elements of the various vectors. The vector summation in Equation 1.2 can be rearranged to produce the form in Equation 1.3, which can be reduced to the traditional form of the collinearity equations (Equation 1.1) after dividing the first two rows by the third one:

$$r_I^m = r_c^m + S_i R_c^m (\omega, \varphi, \kappa) \, r_i^c, \tag{1.2a}$$

$$\begin{bmatrix} X_I \\ Y_I \\ Z_I \end{bmatrix} = \begin{bmatrix} X_o \\ Y_o \\ Z_o \end{bmatrix} + S_i R_c^m (\omega, \varphi, \kappa) \begin{bmatrix} x_i - x_p - dist_x \\ y_i - y_p - dist_y \\ -c \end{bmatrix}, \tag{1.2b}$$

$$\begin{bmatrix} x_i - x_p - dist_x \\ y_i - y_p - dist_y \\ -c \end{bmatrix} = \frac{1}{S_i} R_m^c (\omega, \varphi, \kappa) \begin{bmatrix} X_I - X_o \\ Y_I - Y_o \\ Z_I - Z_o \end{bmatrix}. \tag{1.3}$$

The ground coordinates of a specific object point can be derived by observing the image coordinates of this point in a set of overlapping imagery if the internal characteristics of the implemented camera, which are collectively known as the interior orientation parameters (IOP), as well as the position and attitude, which are known as the exterior orientation parameters (EOP), of the camera at the moments of exposure associated with the involved imagery are available. The IOP are determined through a camera calibration process, and the EOP are determined through a georeferencing procedure. The camera calibration process can be carried out using one of the following options: (a) Laboratory calibration conducted by the camera manufacturer or a certified organization, (b) in-door calibration using a calibration test field, or (c) in situ calibration, in which the IOP are determined under operational conditions. The in-door and in situ calibration procedures are conducted using a bundle adjustment with self-calibration in which the IOP, EOP, and ground coordinates of tie points are estimated using a test field with ground control points (GCP). The coordinates of GCPs are established through an independent survey.

The EOP can be estimated indirectly with the help of a collection of GCPs and identified tie points in a bundle adjustment procedure, directly derived using an integrated GNSS/INS unit onboard the mapping platform and the mounting parameters relating the camera and the GNSS/INS unit, or estimated by incorporating the GNSS/INS-derived position and orientation information in the bundle adjustment (i.e., indirect georeferencing, direct georeferencing, or integrated sensor orientation [ISO], respectively). In the indirect georeferencing, the EOP and ground coordinates of tie points can be estimated simultaneously through the bundle adjustment procedure if sufficient control and suitable imaging configuration are available. For the direct georeferencing and ISO, the mathematical details for the incorporation of GNSS/INS-based position and orientation information for EOP derivation and object space reconstruction will be discussed in a later section of this chapter.

Photogrammetric mapping has relied mainly on large-format metric analog or digital cameras, which have been carefully designed and built to ensure the structural stability of their components. With the increasing resolution of low-cost and medium-format digital cameras, the photogrammetric community has been interested in investigating the possibility of using these cameras for mapping applications, in particular for close-range photogrammetry as well as in conjunction with other mapping tools, such as laser-scanning systems (Fraser, 1997; Habib et al., 2008; Renaudin et al., 2011). Before using low-cost digital cameras for photogrammetric mapping, however, one should ensure accurate estimation of their internal characteristics (IOP) as well as investigate their stability over time (Habib et al., 2008). The following section deals with one of the available techniques for the calibration and stability analysis of medium-format digital cameras.

1.3 CALIBRATION AND STABILITY ANALYSIS OF MEDIUM-FORMAT DIGITAL CAMERAS

Deriving accurate 3D measurements from imagery is contingent on precise knowledge of the IOP of the utilized camera. These characteristics are derived through the process of camera calibration, in which the coordinates of the principal point, principal distance, and distortion parameters are determined (Brown, 1966). The majority of current photogrammetric mapping systems mainly rely on metric large-format analog or digital cameras designed specifically for this purpose. Metric analog camera calibration usually is by a certified government agency (e.g., the U.S. Geological Survey or Natural Resources Canada). For metric large-format digital cameras, the calibration process is usually carried out by the camera manufacturer. The calibration of medium-format digital cameras, on the other hand, is not regulated as well as analog and large-format digital cameras. Because of the various designs of these types of cameras, it has become more practical for users to conduct the calibration procedure. As such, the burden of camera calibration has been shifted to the hands of the mapping data providers. Therefore, there has been an obvious need for the development of simple and automated camera calibration procedures, which can be conducted by the data providers.

Regardless of the utilized calibration procedure, control information is essential for the estimation of the IOP through the bundle adjustment. The control information is often in the form of specially marked targets, whose positions have been precisely determined through surveying techniques. Establishing and maintaining this form of a test field can be quite costly, which might limit the potential use of medium-format digital cameras in mapping and close-range photogrammetric applications. The need for a convenient and efficient calibration technique was addressed by Habib et al. (2002) and Habib and Morgan (2003b), in which the use of linear features for camera calibration was proposed and proved to be a promising alternative. Their approach incorporated the knowledge that in the absence of distortions, object space straight lines should appear as straight lines in the image space. Therefore, deviations from straightness in the image space can be used to allow for an accurate estimation of the distortion parameters that led to such deviations (e.g., radial and decentering lens distortions). The utilization of linear features for camera calibration is not a new concept and has been addressed by the plumb-line method in Brown (1971). The main advantage of the proposed approach by Habib et al. (2002) and Habib and Morgan (2003b) over the plumb-line method is that all the IOP (i.e., principal point coordinates, principal distance, and the polynomial coefficients that describe various distortions) are solved for simultaneously in the bundle adjustment procedure in contrast to the sequential estimation of the lens distortions and principal distance parameters in the plumb-line method. The mathematical details of utilizing linear features for the estimation of the IOP will be discussed in the feature-based photogrammetric triangulation section later in this chapter. Besides the advantages of having an easier-to-construct test field and allowing for accurate estimation of the distortion parameters, image-processing techniques can be utilized to automatically derive the image coordinate measurements along the linear features and point targets (see Figure 1.2), thus leading to a more convenient calibration procedure.

(a)   (b)   (c)

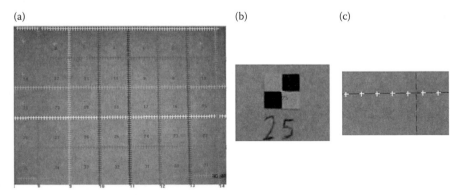

Figure 1.2 Suggested calibration test field with (a) automatically extracted point and linear features, (b) automatically extracted points for distinct targets, and (c) along line features.

After estimating the IOP of a medium-format digital camera and before its use in mapping applications, one should check the stability of these parameters (i.e., ensure that the estimated IOP do not change over time). One of the possible alternatives for testing the camera stability is to conduct several temporal calibrations followed by a statistical test to verify whether the estimated IOP sets are equivalent. The test is conducted by making some assumptions about the statistical distribution governing the available IOP (usually, a normal distribution is assumed). Statistical testing would mainly investigate the changes in the numerical values of the IOP without giving an indication of the impact of these changes on the mapping outcome. Moreover, the statistical testing would not consider possible correlation among the IOP and EOP. Another methodology for testing the camera stability is to evaluate the degree of similarity between the reconstructed bundles using different IOP sets, while providing a meaningful measure of the degree of similarity in terms of whether the changes are significant to the point that impacts the outcome of the mapping process.

Before getting into the details of bundle similarity evaluation, one should note that the objective of camera calibration is to reconstruct a bundle of light rays (which is defined by the perspective center and image points along the focal plane) that is as similar as possible to the incident bundle into the camera at the moment of exposure. Therefore, the stability analysis of two sets of IOP derived from temporally spaced calibration sessions could be based on a quantitative comparison of the reconstructed bundles from these IOP sets (Habib and Morgan, 2005; Habib et al., 2006a). To conduct such a stability analysis, one can start by defining a synthetic grid in the image plane, as shown in Figure 1.3a. Then, the available IOP sets are used to remove various distortions at the grid vertices. By connecting the perspective center (whose position relative to the image plane is describe by the principal point coordinates and the principal distance) to each set of the distortion-free grid vertices, two bundles of light rays are defined (see Figure 1.3b). Finally, the degree of similarity between the reconstructed bundles can be quantified. A measure that can be used to quantify such similarity is a representative estimate of the spatial offsets between corresponding light rays along the image plane. This measure could be derived as the root mean square error of the offsets between corresponding distortion-free grid vertices ($RMSE_{offset}$; see Figure 1.4). Such a measure can be estimated after rotating the two bundles relative to each other, while sharing the same perspective center to ensure the best fit between them. The image planes defined by the involved IOP sets may not coincide. The noncoincidence of these planes could arise from different estimates of the principal distances or the applied rotations to the respective bundles. Therefore, the offsets between corresponding light rays should be estimated along one of the image planes by projecting the image points in one plane onto the other plane (see Figure 1.4). If the representative estimate of the spatial

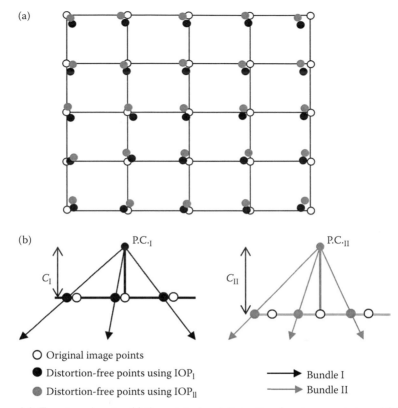

Figure 1.3 Top view showing (a) the original and distortion-free image points and (b) side view of the distortion-free bundles derived using IOP sets from two different calibration sessions.

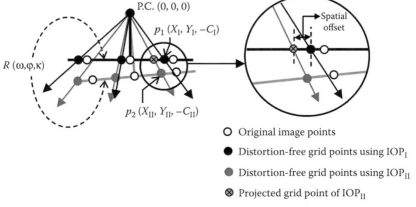

Figure 1.4 Bundles that are rotated to reduce the angular offset between conjugate light rays and a sample of the image-space spatial offset between conjugate light rays.

offsets ($RMSE_{offset}$) lies within the range defined by the noise in the image coordinate measurements (e.g., one-half to two-third of a pixel), then the two bundles are practically equivalent to each other, and the camera can be deemed stable. This similarity measure does not make any assumptions about the statistical properties of the available IOP. Moreover, it considers possible correlations between the elements of the internal and external sensor characteristics, because the bundles are allowed to rotate relative to each other to ensure the best fit between them. In addition, this methodology is independent of the involved distortion models in the two IOP sets (i.e., it can be used regardless of which distortion model is utilized). Once the tools for camera calibration and stability analysis have been settled, standards and specifications should be established to regulate the use of medium-format digital cameras in mapping applications (Habib et al., 2008).

I.4 GNSS/INS-ASSISTED PHOTOGRAMMETRIC MAPPING SYSTEMS

Traditionally, image-based topographic mapping has been performed using a single sensor, more specifically a large-format analogue camera. The object space reconstruction is usually enabled through an indirect georeferencing procedure, in which the image georeferencing parameters and the ground coordinates of object points are determined in a bundle adjustment procedure using corresponding tie points in overlapping images and GCP. Currently, modern airborne mapping systems incorporate a GNSS/INS unit to estimate the position and orientation of the mapping platform (Schwarz, 1995; Toth, 1998, 1999; Cramer et al., 2000; Cramer and Stallmann, 2001; Grejner-Brzezinska, 2001; El-Sheimy et al., 2005; Ip et al., 2007). Theoretically, with the help of a GNSS/INS unit, the georeferencing parameters could be determined without the need for GCP. For photogrammetric georeferencing, we are interested in the position and orientation of the camera coordinate system relative to the mapping reference frame. The integration of the GNSS/INS observations, however, would provide the position and orientation of the inertial measurement unit (IMU) body frame relative to the mapping system. Using the time tags associated with the exposure stations, one can derive the position and orientation of the IMU body frame at that time, $r_b^m(t)$, $R_b^m(t)$, respectively. The GNSS/INS position and orientation can be utilized in the photogrammetric mapping in two different scenarios. In the first scenario, which is denoted as the direct georeferencing, the EOP of the camera at the moments of exposure are derived from the GNSS/INS-derived position and orientation. The EOP are then used to derive the object coordinates of the points of interest through a simple intersection procedure. In the second scenario, which is denoted as the ISO, the GNSS/INS-based position and orientation information are used as prior information in the bundle adjustment procedure. The incorporation of this prior information into the bundle adjustment procedure can be implemented using either additional observations within the bundle adjustment or simultaneous incorporation of such information in the collinearity equations through a modified bundle adjustment. These options will be explained in the following paragraphs.

In the first approach, the observations in Equations 1.4 and 1.5 corresponding to the GNSS/INS-based position and orientation information, respectively, can be added to the collinearity equations within the bundle adjustment procedure. In Equation 1.4, r_b^c represents the lever arm describing the position of the IMU body frame relative to the camera coordinate system. The rotation matrix (R_b^c) in Equation 1.5 represents the boresight matrix (rotation matrix) relating the IMU body and camera coordinate systems. The rotation matrix $(R_b^m(t))$ has nine dependent elements (i.e., they should satisfy six orthogonality conditions). Therefore, only three independent elements should be utilized as part of the additional observations. Therefore, six equations will be added for the GNSS/INS-based position (three as per Equation 1.4) and orientation (three as per Equation 1.5) information for each image within the block:

$$r_b^m(t) = r_c^m(t) + R_c^m(t)\, r_b^c, \tag{1.4}$$

$$R_b^m(t) = R_c^m(t)\, R_b^c. \tag{1.5}$$

An alternative would be the direct incorporation of GNSS/INS-based position and orientation information into the bundle adjustment procedure. In this approach, the point positioning equation (Equation 1.6) can be derived through the summation of three vectors after applying the appropriate rotation matrices and scale factor (see Figure 1.5). The vector (r_c^b) is the lever arm representing the position of the camera relative to the IMU body frame, and the rotation matrix (R_c^b) stands for the boresight rotation matrix relating the camera and IMU coordinate systems. Several authors have investigated the comparative performance of the direct georeferencing and the ISO procedures in terms of the quality of the reconstructed object space (Jacobsen, 2000; Habib and Schenk, 2001; Jacobsen, 2004; Yastikli and Jacobsen, 2005b). Regardless of adopting direct georeferencing or ISO, the utilization of the GNSS/INS-based position and orientation for photogrammetric mapping requires the knowledge of the boresight matrix and the lever-arm between the GNSS/INS and camera units (Jacobsen, 1999). The parameters defining the boresight matrix and lever-arm are known as the system mounting parameters and are determined through a system calibration procedure (Jacobsen, 2001; Honkavaara, 2004):

$$r_I^m = r_b^m(t) + R_b^m(t)\, r_c^b + S_i R_b^m(t)\, R_c^b\, r_i^c. \tag{1.6}$$

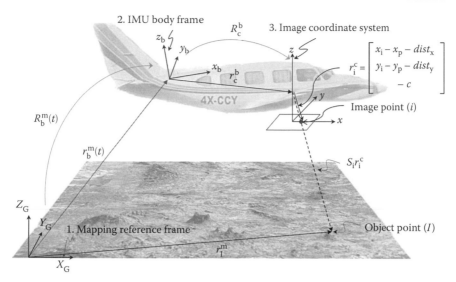

Figure 1.5 Coordinate systems and involved quantities in the point positioning equation based on GNSS/INS-assisted photogrammetric system.

In contrast to indirect georeferencing, in which the photogrammetric system calibration involves camera calibration only, the system calibration of a GNSS/INS-assisted camera involves the estimation of the IOP as well as the system mounting parameters relating the camera and the IMU coordinate systems. The quality of the camera calibration process plays a more important role in the direct sensor orientation than in the indirect sensor orientation. This is mainly due to the fact that for the indirect georeferencing, errors in the IOP can be compensated for by introducing systematic deviations to the EOP while ensuring a good fit in the object space. Such compensation cannot take place in the direct georeferencing or ISO because changes in the camera position and orientation are not allowed due to the GNSS/INS information and the mounting parameters (Habib and Schenk, 2001). Therefore, reliable camera and mounting parameters calibration is essential to ensure accurate object space reconstruction from GNSS/INS-assisted photogrammetric mapping units (Cramer and Stallmann, 2002; Honkavaara, 2003; Honkavaara et al., 2003; Honkavaara, 2004). The system calibration for such units is explained in the next section.

1.5 SYSTEM CALIBRATION

As mentioned, the system calibration of a GNSS/INS-assisted photogrammetric mapping unit encompasses the estimation of the IOP of the involved camera together with the lever arm and boresight matrix relating the camera and IMU coordinate systems (i.e., mounting parameters). Per the photogrammetric literature, two main approaches can be utilized to estimate the system mounting parameters: a two-step or a single-step procedure (Grejner-Brzezinska, 2001; El-Sheimy et al., 2005). In the two-step procedure, the system mounting parameters are estimated by comparing the GNSS/INS-derived position and orientation with the exterior orientation parameters determined from an independent indirect georeferencing process, as in Equations 1.7 and 1.8. These equations would provide several time-dependent estimates of the lever arm and boresight matrix from each image in the block. These estimates are then averaged to produce a single estimate of the system mounting parameters. Because of its simplicity (i.e., any bundle adjustment software can provide EOP values for the system calibration), the two-step procedure has been used extensively by several authors (Toth, 1998, 1999; Cramer, 1999; Jacobsen, 1999; Skaloud, 1999; Cramer and Stallmann, 2001; Yastikli and Jacobsen, 2005a; Casella et al., 2006). The two-step approach has several drawbacks, however. One disadvantage is that correlations among the EOP as well as correlations among the IOP and EOP are ignored (Cramer and Stallmann, 2002). In Jacobsen (1999), high correlation among the EOP was observed because of insufficient flight configuration. In Cramer et al. (2000), correlations among the EOP and IOP resulted in systematic vertical offsets in the derived photogrammetric product. Moreover, the two-step procedure demands a calibration site with well-distributed GCP in the planimetric and vertical directions as well as a geometrically strong imaging configuration:

$$r_c^b (t) = R_m^b (t) \left(r_c^m (t) - r_b^m (t) \right),$$ (1.7)

$$R_c^b (t) = R_m^b (t) \, R_c^m (t).$$ (1.8)

In the single-step procedure, the mounting parameters are estimated in the bundle adjustment through an integrated sensor orientation procedure using the mathematical model in Equation 1.6. Besides having less strict flight and control requirements, the single-step procedure is considered to be a robust approach that can allow for simultaneous camera and mounting parameter calibration. Several authors have highlighted the importance of the estimation of the camera calibration parameters together with the system mounting parameters (Jacobsen, 2001; Cramer and Stallmann, 2002; Honkavara et al., 2003; Jacobsen, 2003; Honkavara, 2004). Some authors have empirically investigated flight and control requirements for the single-step in situ photogrammetric system calibration using real or simulated data sets (Honkavara et al., 2003; Jacobsen, 2003; Honkavara, 2004). The following discussion provides an analysis rationale to determine an optimal flight and control configuration for the estimation of the system calibration parameters.

A rigorous analysis should investigate the possibility of recovering the parameters in question while establishing the minimal and optimal control and flight requirements for the system calibration. The system calibration parameters include the mounting parameters and the camera calibration parameters, which are susceptible to change under operational conditions, such as the principal distance (c) and the principal point coordinates (x_p, y_p) (Jacobsen, 2003). The minimum flight and control requirements can be achieved by establishing the optimum flight configuration that maximizes the impact of biases in the system parameters on the reconstructed object space. Such investigation is carried out through mathematical analysis of the GNSS/INS-assisted camera point positioning equation (Equation 1.6) according to the following rationale:

1. Since photogrammetric surface reconstruction is obtained through the intersection of conjugate light rays from overlapping imagery, a rigorous analysis should first check whether inaccurate or biased system parameters would lead to Y-parallax between conjugate light rays from directly georeferenced stereo-imagery (see Figure 1.6a). System parameters falling in this category can be estimated through the elimination or minimization of the Y-parallax among conjugate light rays in stereo-imagery (i.e., these parameters can be estimated using a stereo-image pair without the need for any GCP).
2. Then, one should investigate whether inaccurate or biased system parameters would lead to biases in the derived object points, whose magnitudes and directions depend on the flight configuration (i.e., the derived surfaces from different flight lines would exhibit discrepancies that depend on the flight configuration—flight direction and height). Using the point positioning equation (Equation 1.6), one can devise a flight configuration that maximizes the impact of biases in the system parameters on the derived object space from the different flight lines. Using such a configuration, the system parameters falling in this category can be estimated while reducing the discrepancy among the derived object points from the overlapping imagery (i.e., achieving the best precision of the derived object points; see Figure 1.6b). Similar to the previous step, system parameters falling in this category can be estimated without the need for any GCP.
3. Finally, for the system parameters, which do not introduce Y-parallax between conjugate light rays or discrepancies between derived points from overlapping imagery in a given flight configuration, control points will be utilized to derive such parameters. In other words, the parameters falling in this category will be estimated while reducing the discrepancy between the derived object space from the GNSS/INS-assisted imagery and the provided control (i.e., achieving the best accuracy of the derived object points; see Figure 1.6c).

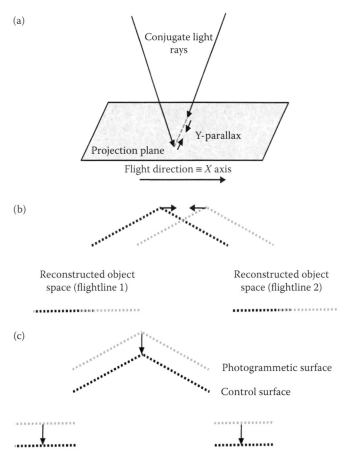

Figure 1.6 Estimation of the system parameters through (a) minimization of the Y-parallax, (b) minimization of the discrepancy between the derived object space surfaces from different flight lines, and (c) minimization of the discrepancy between photogrammetric and control surfaces.

Table 1.1 summarizes the results of the implemented rationale in terms of the possibility of recovering the studied parameters through the minimization of the Y-parallax between conjugate light rays or the minimization of the discrepancies between versions of object space that have been reconstructed from different flight lines. In such cases, control information would not be required. On the basis of this analysis, one can conclude that the vertical bias in the lever-arm components cannot be estimated without the availability of vertical control. Such inability is caused by the fact that a bias in the vertical component of the lever arm does not lead to Y-parallax and would lead to the same effect in the object space regardless of the flying direction or flying height. Therefore, at least one vertical ground control point would be required to estimate this parameter. According to this analysis, the

Table 1.1 Analysis rationale for the derivation of the system parameters (interior orientation and mounting parameters)

Parameter	Y-Parallax	Discrepancies: flying direction/ height dependent	Is control required?
Principal point coordinates	No	Yes/yes	No
Principal distance	No	No/yes	No
Planimetric lever-arm components	No	Yes/no	No
Vertical lever-arm component	No	No/no	Yes
Boresight roll	No	Yes/yes	No
Boresight pitch	Yes	Yes/yes	No
Boresight yaw	Yes	Yes/yes	No

minimum optimal flight and control configuration for the estimation of the mounting parameters and the camera calibration parameters (i.e., principal point coordinates and principal distance) should include five flight lines and one vertical control point as illustrated in Figure 1.7. As demonstrated in this figure, the optimum flight configuration consists of four strips that are captured from two flying heights in opposite directions with 100% side lap,

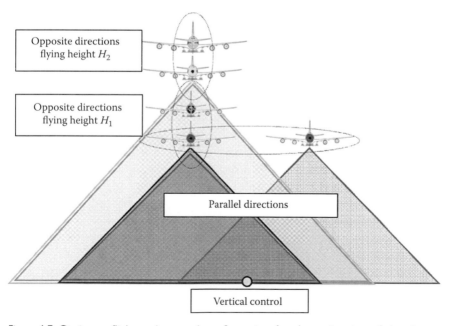

Figure 1.7 Optimum flight and control configuration for the estimation of the photogrammetric system parameters.

and two flight lines that are flown in the same direction with the least side lap possible, while allowing for the identification of a sufficient number of features in the side lap area between these flight lines. Kersting et al. (2011) has shown the feasibility of using such flight and control configuration for the estimation of the mounting parameters, principal point coordinates, and principal distance of the camera. This work has shown the need for considering the camera parameters during the in situ system calibration process.

1.6 PHOTOGRAMMETRIC MAPPING WITH LINE CAMERAS

In spite of the significant improvements in modern large-format frame digital cameras, one can argue that they have not yet provided the same ground coverage and resolution associated with large-format analog cameras. To reduce such a gap, multihead frame cameras and line cameras have been developed and utilized for large-scale mapping of extended areas. Multihead frame cameras (e.g., the digital mapping camera [DMC] developed by Z/I and the UltraCam series developed by Microsoft Vexcel) include several frame cameras, each with its own optical system, within a single enclosure. The different heads are tilted relative to each other and are synchronized to capture several images with some overlap. Using the overlapping images captured at a given time together with accurate knowledge of the internal and relative external characteristics of the different heads, a single virtual image is generated to simulate a perspective image that is captured by a single-head camera. This virtual image can be utilized in existing bundle adjustment packages as if it is captured by a traditional frame camera (Alamus and Kornus, 2008). Line cameras (e.g., the airborne digital sensor [ADS] series developed by Leica Geosystem) implement one or a few scan lines in the focal plane. Figure 1.8 illustrates the dimensionality of the light-sensitive elements in the focal plane of the frame and line cameras. A single exposure of the

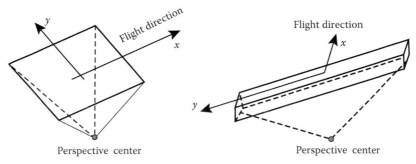

Figure 1.8 Focal plane for a (a) frame camera and (b) line camera.

sensor array in the focal plane of a line camera would result in an image that covers a narrow strip of the object space. To cover an extended area on the ground, several exposures are captured at a high rate while keeping the shutter of the optical system open. The sequence of collected images would compose a scene. In this regard, a scene captured by a line camera is composed of several images, each of which has its own perspective center.

The mathematical model relating the image or scene and the ground coordinates of captured points by a line camera is more complicated than those associated with frame cameras. Rigorous and approximate sensor models are alternative approaches used to describe the mathematics of the involved imaging process for line cameras. The former explicitly involves the image coordinates of points of interest together with the internal and external characteristics of the imaging sensor to faithfully represent the geometry of the scene formation. Approximate modeling directly utilizes the scene coordinates and can be divided into two categories. The first category simplifies the rigorous model after making some assumptions about the system's trajectory or object space (El-Manadili and Novak, 1996; Gupta and Hartley, 1997; Wang, 1999; Ono et al., 2000; Habib et al., 2004). Parallel projection and modified parallel projection are examples of such category. Other approximate models are based on empirical formulation of the scene-to-ground mathematical relationship (Dowman and Dolloff, 2000; Tao and Hu, 2001; Fraser et al., 2002; Hanley et al., 2002; Fraser and Hanley, 2003; Tao et al., 2004). This category includes, among others, the well-known rational function model (RFM). The selection whether to use rigorous or approximate sensor model alternatives should be based on the achievable accuracy from a given sensor model, available number of ground control, availability of the internal or external characteristics of the imaging system, validity of the assumptions associated with the approximate sensor model, and numerical stability of the adjustment procedure. Table 1.2 summarizes the main characteristics of sensor modeling alternatives for line camera scenes.

Currently, the majority of line cameras are utilized in spaceborne mapping systems with some being used onboard airborne systems (e.g., the ADS series from Leica Geosystems). The manipulation of scenes captured by airborne platforms is mainly based on a rigorous model that is aided by the provided GNSS/INS position and orientation information along the system's trajectory. Those systems have shown a performance that is quite similar to frame sensors. For line cameras onboard spaceborne platform, rigorous or approximate sensor models have been implemented (although the RFM-based approach is more common). Collected scenes from those satellites are extensively used for a variety of mapping and geographic information system (GIS) applications, such as topographic mapping, map updating, orthophoto generation, environmental monitoring, and change detection. A detailed discussion about the comparative performance of

Table 1.2 Comparative analysis of the characteristics of some of the available sensor modeling alternatives for line cameras

Model	Linear/ nonlinear	Derivable from the actual imaging geometry	Number of parameters	Stability of the adjustment	Coordinates (image/scene)
Collineartiy	Nonlinear	Yes	IOP + 6 EOPs (per scan line)	Low	Image
Parallel	Linear	Yes	8	High	Scene
Modified parallel	Nonlinear	Yes	9	High	Scene
RFM	Linear	No	78 or 59 (third order)	Low	Scene
DLT	Linear	Yes (frame camera) No (line camera)	11	High	Image (frame camera) Scene (line camera)

DLT = direct linear transformation; EOP = exterior orientation parameters; IOP = interior orientation parameters; RFM = rational function model.

available sensor models can be found in Habib et al. (2007a), which deals with the expected accuracy obtained from rigorous and approximate modeling of line camera scenes as related to the number of available GCP, robustness of the reconstruction process against biases in the available sensor characteristics, and impact of incorporating multisource imagery in a single triangulation mechanism.

1.7 FEATURE-BASED PHOTOGRAMMETRY

So far, the discussed models for photogrammetric mapping are based on the utilization of distinct points that can be identified in overlapping imagery. With the increased interest in the automation of the photogrammetric mapping process, the research community has been motivated to investigate the possibility of incorporating higher order primitives (e.g., linear and areal features) in the photogrammetric triangulation process (Zhang et al., 2008; Tommaselli and Medeiros, 2010). Advances in the digital image–processing techniques, such as edge detection and region segmentation operators, allow for the automated derivation of these primitives (Canny, 1986; Pratt, 2007). Moreover, linear and areal features would play a more effective role, when compared with point primitives, in the automated map compilation process. Another advantage of dealing with linear and areal features is the increased flexibility of integrating photogrammetry with other sources, such as GIS databases and LiDAR data. Before incorporating these primitives in the bundle adjustment, decisions should be made regarding their representation in the image and object space as well as the mathematical model used to relate the parameters describing the image and object space features. The following subsections deal with these prerequisites for the utilization of linear and areal features in the photogrammetric bundle adjustment.

1.7.1 Linear features

When deciding on the representation scheme of image and object space linear features, several factors should be considered. The most convenient representation scheme should not have any singularities (i.e., it should be capable of representing all 2D and 3D lines in the image space and object space, respectively), should provide well-defined segments rather than infinite lines, should be able to deal with several imaging modalities (i.e., imagery captured by frame and line cameras), should be able to deal with the presence of distortions in the captured imagery, and should lead to a simple model for its incorporation into the bundle adjustment procedure (Habib and Morgan, 2003a). Considering these objectives, image space lines can be represented by a sequence of image points along the feature. This is an appealing representation because it can handle image space linear features in the presence of distortions as they will cause deviations from straightness in the image space. Furthermore, such a representation will allow for the inclusion of linear features in scenes captured by line cameras because perturbations in the flight trajectory during the scene acquisition would lead to deviations from straightness in image space linear features corresponding to object space straight lines (Habib et al., 2002). The selected intermediate points along corresponding line segments in overlapping scenes need not be conjugate. As for the object space, 3D lines can be represented by their endpoints. The points defining the object line need not be visible or identifiable in all the captured imagery.

Having settled on the representation scheme, one can focus on the mathematical model for relating image and object space linear features. When dealing with tie lines, the endpoints that will be used to define the beginning and end of the object line are monoscopically measured in one or two of the overlapping images within which the line in question appears (in Figure 1.9, the endpoints are identified in a single image). For the endpoints, the collinearity equations will be used to describe the mathematical relationship between image and object coordinates of these points. A total of four collinearity equations (two per each endpoint) are introduced. In addition to the endpoints, a sequence of intermediate points, which are also monoscopically defined, will be used to represent the linear feature in the image space. The image coordinates of these intermediate points can be incorporated in the coplanarity constraint in Equation 1.9. This constraint indicates that the vector from the perspective center to any intermediate image point along

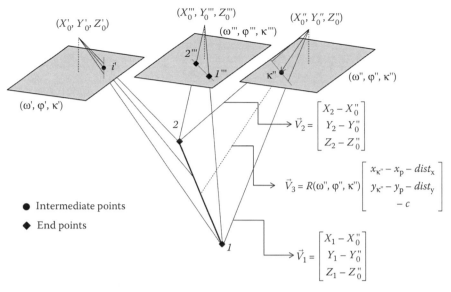

Figure 1.9 Perspective transformation between image and object space straight lines and the coplanarity constraint for intermediate points along the lines.

the image line is contained within the plane defined by the perspective center of that image and the two points defining endpoints of the object line. In other words, for a given intermediate point, k'', the points $\{(X_1, Y_1, Z_1), (X_2, Y_2, Z_2), (X_0'', Y_0'', Z_0'')\}$ and the vector $(x_{k''} - x_p - dist_x, y_{k''} - y_p - dist_y, 0)$ are coplanar (see Figure 1.9):

$$(\vec{V_1} \times \vec{V_2}) \odot \vec{V_3} = 0, \tag{1.9}$$

where

$\vec{V_1}$ = the vector connecting the perspective center to the first endpoint along the object line

$\vec{V_2}$ = the vector connecting the perspective center to the second endpoint along the object line

$\vec{V_3}$ = the vector connecting the perspective center to an intermediate point along the corresponding image line

The coplanarity constraint can be introduced for all of the intermediate points along the image space linear feature. Moreover, this constraint is valid for both frame and line cameras. For scenes captured by line cameras, the involved EOP should correspond to the image associated with the intermediate point under consideration. For frame cameras with known IOP, a maximum of two independent constraints can be defined for a given image line. For self-calibration procedures, however, more intermediate points will provide additional constraints, which will help in better recovery of the IOP because the distortion pattern will change from one intermediate point to the next along the image space linear feature (Habib and Morgan, 2003a). For line cameras, the coplanarity constraint would help in better recovery of the EOP associated with line cameras (Lee and Habib, 2002). Such a contribution is attributed to the fact that the system's trajectory will affect the shape of the linear feature in the final scene. This approach can be utilized for control linear features with minor modification. For such features, the points defining the beginning and the end of the linear features are already available. Therefore, there is no need to identify these points in the imagery, and consequently the associated collinearity equations will not be required.

1.7.2 Areal features

Planar patches from existing 3D GIS databases and LiDAR data can be incorporated in the photogrammetric bundle adjustment instead of the traditional GCP. Using control areal features will be beneficial in reducing the cost associated with establishing GCP that are distinguishable in the imagery. For the proposed approach in this section, the object space areal features will be represented by a set of points that can be derived, for example, through automated segmentation of areal features from LiDAR data. Image space areal feature, on the other hand, will be represented by three points that have to be identified in all of the overlapping imagery covering such feature. As an example, one can consider a planar feature that is represented by two sets of points, namely, the photogrammetric set $S_{PH} = \{A, B, C\}$ and the control set $S_C = \{(X_P, Y_P, Z_P), P = 1 \text{ to } n\}$ (see Figure 1.10). The mathematical model should be capable of incorporating these sets of points without assuming one-to-one correspondence between the photogrammetric and control data. For the photogrammetric set, the image and object space coordinates will be related to each other through the collinearity equations. Conversely, the individual points within the control set can be incorporated through a constraint that mathematically describes the coplanarity of each point and the photogrammetric set (see Figure 1.10). The coplanarity of the individual points in the control set and the photogrammetric points can be mathematically expressed through Equation 1.10:

$$
\begin{vmatrix} X_P & Y_P & Z_P & 1 \\ X_A & Y_A & Z_A & 1 \\ X_B & Y_B & Z_B & 1 \\ X_C & Y_C & Z_C & 1 \end{vmatrix} = \begin{vmatrix} X_P - X_A & Y_P - Y_A & Z_P - Z_A \\ X_B - X_A & Y_B - Y_A & Z_B - Z_A \\ X_C - X_A & Y_C - Y_A & Z_C - Z_A \end{vmatrix} = 0. \tag{1.10}
$$

Previous research has shown the benefit of using linear and areal features for various activities, such as frame-camera calibration, indirect georeferencing of

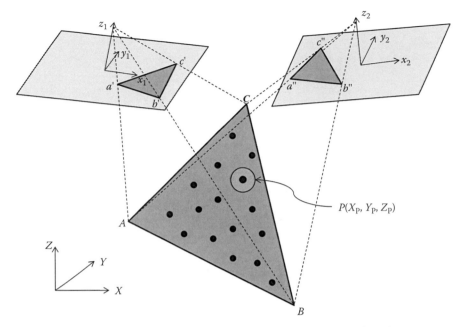

Figure 1.10 Coplanarity of photogrammetric and object space control patches.

line-camera scenes, LiDAR data for photogrammetric georeferencing, coregistration of photogrammetric and terrestrial laser scans, and building model generation (Lee and Habib, 2002; Habib and Morgan, 2003a; Habib et al., 2005; Habib et al., 2006b; Kim and Habib, 2009; Renaudin et al., 2011).

1.8 INTEGRATION OF PHOTOGRAMMETRIC AND LiDAR DATA

Over the last few years, LiDAR systems have emerged as effective tools for the acquisition of accurate topographic data with high point density. The high quality of the collected data is mainly attributed to significant advances in the GNSS/INS georeferencing systems. In this regard, one might ask if LiDAR and photogrammetric systems are viewed as competing mapping technologies. Considering the characteristics of acquired spatial data from imaging and LiDAR systems, one can argue that they are complementary systems. In other words, the integration of photogrammetric and LiDAR data will be beneficial for accurate and complete descriptions of the object space. Highlighting the complementary characteristics of imaging and LiDAR systems, Tables 1.3 and 1.4 list the advantages and disadvantages of each system as contrasted by the disadvantages and the advantages of the other system, respectively. As seen in these tables, the disadvantages of one system can be

Table 1.3 Photogrammetric weaknesses as contrasted by LiDAR strengths

LiDAR pros	Photogrammetric cons
Dense information along homogeneous surfaces	Almost no positional information along homogeneous surfaces
Day or night data collection	Daytime data collection
Direct acquisition of 3D coordinates	Complicated and sometimes unreliable matching procedures
Vertical accuracy better than the planimetric accuracy	Vertical accuracy worse than the planimetric accuracy

compensated for by the advantages of the other system (Baltsavias, 1999). The synergistic characteristics of both systems, however, can be fully utilized only after ensuring that both data sets are georeferenced relative to the same reference frame (Habib and Schenk, 1999; Chen et al., 2004). The introduced models in the feature-based photogrammetry section can be used to ensure the coregistration of the photogrammetric and LiDAR data to a common reference frame. More specifically, LiDAR linear and areal features have been used as control information for the photogrammetric georeferencing (Habib et al., 2005; Habib et al., 2006b; Shin et al., 2007). Automated segmentation techniques can be utilized to derive control areal and linear features from LiDAR data (e.g., segmentation of planar patches and intersection of neighboring planar patches, respectively). Once the photogrammetric and LiDAR data have been registered to a common reference frame, they can be integrated to generate different products, such as orthorectification, digital building model (DBM) generation, and 3D visualization (Chen et al., 2004; Brenner, 2005; Cheng et al., 2008; Demir et al., 2009; Awrangjeb et al., 2010). The discussion in the remaining portion of this chapter deals with some of the issues that need to be addressed when integrating photogrammetric and LiDAR data for the applications discussed in this chapter.

LiDAR data are excellent sources for the generation a digital surface model (DSM), which can be used for orthophoto generation, in which the relief displacement and scale variation in perspective imagery (see Figure 1.11a) are removed (see Figure 1.11b). When dealing with large-scale data

Table 1.4 LiDAR weaknesses as contrasted by photogrammetric strengths

Photogrammetric pros	LiDAR cons
High redundancy	No inherent redundancy
Rich in semantic information	Positional; difficult to derive semantic information
Dense positional information along object space break lines	Almost no information along break lines
Planimetric accuracy better than the vertical accuracy	Planimetric accuracy worse than the vertical accuracy

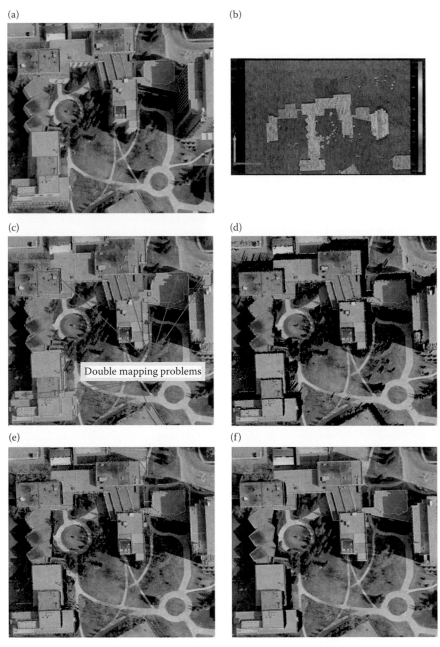

Figure 1.11 (**See color insert.**) Integration of photogrammetric and LiDAR data for orthophoto generation. (a) Perspective image. (b) Corresponding LiDAR data. (c) Orthophoto with the double mapping problems. (d) True orthophoto after occlusion detection. (e) True orthophoto after occlusion filling. (f) True orthophoto with enhanced building boundaries.

over urban areas, however, the traditional orthophoto-generation procedure will lead to what is known as the double mapping problem (see Figure
1.11c). The double mapping problem is attributed to improper projection of
the image space color information into occluded areas in the object space as
a result of object space relief and the perspective image-generation process.
In this regard, the differential rectification should be preceded by a visibility analysis process to automatically identify occluded areas in the object
space, leading to what is known as true orthophoto-generation techniques
(Amhar et al., 1998; Habib and Kim, 2006; Habib et al., 2007b). Figures
1.11c and 1.11d illustrate a portion of an orthophoto with the double mapping problem and the resulting true orthophoto after the visibility analysis
and the differential rectification process, respectively. The blank areas (see
Figure 1.11d) correspond to the occluded object space areas in the perspective image utilized for the orthophoto-generation process. These blank
areas can be filled with spectral information from overlapping imagery
(see Figure 1.11e). Smooth continuity of features when moving in between
projected spectral information from different imagery can be ensured only
after accurate calibration of the involved camera stations and registration
of the photogrammetric and LiDAR data to a common reference frame. A
closer look at Figure 1.11e reveals the lower quality of the building boundaries as manifested by irregularities of these boundaries. The reason for
such irregularity is the fact that the LiDAR-based DSM does not portray
accurate building boundaries because of the sparse nature of the LiDAR
footprints. A solution to this problem can be attained through the enhancement of the LiDAR-based DSM by incorporating a DBM of the various
building boundaries (see Figure 1.11f). In this regard, current research has

Figure 1.12 (**See color insert.**) 3D visualization achieved by draping a true orthophoto
on top of an enhanced LiDAR DSM with a DBM wireframe.

shown that the integration of photogrammetric and LiDAR data is one of the most promising alternatives for the automated generation of DBM (Rottensteiner, 2005; Lee et al., 2008; Kim and Habib, 2009; Habib et al., 2010). After generating true orthophotos with accurate building boundaries, the orthophotos can be draped on top of the enhanced DSM to produce realistic views of 3D environments. Figure 1.12 illustrates a perspective view of the 3D visualized model of the area depicted in Figure 1.11.

1.9 CONCLUSIONS

This chapter provides snapshots of the impacts of recent technical advances in the geospatial data acquisition and processing systems on photogrammetric mapping. The following list provides the main conclusions that can be drawn from the discussion in this chapter.

1. The widespread adoption of new imaging systems (e.g., medium-format digital cameras) would lead to an expansion in the photogrammetric data provider or user sector. To ensure accurate delivery of mapping products from such an expansion, one should ensure accurate calibration and meaningful investigation of the stability of the internal characteristics of the implemented cameras.
2. The improved performance of GNSS/INSS technology is allowing for the possibility of carrying out the photogrammetric mapping with minimal or no ground control requirement. Such an advantage, however, can be attained only after accurate calibration or estimation of the mounting parameters relating the camera and GNSS/INS unit. Moreover, in situ calibration techniques should allow for simultaneous estimation of the mounting parameters and the internal characteristics of the utilized camera.
3. Because of the limited number of light-sensitive elements in current digital frame cameras, line cameras have been developed to ensure large ground coverage and good resolution. Such a development, however, led to more complexity and wider variety in the sensor modeling process. The choice of the sensor modeling should be based on the desired accuracy, available control, and accessibility to the system's internal and external characteristics.
4. With the increased need for more automation in the mapping process, higher order primitives that could be automatically extracted from the imagery should be incorporated in the photogrammetric bundle adjustment. In this regard, having flexible models that can deal with various imaging modalities and primitives that can be easily included in existing bundle adjustment procedures would be quite advantageous.

5. New technologies, such as LiDAR systems, should not be viewed as competing with photogrammetric mapping. It has been argued that the integration of photogrammetric and LiDAR mapping systems will be beneficial in generating higher quality products, such as true orthophotos, automatically generated DBM, and 3D visualization of urban environments.

REFERENCES

Alamus, R., and Kornus, W. 2008. DMC geometry analysis and virtual image characterisation. *Photogrammetric Record* 23(124): 353–371.

Amhar, F., J. Jansa, and C. Ries. 1998. The generation of true orthophotos using a 3D building model in conjunction with a conventional DTM. *International Archives of Photogrammetry and Remote Sensing* 32(Part 4): 16–22.

Awrangjeb, M., M. Ravanbakhsh, and C. S. Fraser. 2010. Automatic detection of residential buildings using LiDAR data and multispectral imagery. *ISPRS Journal of Photogrammetry & Remote Sensing* 65: 457–467.

Baltsavias, E. 1999. A comparison between photogrammetry and laser scanning. *ISPRS Journal of Photogrammetry & Remote Sensing* 54(1): 83–94.

Brenner, C. 2005. Building reconstruction from images and laser scanning. *International Journal of Applied Earth Observation and Geo-information* 6(3–4): 187–198.

Brown, D. 1966. Decentric distortion of lenses. *Journal of Photogrammetric Engineering & Remote Sensing* 32(3): 444–462.

Brown, D. 1971. Close range camera calibration. *Journal of Photogrammetric Engineering & Remote Sensing* 37(8): 855–866.

Canny, J. 1986. A computational approach to edge detection. *IEEE Transactions on Pattern Analysis and Machine Intelligence* 8: 679–690.

Casella, V., R. Galetto, and M. Franzini. 2006. An Italian project on the evaluation of direct geo-referencing in photogrammetry. In *European Calibration and Orientation Workshop (EuroCOW)*. http://www.isprs.org/proceedings/2006/euroCOW06/euroCOW06_files/papers/Eurocow2006_casella_galetto_franzini.pdf.

Chen, L., T. Teo, Y. Shao, Y. Lai., and J. Rau. 2004. Fusion of LiDAR data and optical imagery for building modeling. *International Archives of Photogrammetry and Remote Sensing* 35(B4): 732–737.

Cheng, L., J. Gong, X. Chen, and P. Han. 2008. Building boundary extraction from high resolution imagery and LiDAR data. *International Archives of the Photogrammetry, Remote Sensing and Spatial Information Sciences* 37(Part B3): 693– 698.

Cramer, M. 1999. Direct geo-coding: Is aerial triangulation obsolete? *Photogrammetric Week 1999*, 59–70.

Cramer, M., and D. Stallmann. 2001. On the use of GPS/inertial exterior orientation parameters in airborne photogrammetry. In *ISPRS Workshop High Resolution Mapping from Space 2001*, Hanover, Germany, 32–44.

Cramer, M., and D. Stallmann. 2002. System calibration for direct geo-referencing. Paper from ISPRS Comm. III Symposium Photogrammetric Computer Vision, September 9–13, Graz, Austria.

Cramer, M., D. Stallmann, and N. Haala. 2000. Direct geo-referencing using GPS/ inertial exterior orientations for photogrammetric applications. *International Archives of Photogrammetry and Remote Sensing* 33(B3): 198–205.

Demir, N., D. Poli, and E. Baltsavias. 2009. Extraction of buildings using images & LiDAR data and a combination of various methods. *International Archives of the Photogrammetry, Remote Sensing and Spatial Information Sciences*, 38(Part 3/W4): 71–76.

Dowman, I., and J. Dolloff. 2000. An evaluation of rational functions for photogrammetric restitutions. *International Archives of Photogrammetry and Remote Sensing*, 33(B3): 254–266.

El-Manadili, Y., and K. Novak. 1996. Precision rectification of SPOT imagery using the direct linear transformation model. *Photogrammetric Engineering and Remote Sensing* 62(1): 67–72.

El-Sheimy, N., C. Valeo, and A. Habib. 2005. *Digital Terrain Modeling: Acquisition, Manipulation and Applications.* Boston: Artech House Remote Sensing Library.

Fraser, C. 1997. Digital camera self-calibration. *ISPRS Journal of Photogrammetry and Remote Sensing* 52(August): 149–159.

Fraser, C., and H. Hanley. 2003. Bias compensation in rational functions for IKONOS satellite imagery. *Photogrammetric Engineering & Remote Sensing* 69(1): 53–57.

Fraser, C., H. Hanley, and T. Yamakawa. 2002. High-precision geo-positioning from IKONOS satellite imagery. In *Proceeding of ACSM-ASPRS 2002*, Washington, DC, CD-ROM.

Grejner-Brzezinska, D. 2001. Direct sensor orientation in airborne and land-based mapping applications. *Report No. 461, Geodetic Geo-Information Science.* Department of Civil and Environmental Engineering and Geodetic Science, The Ohio State University.

Gupta, R., and R. Hartley. 1997. Linear push-broom cameras. *IEEE Transactions on Pattern Analysis and Machine Intelligence* 19(9): 963–975.

Habib, A., M. Ghanma, M. Morgan, and R. Al-Ruzouq. 2005. Photogrammetric and LiDAR data registration using linear features. *Photogrammetric Engineering and Remote Sensing* 71(6): 699–707.

Habib, A., A. Jarvis, I. Detchev, G. Stensaas, D. Moe, and J. Christopherson. 2008. Standards and specifications for the calibration and stability of amateur digital cameras for close-range mapping applications. *The International Archives of Photogrammetry, Remote Sensing and Spatial Information Sciences—ISPRS Congress Beijing* 37(B1): 1059–1064.

Habib, A., and C. Kim. 2006. LiDAR-aided true orthophoto and DBM generation system. In *Innovations in 3D Geo-Information Systems, Lecture Notes in Geo Information and Cartography*, edited by A. Abdul-Rahman, S. Zlatanova, and V. Coors, 47–65. Berlin, Germany: Springer-Verlag.

Habib, A., E. Kim, and C. Kim. 2007b. New methodologies for true orthophoto generation. *Photogrammetric Engineering and Remote Sensing* 73(1): 25–36.

Habib, A., E. Kim, M. Morgan, and I. Couloigner. 2004. DEM generation from high resolution satellite imagery using parallel projection model. In *Proceeding of the 20th ISPRS Congress, Commission 1, TS: HRS DEM Generation from SPOT-5 HRS Data, July 12–23, Istanbul, Turkey*, 393–398.

Habib, A., and M. Morgan. 2003a. Linear features in photogrammetry, invited paper. *Geodetic Science Bulletin* 9(1): 3.

Habib, A., and M. Morgan. 2003b. Automatic calibration of low-cost digital cameras. *Optical Engineering* 42(4): 948–955.

Habib, A., and M. Morgan. 2005. Stability analysis and geometric calibration of off-the-shelf digital cameras. *Photogrammetric Engineering and Remote Sensing* 71(6): 733–741.

Habib, A., M. Morgan, and Y. Lee. 2002. Bundle adjustment with self-calibration using straight lines. *Photogrammetric Record* 17(100): 635–650.

Habib, A., A. Pullivelli, E. Mitishita, M. Ghanma, and E. Kim. 2006a. Stability analysis of low-cost digital cameras for aerial mapping using different geo-referencing techniques. *Journal of Photogrammetric Record* 21(113): 29–43.

Habib, A., and T. Schenk. 1999. A new approach for matching surfaces from laser scanners and optical sensors. *International Archives of Photogrammetry and Remote Sensing* 32(3W14): 55–61.

Habib, A., and T. Schenk. 2001. Accuracy analysis of reconstructed points in object space from direct and indirect exterior orientation methods. Paper from OEEPE Workshop on Integrated Sensor Orientation. Institute for Photogrammetry and Engineering Surveying, University of Hanover, September 17–18, Hanover, Germany.

Habib, A., S. Shin, C. Kim, and M. Al-Durgham. 2006b. Integration of photogrammetric and LiDAR data in a multi-primitive triangulation environment. In *Innovations in 3D Geo Information Systems, Lecture Notes in Geo-Information and Cartography*, edited by A. Abdul-Rahman, S. Zlatanova, and V. Coors, 29–45. Berlin, Germany: Springer.

Habib, A., S. Shin, C. Kim, K. Bang, E. Kim, and D. Lee. 2007a. Comprehensive analysis of sensor modeling alternatives for high resolution imaging satellites. *Photogrammetric Engineering and Remote Sensing* 73(11): 1241–125.

Habib, A., R. Zhai, and C. Kim. 2010. Generation of complex polyhedral building models by integrating stereo-aerial imagery and LiDAR data. *Photogrammetric Engineering and Remote Sensing* 76: 609–623.

Hanley, H., T. Yamakawa, and C. Fraser. 2002. Sensor orientation for high resolution satellite imagery. In *Pecora 15/Land Satellite Information IV/ISPRS Commission I/FIEOS, November 10–15, 2002, Denver, CO*, CD-ROM.

Honkavaara, E. 2003. Calibration field structures for GPS/IMU/camera-system calibration. *The Photogrammetric Journal of Finland* 18(2): 3–15.

Honkavaara, E. 2004. In-flight camera calibration for direct geo-referencing. *International Archives of Photogrammetry. Remote Sensing and Spatial Information Sciences* 35(1): 166–171.

Honkavara, E., R. Ilves, and J. Jaakkola. 2003. Practical results of GPS/IMU camera system calibration. In *International Workshop, Theory, Technology and Realities of Inertial/GPS Sensor Orientation, ISPRS WG I/5, Castelldefels, Spain*, CD-ROM.

Ip, A., N. El-Sheimy, and M. Mostafa. 2007. Performance analysis of integrated sensor orientation. *Photogrammetric Engineering and Remote Sensing* 73(1): 89–97.

Jacobsen, K. 1999. Determination of image orientation supported by IMU and GPS. In *Joint Workshop of ISPRS Working Groups I/1, I/3 and IV/4 on Sensors and Mapping from Space, September 27–30, 1999, Hanover, Germany*, CD-ROM.

Jacobsen, K. 2000. Combined bundle block adjustment versus direct sensor orientation. Paper from ASPRS Annual Convention 2000, May 21–26, Washington, DC.

Jacobsen, K. 2001. Aspects of handling image orientation by direct sensor orientation. In *ASPRS Annual Convention St. Louis 2001*, CD-ROM.

Jacobsen, K. 2003. Issues and method for in-flight and on-orbit calibration. Paper from Workshop on Radiometric and Geometric Calibration, December 3–5, Gulfport, LA.

Jacobsen, K. 2004. Direct/integrated sensor orientation pros and cons. *The International Archives of Photogrammetry and Remote Sensing* 25(B3): 829–835.

Kersting, K., A. Habib, and K. Bang. 2011. Mounting parameters calibration of GPS/INS-assisted photogrammetric systems. In *Proceedings of the 2011 International Workshop on Multi-Platform/Multi-Sensor Remote Sensing and Mapping, Xiamen, China, January 10–12, 2011*.

Kim, C., and A. Habib. 2009. Object-based integration of photogrammetric and LiDAR data for automated generation of complex polyhedral building models. *Sensors* 9: 5679–5701.

Lee, Y., and A. Habib. 2002. Pose estimation of line cameras using linear features. In *Proceedings of ISPRS Commission III Symposium Photogrammetric Computer Vision, Graz, Austria, September 9–13, 2002*.

Lee, D., K. Lee, and S. Lee. 2008. Fusion of LiDAR and imagery for reliable building extraction. *Photogrammetric Engineering and Remote Sensing* 74(2): 215–225.

Ono, T., S. Hattori, H. Hasegawa, and S. Akamatsu. 2000. Digital mapping using high resolution satellite imagery based on 2D affine projection model. *International Archives of Photogrammetry and Remote Sensing* 33(B3): 672–677.

Pratt, W. 2007. *Digital Image Processing*. 4th ed. Los Altos, CA: Wiley & Sons.

Renaudin, E., A. Habib, and A. Kersting. 2011. Feature-based registration of terrestrial laser scans with minimum overlap using photogrammetric data. *ETRI Journal* 33(4): 517–527.

Rottensteiner, F., J. Trinder, S. Clode, and K. Kubik. 2005. Using the Dempster Shafer Method for the fusion of LiDAR data and multi-spectral images for building detection. *Information Fusion* 6(4): 283–300.

Schwarz, K. 1995. Integrated airborne navigation systems for photogrammetry. *Photogrammetric Week 1995* Wichmann Verlag: 139–154.

Shin, S., A. Habib, M. Ghanma, C. Kim, and E. Kim. 2007. Algorithms for multi-sensor and multi-primitive photogrammetric triangulation. *ETRI Journal* 29(4): 411–420.

Skaloud, J. 1999. Optimizing geo-referencing of airborne survey systems by INS/DGPS, PhD dissertation. Department of Geomatics Engineering, University of Calgary.

Tao, V., and Y. Hu. 2001. A comprehensive study of rational function model for photogrammetric processing. *Photogrammetric Engineering & Remote Sensing* 67(12): 1347–1357.

Tao, V., Y. Hu, and W. Jiang. 2004. Photogrammetric exploitation of IKONOS imagery for mapping applications. *International Journal of Remote Sensing* 25(14): 2833–2853.

Tommaselli, A., and N. Medeiros. 2010. Determination of the indirect orientation of orbital pushbroom images using control straight lines. *Photogrammetric Record* 25(130): 159–179.

Toth, C. 1998. Direct platform orientation of multi-sensor data acquisition systems. *International Archives of Photogrammetry and Remote Sensing* 32(4): 629–634.

Toth, C. 1999. Experiences with frame CCD arrays and direct geo-referencing. *Photogrammetric Week 1999* Wichmann Verlag: 95–109.

Wang, Y. 1999. Automated triangulation of linear scanner imagery. In *Joint Workshop of ISPRS WG I/1, I/3 and IV/4 on Sensors and Mapping from Space, September 27–30, 1999, Hanover, Germany*, CD-ROM.

Yastikli, N., and K. Jacobsen. 2005a. Direct sensor orientation for large scale mapping-potential, problems, and solutions. *The Photogrammetric Record* 20(111): 274–284.

Yastikli, N., and K. Jacobsen. 2005b. Influence of system calibration on direct sensor orientation. *Photogrammetric Engineering and Remote Sensing* 71(5): 629–633.

Zhang, Z., Y. Zhang, J. Zhang, and H. Zhang. 2008. Photogrammetric modeling of linear features with generalized point photogrammetry. *Photogrammetric Engineering and Remote Sensing* 74(9): 1119–1129.

Chapter 2

Airborne LiDAR remote sensing and its applications

Keqi Zhang, Zheng Cui, and Patricia A. Houle

CONTENTS

The airborne light detection and ranging (LiDAR) technology provides highly accurate measurements of objects on the Earth's surface. In contrast to optical remote-sensing technology, a LiDAR remote-sensing system derives voluminous three-dimensional point measurements of the Earth's surface. The major challenge for the utilization of the airborne LiDAR

33

technology is to develop appropriate methods to extract the desired features such as topography, vegetation, and buildings from LiDAR measurements. Numerous methods on feature extraction from LiDAR measurements and the applications of LiDAR technology to measuring and monitoring the Earth's surface processes have been developed in the past 20 years. To appropriately use the LiDAR data, it is necessary to understand the characteristics of airborne LiDAR technology and feature extraction algorithms related to processing LiDAR data. This chapter briefly introduces the airborne LiDAR technology, presents the filtering algorithms for separating ground and nonground LiDAR points and discusses the issues related to filtering processes, presents a framework to construct building models from LiDAR measurements, and describes the applications of LiDAR technology on mapping vegetation.

2.1 INTRODUCTION

Airborne light detection and ranging (LiDAR) is an active remote-sensing technology that allows accurate and cost-effective measurements of topography, vegetation, and buildings over large areas. The use of LiDAR for topographic mapping began in the late 1970s (Ackermann, 1999), but its accuracy was limited by poor determination of aircraft position and orientation. During the past 20 years, advances in global positioning systems (GPS), inertial navigation units (IMU), laser ranging, and microcomputers led to the development of highly accurate commercial airborne LiDAR mapping systems (Petrie and Toth, 2008). LiDAR technology is quickly becoming the primary method to acquire high-resolution elevation data of ground objects. The applications of LiDAR technology have increased exponentially in the past two decades, including mapping freshwater and storm surge flood areas, detecting landslides and erosion, constructing building models, and measuring individual trees and canopy structures (Whitman et al., 2003; Zhang et al., 2005; Hyyppä et al., 2008; Zhang et al., 2008; Roering et al., 2009; Zhao et al., 2011). The airborne LiDAR systems collect voluminous irregularly spaced, three-dimensional (3D) point measurements of terrain (ground) and nonterrain (nonground) objects scanned by the laser beneath the aircraft. On the one hand, the sheer volume of point measurements provides researchers with unprecedented information for studying Earth surface processes. On the other hand, voluminous data raise serious challenges for users to extract desired information from LiDAR data sets for specific applications. Numerous methods have been developed to extract topographic, vegetation, and building information from LiDAR data (Shan and Toth, 2008; Slatton et al., 2007), but much remains to be done in the future. The objectives of this chapter are first to briefly introduce the airborne LiDAR technology and how to extract ground and nonground features using LiDAR data. The remainder of the chapter is arranged as follows: Section 2.2 introduces the airborne LiDAR technology, Section 2.3 describes the filtering algorithms for separating ground and non-ground LiDAR points, Section 2.4 presents a framework to construct building models from LiDAR data, Section 2.5 describes the applications of LiDAR technology on mapping vegetation, and Section 2.6 presents the conclusions to the chapter.

2.2 AIRBORNE LiDAR SYSTEM

Most current airborne laser systems use the travel time of a laser pulse to detect the range; therefore, these airborne LiDAR systems are called "pulsed laser systems." A pulsed airborne LiDAR system has three basic components: a laser scanner, a GPS, and an IMU. The laser scanner mainly consists of a laser-ranging unit for sending and receiving laser signals and a swing mirror unit to direct emitting laser pulses. The laser-ranging unit determines the range from the aircraft to an object on the ground using the travel time of an emitted and reflected laser pulse and the shooting angle of the laser beam. The oscillating or spinning movement of the mirror in front of the laser unit allows laser pulses to be emitted toward both sides of the nadir position, which is perpendicular to the Earth's surface. As the aircraft flies forward, the laser scanner generates a swath of laser points with zigzag or elliptical ground scanning patterns (see Figure 2.1).

The reflected laser from an emitted pulse can have several peak intensities because of partial penetration and multiple reflections from objects such as trees and the ground underneath (see Figure 2.2). The laser intensity curve with multiple peaks provides multiple range values that are critical for separating trees from the topographic features and for measuring canopy structures. Because of computer and storage technology limitations, the earlier airborne LiDAR systems recorded the range value of only the first return in cases in which the reflected intensity was above a given threshold. The later LiDAR systems can digitize both first and last range values corresponding to the positions of the first and last above-threshold intensities in a curve for returned laser intensity. More recent systems allow digitizing several intermediate returns in addition to first and last returns and even can record the entire return intensity curve with a very high sampling rate, resulting in several hundred points for an intensity curve.

One limitation of the pulse laser systems is that an emitted laser pulse has a width of about 5–10 nanoseconds (Carter, Shrestha, and Slatton, 2007), which leads to a minimum distance for discerning two close objects. The receiver cannot differentiate which portion of the emitted pulse returns to the sensor first because of a complicated interaction between the laser pulse

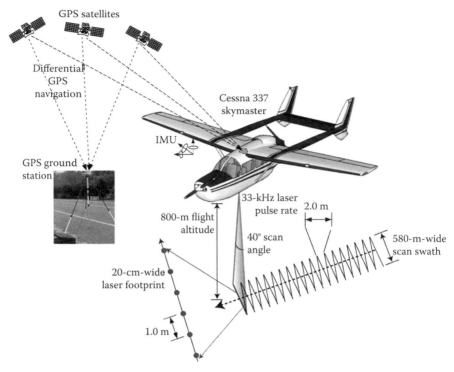

Figure 2.1 Schematic diagram showing data acquisition parameters used for the LiDAR survey for Big Pine Key in Florida.

and objects on the Earth's surface. The minimum distance ΔR between two discernible objects are computed by (Wehr, 2008)

$$\Delta R = \frac{C}{2} T_P, \tag{2.1}$$

where C is the speed of light, about 300,000,000 m/s, and T_p is the width of a laser pulse. For a 5 nanosecond laser pulse, ΔR is equal to 0.75 m. In other words, it is difficult to reliably separate first and last returns from the same laser pulse when the elevation differences of two objects are less than 0.75 meters (m). This indicates that the pulsed laser system has limitations in detecting heights of low vegetation in places such as marshes.

The GPS and IMU units on the aircraft measure the aircraft position and orientation. The 3D coordinates (x, y, and z) of an object referenced to the Earth's center can be determined by combing the laser range, GPS, and IMU data. To generate accurate point measurements of ground objects, ground GPS stations within 30–50 kilometer (km) distances from the aircraft record the GPS signals simultaneously with the aircraft GPS unit

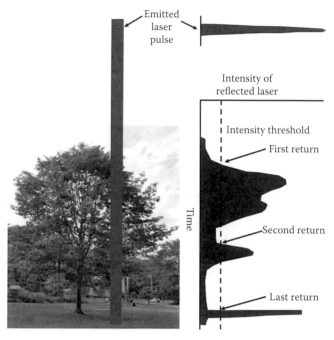

Figure 2.2 Schematic reflective intensity curve of a laser pulse for a tree and the terrain beneath the tree.

during a survey to provide an accurate determination of the aircraft trajectory. Typically, the software accompanying an airborne LiDAR system provides functions to read recorded laser ranging, GPS, and IMU measurements and produces georeferenced LiDAR point data by integrating these data. Because every flight only covers a strip of ground beneath the aircraft, stitching of flight strips is performed on the basis of the overlapping portions between strips to produce a complete LiDAR coverage for a survey area.

The laser scanner, GPS, and IMU are mounted at the different positions on the aircraft. To derive consistent LiDAR data, it is required to ensure that these units work properly and maintain their mounted position unchanged during survey flights. Therefore, the airborne LiDAR system usually needs to be calibrated before and after a survey against unchanging ground objects with distinct facets, such as large flat-roof buildings. Intermediate calibrations may be needed, depending on the duration of a survey mission. The systematic calibration of the LiDAR system is only one of the steps required to derive high-quality LiDAR data. Additional quality control and assurance measures are also needed. The vertical accuracy of LiDAR measurements is relatively easy to verify using two different methods. One method compares the LiDAR elevations with measurements of higher accuracy, such as those from GPS surveys, for the same ground features such as roads. The root square mean errors of LiDAR elevations are routinely less than 0.15 m for smooth ground surfaces. The vertical accuracy of ground elevation is decreased in vegetated areas because few LiDAR measurements can reach the ground due to vegetation blocking the penetration of the laser beams. The second method compares the elevation differences between objects, such as parking lots and road surfaces, that do not change between two surveys. This method is particularly useful if the purpose of analyzing the LiDAR data is to perform change detection (Zhang et al., 2005). It is much more difficult to estimate the horizontal accuracy of LiDAR measurements because LiDAR does not detect the edges of objects well because of its irregularly spaced point measurements. Typically, the horizontal errors are three times that of the vertical errors (Vosselman, 2008).

2.3 FILTERING LiDAR MEASUREMENTS

Airborne LiDAR systems generate a 3D cloud of point measurements for reflective objects scanned by the laser beneath the flight path (see Figure 2.3). These measurements include points for the terrain and nonterrain objects, such as buildings, trees, and vehicles. Topography is often represented by digital terrain models (DTMs), also called "bare earth" digital elevation models. Generating a DTM from LiDAR measurements involves two tasks: filtering out the nonterrain points and the interpolation of the irregularly spaced remaining terrain points. Because LiDAR surveys often involve voluminous point measurements, it is cost prohibitive and time consuming

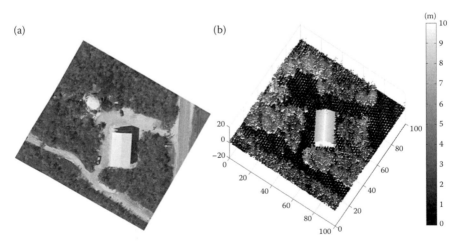

Figure 2.3 (a) Aerial photograph collected on February 27, 2006, from Google Earth for a building and surrounding vegetation and roads in Big Pine Key in Florida. (b) LiDAR points collected on January 19, 2007.

to manually remove nonterrain points. A highly automated filtering algorithm is needed for practical reasons.

A number of algorithms have been developed to filter LiDAR data (Sithole and Vosselman, 2004; Zhang and Whitman, 2005; Liu, 2008; Meng, Currit, and Zhao, 2010). Most LiDAR filtering algorithms are based on the following observation: topographic changes occur in large scales compared with the height changes of buildings and vegetation. The elevation change of terrain is usually gradual and exhibits a large degree of spatial autocorrelation in its neighborhood, whereas the elevation change between buildings or trees and the ground is drastic. Therefore, the difference in elevation change in a neighborhood (window) can be used to construct filtering measures to classify LiDAR data. The often-used filtering measures include a local slope (Vosselman, 2000), height range (Whitman et al., 2003), and distance to a locally fitted surface (Kraus and Pfeifer, 1998; Haugerud and Harding, 2001; Pfeifer, Stadler, and Briese, 2001; Meng et al., 2009) (see Figure 2.4). Whether a LiDAR point is a terrain measurement is determined by comparing the filtering measures with predefined thresholds. Several commercial computer programs such as TerraScan (http://www.terrasolid.fi) have been developed to filter the LiDAR data. Zhang and Cui (2007) developed a public domain, airborne LiDAR data-processing and analysis tool kit (ALDPAT), which can be downloaded from http://www.ihrc.lidar. In this section, the following five filters used in ALDPAT to separate ground and nonground point measurements are presented:

- Elevation threshold with expand window (ETEW)
- Maximum local slope (MLS)
- Progressive morphology (PM)
- Iterative local polynomial fitting (ILPF)
- Adaptive triangulated irregular network (ATIN)

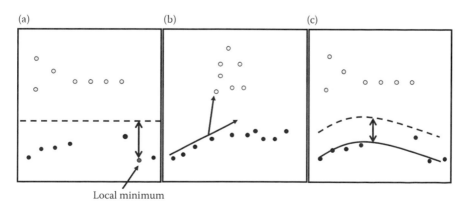

Figure 2.4 Measures for filtering LiDAR points, including (a) height difference, (b) slope, and (c) the distance to the surface in a local window.

2.3.1 Elevation threshold with expand window filter

Elevation differences between neighboring ground measurements usually are distinct from those between the ground and the tops of trees and buildings in an area of limited size (Zhang and Whitman, 2005). Therefore, elevation differences in a window can be used to separate ground and nonground LiDAR measurements. The ETEW method uses an expanding search window to identify and remove nonground points:

- The data set is subdivided into an array of square cells, and all other points, except the minimum elevation point, are discarded.
- For the next iteration, the cells are increased in size and the minimum elevation in each cell is determined. Then, all points with elevations greater than a threshold above the minimum are discarded.
- The process is repeated with the cells and thresholds increasing in size until no points from the previous iteration are discarded.

For ith iteration, a point $p_{i,j}$ is removed if

$$Z_{i,j} - Z_{i,min} > h_{i,T}, \tag{2.2}$$

where $Z_{i,j}$ represents the elevation of jth point $(p_{i,j})$ in a cell for ith iteration, $Z_{i,min}$ is the minimum elevation in this cell, and $h_{i,T}$ is the height threshold. The $h_{i,T}$ is related to the cell size and defined by

$$h_{i,T} = sc_i, \tag{2.3}$$

where s is a predefined maximum terrain slope and c_i is the cell size for ith iteration. In the current implementation of the algorithm, the cell size c_i is doubled each iteration such that

$$c_i = 2c_{i-1} \qquad i = 2, 3, \dots M, \tag{2.4}$$

where M is the total number of iterations.

This filter works fine in areas with gentle slopes, in areas without large variations of slopes, and in areas where the sizes of nonground objects are small. For an area with a mixture of buildings, vegetation, and steep topography, it is difficult for the ETEW to effectively separate ground and nonground measurements.

2.3.2 Maximum local slope filter

Because the slopes of terrain are usually different from those between the ground and the tops of trees and buildings, the slope difference can be used to separate ground and nonground measurements from a LiDAR data set (Zhang and Whitman, 2005). Vosselman (2000) developed a filter that identifies ground measurements by comparing local slopes between a LiDAR point and its neighbors. The method implemented here is similar to Vosselman's filter:

- The data set is divided into an array of square cells. Each point measurement $p_j(x_j, y_j, z_j)$ from the LiDAR data set is assigned into a cell in terms of its x and y coordinates. If more than one point falls in the same cell, one with the lowest elevation is selected as the array element.
- A LiDAR point, $p_0(x_0, y_0, z_0)$, is classified as a ground measurement if the maximum value ($s_{0,max}$) of slopes between this point and any other point (p_j) within a given radius is less than the predefined threshold (s):

$$\left\{ \begin{array}{l} s_{0,j} = \dfrac{z_0 - z_j}{\sqrt{(x_0 - x_j)^2 + (y_0 - y_j)^2}} \\ p_0 \in \text{ground measurements} \mid \text{if } s_{0,max} < s \end{array} \right\}, \qquad (2.5)$$

where $s_{0,j}$ is slope between p_0 and p_j, x_j and y_j represent horizontal coordinates of p_j and z_j is its elevation.

This method is effective in removing the vegetation in mountainous areas without large buildings. For forests with open spaces, a small search radius often is used to remove trees without much influence on the ground points inside the window. For an area with a mixture of vegetation and buildings, however, a large search radius is needed to remove the large size of buildings, which could lead to an incorrect removal of small topographic high features.

2.3.3 Progressive morphology filter

Mathematical morphology uses operations based on set theory to extract features from images. Zhang et al. (2003) developed the PM filter to remove nonground measurements from a LiDAR data set. By gradually increasing the window size and using elevation difference thresholds, the PM filter removes the measurements for different-size nonground objects while preserving ground data. The procedure of the PM filter is listed as follows:

- Overlaying a rectangular mesh on the LiDAR data set. A minimum elevation point $p_j(x_j, y_j, z_j)$ is selected for the cell of the mesh if at least one point falls within the cell. If no measurements exist in a cell, it is assigned the point of its nearest neighbor. Elevations of points in the cells make up an initial approximate surface. The cell size is usually selected to be smaller than the average spacing between LiDAR measurements so that most LiDAR points are preserved.

- Performing an opening morphological operation on the initial surface to derive a secondary surface. The elevation difference ($dh_{i,j}$) of a cell j between the previous (initial or $i-1$) and current (secondary or i) surfaces is compared with a threshold $dh_{i,T}$ to determine whether the point p_j in this cell is a nonground measurement. The threshold $dh_{i,T}$ is determined by

$$dh_{i,T} = \begin{cases} dh_0 & \text{if } w_i \le 3 \\ s\,(w_i - w_{i-1})\,c + dh_0 & \text{if } w_i > 3 \\ dh_{max} & \text{if } dh_{i,T} > dh_{max} \end{cases}, \tag{2.6}$$

where dh_0 is the initial elevation difference threshold, which approximates the error of LiDAR measurements (0.2–0.3 m); dh_{max} is the maximum elevation difference threshold; s is the predefined maximum terrain slope; c is the cell size of the mesh; and w_i is the filtering window size (in number of cells) at ith iteration.

- Increasing the size of filtering window and the current surface in the second step is used as the input for the next opening operation. The second and third steps are repeated until the size of the filtering window is larger than the predefined maximum size of nonground objects.

The maximum elevation difference threshold can be set either to a fixed value to ensure the removal of large and low buildings in an urban area or to the largest elevation difference in a study area. The filtering window can be a one-dimensional (1D) line or two-dimensional (2D) rectangle or any other shape. When a line window was used, the opening operation was applied to both x and y directions at each step, except for the data set for linear geomorphic features such as coastal barrier islands, to ensure that the nonground objects were removed. This filter is effective in removing nonground objects and preserves topographic features in gently sloped areas. Figure 2.5 shows that the aerial photograph and digital surface model generated using unfiltered LiDAR points for the Potomac River area in Maryland, United States, where there are many buildings and trees. A DTM generated using the filtered LiDAR points indicates that the PM filter successfully removes the vegetation and building points (see Figure 2.6a). The advantage of the PM filter is that it preserves the shapes of special geomorphic features, such as cliffs, with sizes that are larger than the window sizes and that the speed of processing data is fast. Figure 2.7 shows that geomorphic features, such as banks of small rivers and road edges, are

(a)   (b)

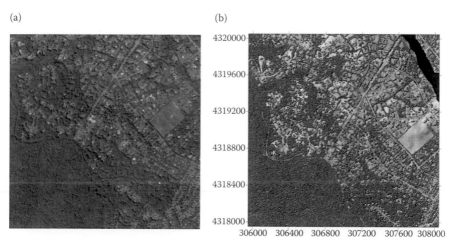

Figure 2.5 (a) Aerial photograph collected on August 28, 2010, from Google Earth for the Potomac area in Montgomery County, Maryland. (b) LiDAR image generated using first return points.

clearly identified and preserved by the PM filter, whereas the ILPF filter smooths some of these features.

The disadvantage of the PM filter is that the tops of mountains can be incorrectly removed because of the use of a constant slope to determine the threshold for each window. This can be alleviated by increasing the window sizes nonlinearly, such as by the power of 2, and changing the slope parameter of the threshold equation as window size increases. Small initial thresholds derived using the low slope value in Equation 2.6 are often selected to remove small nonground objects, such as vegetation, whereas

(a)   (b)

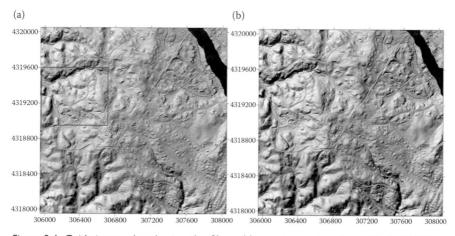

Figure 2.6 Grids interpolated using the filtered last return points from (a) the PM filter and (b) from ILPF filter. The extent of Figure 2.7 is indicated by a rectangle.

(a)   (b)

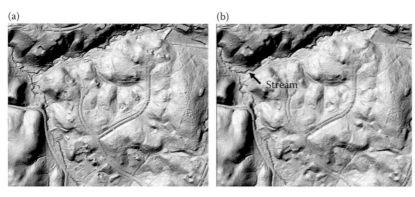

Figure 2.7 Enlarged left portions of grids in Figure 2.6. The streams are well preserved by the (a) PM filter, whereas the streams are "blurred" (dilated) by the (b) ILPF filter.

large subsequent thresholds derived using large slopes are used to remove large nonground objects.

2.3.4 Iterative local polynomial fitting filter

Previous algorithms separated ground and nonground measurements by removing nonground points from a LiDAR data set. Alternatively, LiDAR data can be classified by selecting ground measurements iteratively from the original data set. The ILPF algorithm adopts this strategy.

- Selecting the lowest points within a large moving window (e.g., 40 m) over a grid with a small spacing (e.g., 2 m). The large moving window is centered over each grid node, and its initial size is usually larger than nonground objects in the study area. These lowest points consist of an initial set of ground measurements.
- Reducing the moving window size and lowest point within the window is selected as a candidate of ground measurement. The candidate is added to the set of ground measurements if the elevation difference between the candidate and the interpolated surface at the grid node is less than a predefined tolerance. The interpolated surface is produced in terms of ground measurements identified in the previous step. This process is repeated until the moving window size (e.g., 1 m) is smaller than the grid spacing.

The ILPF method works well for most mountainous areas, but it fails to identify the top of a mountain with a very steep slope. Figures 2.6b and 2.7b show that the DTM generated by the ILPF filter is smoother than the DTM from the PM filter; thus, it is more suitable for producing contour maps.

2.3.5 Adaptive triangulated irregular network filter

The ATIN filter employs the distance of points on the surface of a triangulated irregular network (TIN) to select ground points from a LiDAR data set. This filter was developed by Axelsson (2000) and implemented in the commercial LiDAR data-processing software, TerraScan. The algorithm has been modified slightly and implemented as follows:

- Subdividing the data set into an array of square cells and selecting points within a cell with the minimum elevation to be seeds of a ground point data set. The size of a square cell is set to be larger than the maximum size of nonground objects in the study area. A TIN is built using seed ground points based on the Delauney triangulation algorithm.
- Examining points above each triangle of the TIN in terms of their distances to the triangle surface. If the distance of a point is less than the predefined threshold, the point is added to the ground point data set. To include the measurements for steep terrains such as cliffs, the distance of a mirror point to the corresponding surface is also employed in the process of selecting ground points (see Figure 2.8).

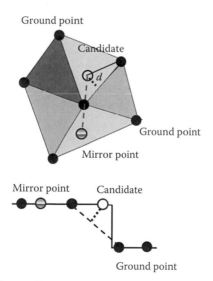

Figure 2.8 Selection of ground points in the ATIN filter. The distance (*d*) from a candidate point to the triangle surface is compared with a predefined threshold to determine whether the point is a ground point. If the distance is less than the threshold, the candidate is classified as a ground measurement. If the distance is greater than the threshold, but the distance of the mirror point is less than the threshold, the candidate is also classified as a ground point. A mirror point is the reflection of the candidate point against a vertex of the ground triangle along the line connecting the candidate and the vertex. In such a way, the points for steep terrains are included into ground data sets.

- Constructing a new TIN using the ground point data set. The second and third steps are repeated until no points can be added to the ground point data set.

This filter works for both low and high relief topography as long as the seed ground points are selected appropriately. The disadvantage of the ATIN filter is that the computation time is relatively long compared with other filters such as the PM filter.

2.3.6 Issues with filtering of LiDAR data

A LiDAR data set may contain a few points with elevations remarkably lower than those of their neighbors because of the multipath reflection of an emitted laser pulse (see Figure 2.9a). These low elevation measurements that are called negative blunders often remain in the filtered data set because most filters classify low elevation points as ground points. These blunders can cause incorrect removal of neighboring ground points because of the large elevation differences between the blunders and surrounding ground points (see Figure 2.9b). Additionally, when a DTM is generated by inter-polating the ground measurements including negative blunders, conical pits similar to "bomb craters" are generated (see Figure 2.10a). Fortunately, negative blunders are distinct from their surroundings in elevation and, in most cases, occupy a small area in space. Therefore, these points can be removed by performing a closing operation with a small window immediately following an opening operation with a large window. A point is classified

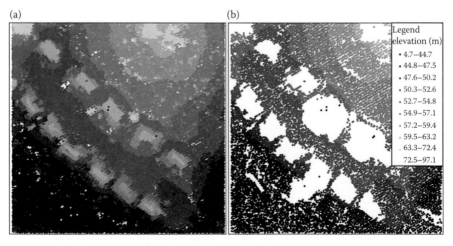

(a)   (b)

Legend
elevation (m)
- 4.7–44.7
- 44.8–47.5
- 47.6–50.2
- 50.3–52.6
- 52.7–54.8
- 54.9–57.1
- 57.2–59.4
- 59.5–63.2
- 63.3–72.4
- 72.5–97.1

Figure 2.9 (a) Raw LiDAR points. (b) Filtered points derived using the PM filter for the Potomac area in Maryland. The blunders of low elevations remain in the filtered data set and cause incorrect removal of neighboring ground points.

(a)　　　　　　　　　　　　　　　　　(b)

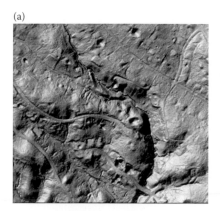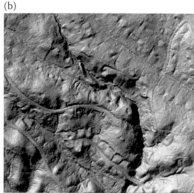

Figure 2.10 Grids generated (a) using the filtered points with low elevation blunders at Potomac in Maryland, United States, and (b) using the filtered points after removal of low elevation blunders. Morphological opening and closing operations were applied to the last return measurements to remove "blunders" before using the ILPF filter. If the elevation of a point is lower than the closed surface and the difference between the point and the surface is less than 5 m, the point was classified as a blunder.

as a blunder and removed from the raw LiDAR data set before filtering if the elevation difference between a point and the surface from the closing operation is less than a given threshold. Figure 2.10b shows that the large "bomb craters" caused by individual multipath low points are removed by the morphological closing operation. The side effect of the closing operation is that it could diminish the details of the terrain by removing true low-elevation points less than the window size. An alternative method to remove the blunders is to use the elevation difference between a LiDAR point and the surface constructed through Delaunay triangulation (Silván and Wang, 2006; Meng et al., 2009).

The filters presented in the previous section perform well in landscapes of low complexity, but they are vulnerable to commit large errors of either omission or commission over landscapes with dense vegetation, complex shape and configuration (terraces), disconnected terrain (courtyards), and special geomorphic features, such as coastal sand dunes. When LiDAR measurements for nonterrain objects display spatial characteristics similar to the terrain, commission errors occur. On the contrary, when the terrain exhibits spatial characteristics similar to the buildings, omission errors occur. All existing filtering algorithms suffer commission and omission errors in various degrees, depending on their filtering measures and the types of topography, vegetation, and buildings in a study area. No magic filter can identify all landforms without errors because of the complexity of the Earth's surface. Therefore, a manual adjustment is often needed to produce high-quality DTMs after an automated filter is applied. It is preferred for a filter to produce more commission errors than omission errors for editing because the removal of commission errors is easier than searching for omission errors. Manual classification and quality control are expensive and often consume 60%–80% of data-processing time when producing DTMs (Flood, 2001). Therefore, it is very helpful to develop automated classification algorithms that commit a minimum number of errors.

Three common factors contribute to the errors for LiDAR data filtering. First, special geomorphic forms, such as riverbanks and cliffs, exhibit a large elevation change in a short distance similar to buildings and trees. These features are often mistakenly removed by many existing filters. Second, the size of individual or clusters of nonterrain objects often varies for a study area. This poses special challenges for the filters with a fixed window size, such as the MLS filter. This problem has been solved by adopting multisize windows during a filtering process. For example, the PM filter removes various sizes of nonterrain objects by gradually increasing window size from the minimum to a maximum, which is greater than the largest nonterrain objects in a study area. In contrast, the ATIN and ILPF filters select terrain measurements by decreasing window size from the maximum to the minimum.

Third, the threshold used to compare with the filtering measures is not adaptive. The topography, vegetation, and buildings in a study area could change considerably. An adaptive threshold requires that the threshold changes its value as topographic slope and the characteristics of nonterrain objects change. Many filters can only partially satisfy this requirement. Filters such as PM change the threshold correspondingly as window sizes increase or decrease. The topographic slope used to determine the threshold value, however, is globally set to be a constant for a study area. This works for an area with gentle slopes but not for steep slope areas. Hu and Tao (2005) attempted to design an adaptive threshold by incorporating both the range and slope measures for a filtering window. Their method is not fully adaptive because a constant maximum slope value was used in the calculation of the threshold. Kampa and Slatton (2004) developed an adaptive filter to classify LiDAR points in a vegetated mountain area by fitting two Gaussian probability density functions for vegetation points and ground points within a window. The problem with this method is that in a vegetated area of steep slopes, the boundary between ground and vegetation measurements is not distinct in probability density curves. These areas can be identified by this method, however, by calculating the separation between ground and vegetation density functions.

The construction of a fully adaptive filtering threshold when ground and nonground points are not separated is a chicken–egg problem that is difficult to solve. An alternative way to filter LiDAR data with large changes in topographic slopes is to construct interpolated surfaces to gradually fit the topographic changes in the study area. Whether a point represents a ground measurement is determined by the distance of the point to the fitted surface (see Figure 2.4). If an interpolated surface can fit the change of topography, the points within an elevation buffer zone can be classified as ground measurements. The advantage of this method is that a height threshold for the elevation buffer zone can be set to be a constant if the fitting surface follows the changes of topography. The key step for this algorithm is to generate the surface from LiDAR measurements to fit topographic changes.

The ILPF method can be used to generate the surface fitting topographic changes. In some cases, however, the top of a steep mountain may be removed incorrectly because the interpolated surface based on previously identified ground points is too low. This problem can be overcome by considering the fitness of the surface derived using minimum elevation points from the current window to previously identified ground measurements. Figure 2.11 shows how the ILPF method is improved based on LiDAR points (circles) along a profile. The set of ground measurements consists of all minimum elevation points for a window of 40 m moving over the x-axis with 2 m spacing. A ground (previous) surface is generated by interpolating this set of ground points (black dots). The rectangle symbol represents minimum elevation points (ground candidates) within a window of 10 m. In the ILPF filter, the elevation difference of a candidate to the previous surface determines whether it is added to the set of ground measurements. On the basis of this elevation difference rule, candidate points close to the previous surface around 225 m at the x-axis are added into the ground set.

Three ground measurements at the top of a small mountain (around 200 m at the x-axis) are missed, however, because the previous surface is too

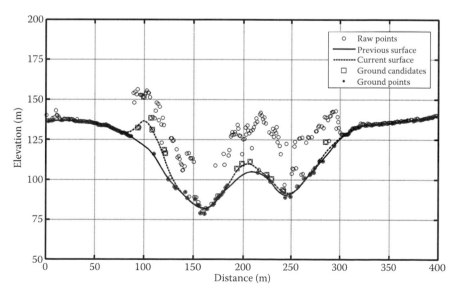

Figure 2.11 LiDAR points along a profile with a width of 2 m for Puget Sound in Washington, to illustrate how the ILPF filter is improved.

low due to a lack of previously identified ground points at the mountaintop. Therefore, the elevation difference between a candidate and the current surface is employed to recover the missed ground points. The current surface, represented by a dashed line, is derived by interpolating both candidate and ground points. The three points at the top of a mountain were identified as ground points because their elevation differences from current surface are less than a predefined threshold. Nonground points (e.g., those around 100 m at the x-axis) are included mistakenly when the elevation difference from the current surface is used to recover the missed ground points. To remove these commission errors, the fitness of the previous and current surfaces to the ground points within a local window is introduced as another criterion. If the fitness of the current surface is better than that of the previous surface or is less than a given threshold, a candidate is selected as a ground point (e.g., points around 200 m at the x-axis); otherwise, it is not included (e.g., points around 100 m at the x-axis). The fitness of the surface to the ground points is measured by

$$\sigma_s = \sum_{j=1}^{N} (z_g - z_i)^2, \tag{2.7}$$

where N is the total number of ground points within a local window, z_g is the elevation of a ground point, and z_i the elevation of the current surface at the same location as a ground point.

The DTM generated from the filtered LiDAR points for a steep mountain area that are derived using the improved ILPF method indicates that many small ridges between the valleys in densely vegetated areas are clearly identified (see Figure 2.12).

2.4 CONSTRUCTION OF BUILDING MODELS

Building models are essential for constructing digital city visualization systems and urban landscape models (Jensen, 2000). Building models can be divided into two categories: two-and-a-half-dimensional (2.5D; simple) and 3D (sophisticated) models (Zhang, Yan, and Chen, 2008). Geometric attributes of a 2.5D building model consist of a footprint polygon and a height value (see Figure 2.13b), whereas geometric attributes of a 3D building model include a footprint polygon, planes or other types of surfaces for various parts of the roof, and their projections (polygons) on the ground plane (see Figure 2.13c). The advantage of the 2.5D building model is that the buildings require less geometric attributes and are represented by various types of "boxes," and thus, the 3D rendering is fast. The 2.5D building model is sufficient for many applications that do not require the details of buildings, such as numerical modeling of urban flooding and heat island

(a)   (b)

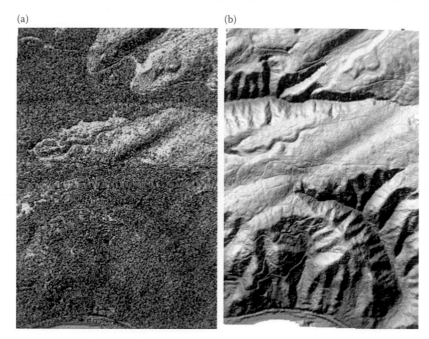

Figure 2.12 Grids for an area at Puget Sound in Washington, generated (a) using raw LiDAR points and (b) using filtered points by the improved ILPF filter.

effect, urban population and energy demand estimation, and large-scale 3D visualization. The 3D models offer more details and roof shapes of buildings in comparison with 2.5D models. The 3D models are required by applications such as hurricane wind damage models for individual properties (Zhang et al., 2006), property tax estimation, and detailed urban landscape modeling. The disadvantage of 3D models is that rendering is slow.

Building models can be extracted from high-resolution remote-sensing data such as aerial photographs, high-resolution satellite images, and LiDAR

(a)   (b)   (c)

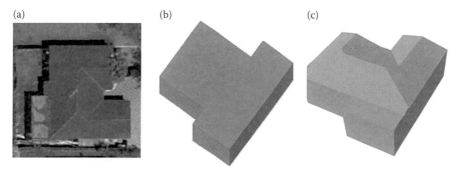

Figure 2.13 (a) Aerial photo. (b) 2.5D model. (c) 3D model for a residential building.

measurements. Compared with aerial photographs and satellite images obtained by optical sensors, LiDAR measurements are not influenced by sun shadow and relief displacement and provide direct measurements of building heights. The major step to construct a 2.5D building model is to extract footprints from LiDAR measurements. The building height for a 2.5D model can be derived using a height statistics for a footprint.

The methods used to derive 3D building models fall into two categories: model-driven and data-driven methods (Schwalbe, Maas, and Seidel, 2005). In the model-driven method, building models are identified by fitting predefined models into the LiDAR measurements. This method is robust, but building models for a study area are not always available in advance, which limits the application of the model-driven method. In the data-driven method, building measurements are grouped first for different roof planes. Then, a 2D topology of each building, which is represented by a set of connected planar roof surfaces projected onto a horizontal plane, is derived. Finally, 3D building models are constructed based on 2D topology and associated roof planes. Many algorithms have been developed to extract building models from LiDAR measurements (Shan and Toth, 2008; Meng, Wang, and Currit, 2009). Zhang et al. (2006, 2008) developed a framework for the construction of building models from LiDAR measurements, which includes six major components: (a) separating ground and nonground measurements, (b) segmenting building points in nonground measurements based on a plane-fitting technique, (c) deriving initial footprints and 2D topology of roof facets, (d) estimating the dominant directions of a building, (e) adjusting footprints and the 2D topology using a snake-based algorithm, and (f) generating 2.5D building models based on footprints and 3D building models based on adjusted 2D topology and plane parameters derived from building measurements.

2.4.1 Separating ground and nonground measurements

The first step for constructing a building model is to identify nonground measurements that include building points in a LiDAR data set. Numerous algorithms have been developed to identify ground measurements in a LiDAR data set. The remaining points in the LiDAR data set after ground points are removed are nonground points. Thus, any filter such as the PM method can be utilized in this step as long as the filter can effectively separate ground and nonground measurements.

2.4.2 Segmenting nonground measurements

The nonground objects mainly consist of buildings and trees. The distinct difference between buildings and trees is that the roof surfaces are approximately planar, whereas canopy surfaces are irregular. This difference allows us to construct measures to separate building and vegetation measurements. For example, Figure 2.14 shows the normalized sum of squares due to deviations (SSD) in elevations of points from an associated plane. The plane is created by fitting the equation of a plane to the point and its eight neighbors. Given nonground point $p_k(x_0, y_0, z_0)$, a Cartesian coordinate system (x, y, and z) is established using p_k as the origin. Assume that the plane is defined as follows:

$$z = ax + by + c. \tag{2.8}$$

The parameters (a, b, c) of the best fitting plane for p_k and its neighbors are derived by minimizing the SSD, which is represented by ssd_k:

$$ssd_k = \sum_{(p_i) \in K} (z_i - h_i)^2, \tag{2.9}$$

where K is a set for p_k and its neighbors, and h_i and z_i are observed and plane fitted surface elevations, respectively.

The SSD value for each LiDAR point is calculated in terms of Equation 2.9 after the parameters for Equation 2.8 are estimated. The SSD values for LiDAR points on the roof are small and consistent, whereas SSD values for nonbuilding points are large and variable as shown in Figure 2.14b. On the basis of this observation, a region-growing algorithm using a local plane-fitting technique was developed to segment building points in nonground

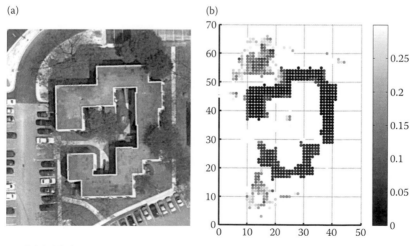

(a)

(b)

Figure 2.14 (a) Aerial photograph for a flat roof building and the surrounding area. (b) The SSDs of roof points that are normalized by dividing the total number of points used to calculate a SSD.

measurements (Zhang, Yan, and Chen, 2006; Yan et al., 2007;). First, the region-growing algorithm starts with a seed point inside the building area with a minimum SSD value and groups nonground measurements located on the same planes into patches (i.e., roof facets) with the same identification number. Then, connected patches with shared edges are merged, and building measurements are identified in terms of patch sizes. Building patch sizes are relatively larger because all measurements for a roof facet are located in the same plane, whereas vegetation patch sizes are small because of large changes in elevations for an irregular canopy. Therefore, building and vegetation patches are separated in terms of a predefined minimum area threshold that typically is the roof area of a minimal building in the study area.

2.4.3 Deriving initial footprints and 2D topology of roof edges

For 2.5D building models, raw footprints are derived by connecting the boundary points for the merged patches. For 3D building models, the equations for roof facets and their edges for each unmerged building patch have to be derived to construct a 3D building model. It is assumed that a building roof consists of various roof facets with shared edges, and roof facets follow plane equations. Roof facet edges and their spatial relationships are characterized by the 2D topology, which is represented by a set of connected polygons that are the projections of the roof facets onto a horizontal plane. The edges and vertices that form the topology are derived from grouped LiDAR measurements for various facets of a roof by comparing group numbers of neighboring points. The raw footprints for 2.5D building models and the topology for 3D building models are simplified using the Douglas–Peucker algorithm (Douglas and Peucker, 1973) to reduce the noise of the edges resulting from irregularly spaced LiDAR points and measurement errors. Parallel edges of a building can be distorted in some cases during the simplification because no geometric constraints are applied. Therefore, refinement of the footprint and the topology is performed based on the dominant directions of a building.

2.4.4 Estimating dominant directions of a building

To estimate the dominant directions of buildings, it is assumed that a building model has two dominant horizontal directions X and Y that are perpendicular to each other (see Figure 2.15a). It is also assumed that most of footprint edges are parallel to the dominant directions, but a few edges can be oblique to the dominant directions. The X and Y directions for a building with a flat roof are estimated by using lengths of footprint edges and directions. The dominant directions of a building are derived by optimizing

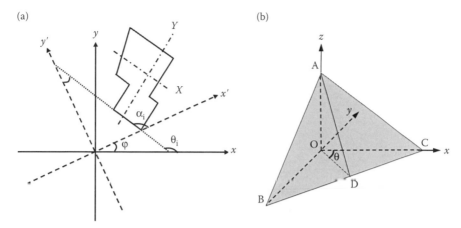

Figure 2.15 (a) Relationship between angles β_i, θ_i, α_i, and φ. The coordinate system x' and y' is a counterclockwise rotation of the coordinate system x and y by an angle φ. θ_i is the counterclockwise intersection angle between a segment of a building footprint and the axis x. α_i is the counterclockwise intersection angle between a segment and the axis x'. β_i is the minimum intersection angle between a segment and the axes x' and y'. X and Y represent the dominant directions of the building. (b) The θ angle of a plane ABC which is equal to angle COD. The line OD is perpendicular to line BC.

a rotation angle φ using a function SL, which is the sum of the product of the segment length L_i and associated angle θ_i (Zhang, Yan, and Chen 2006):

$$SL = \frac{1}{\sum\limits_{i=1}^{N} L_i} \sum_{i=1}^{N} \beta_i L_i$$

$$\alpha_i = \begin{cases} \theta_i - \varphi & \theta_i \geq \varphi \\ 180 + \theta_i - \varphi & \theta_i < \varphi \end{cases}$$

$$\beta_i = \begin{cases} \min(\alpha_i, 90 - \alpha_i) & \alpha_i \leq 90 \\ \min(180 - \alpha_i, \alpha_i - 90) & \alpha_i > 90 \end{cases},$$

(2.10)

where N is the total number of segments of a footprint polygon. It has been proven that the estimated dominant directions are the same as the directions of the parallel and perpendicular segments as long as the total length of the oblique lines is less than the total length of the parallel and perpendicular segments of a footprint (Zhang, Yan, and Chen, 2006).

The method works fine for buildings with large flat roofs. The robustness of the method, however, is deceased for buildings with many small non–flat roof facets and oblique edges when the point density of the LiDAR measurements is less than 1 per square meter. This problem can be alleviated by replacing segment lengths in Equation 2.10 with the areas of roof facets.

The parameter θ is the angle between a roof facet plane, $z = Ax + By + C$, and the plane XOZ and is determined by the parameters of the plane equations for roof facets (see Figure 2.15b):

$$
\theta = \begin{cases} \arctan(B/A) & \arctan(B/A) \geq 0 \\ \arctan(B/A) + 180 & \arctan(B/A) < 0 \end{cases}.
\tag{2.11}
$$

2.4.5 Adjusting footprints and 2D topology

A snake-based (active contour) algorithm was developed to adjust the 2D topology of a 3D building model based on dominant building directions (Yan et al., 2007; Zhang, Yan, and Chen, 2008). Because the footprint of a 2.5D building model is just a special case of the 2D topology without inside line segments separating various roof facet polygons, the snake-based algorithm can be used to adjust footprints, too. This algorithm refines the 2D topology by enforcing the requirement that most edges be parallel to the dominant directions as well as by keeping the adjusted topology as close to its original location as possible. The total energy E_{Total} of a contour with a parametric representation $v(s) = (x(s), y(s))$ consists of the direction energy (E_{Dir}) and the deviation distance energy (E_{Dis}):

$$
E_{\text{Total}} = \int_0^1 (E_{\text{Dir}}(v(s)) + E_{\text{Dis}}(v(s))\, ds
\tag{2.12}
$$

Direction energy E_{Dir}, which relates the deviation of an edge to the dominant direction is used to enforce the parallel constraint. Distance energy E_{Dis}, which computes distances between the original and transformed topologies is used to limit the deviation of the adjusted topology from its original position.

The 2D topology is refined by minimizing the above energy function. Unfortunately, the global minimization for a general 2D topology is a non-deterministic polynomial (NP) time completeness problem and involves a time-consuming computation. The energy function for a subset of the 2D topology can be minimized globally in a polynomial time using an algorithm based on graph theory (Yan et al., 2007). Fortunately, most building topologies belong to this subset, and our experiments indicate that more than 95% of building topologies in the study area can be refined by the graph-based algorithm.

2.4.6 Generating 2.5D and 3D building models

The proposed framework has been used to extract 2.5D and 3D building models from LiDAR measurements. Figure 2.16 shows 2.5D and 3D

(a) (b)

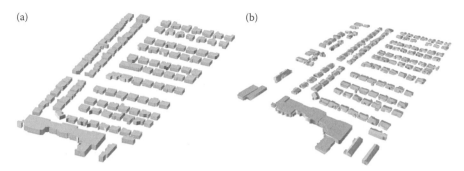

Figure 2.16 (a) Constructed 2.5D and (b) 3D building models using LiDAR data for an area with both commercial and residential buildings.

models for both residential and commercial buildings reconstructed through the proposed framework. The application of the framework to the Florida International University campus and adjacent residential areas shows that the region-growing algorithm identified the building patches well, the snake algorithm adjusted most of the 2D topologies properly, and the 3D building models were reconstructed effectively. The quantitative accuracy analysis indicates that all buildings were extracted. Errors of about 12% in the total area of footprints were committed by the proposed algorithms, despite the fact that there are several complex buildings on the Florida International University campus (Zhang, Yan, and Chen, 2006). The entire process of 3D building model reconstruction is highly automated and requires little human assistance, which is very useful for processing voluminous LiDAR measurements.

2.5 EXTRACTION OF VEGETATION INFORMATION

In addition to the production of high-quality digital terrain and building models, airborne LiDAR is used extensively for the analysis of vegetation structure (Omasa, Hosoi, and Konishi, 2007; Vierling et al., 2008; Zhao et al., 2011). LiDAR provides a 3D view of the canopy structure, adding an invaluable tool for the preservation, restoration, and management of forested areas. The use of LiDAR has been applied to forests in two general areas: stand or patch analysis and individual tree analysis.

2.5.1 Extraction of individual trees and estimation of biophysical variables

The key to the identification of individual tree species is the accurate delineation of tree crowns. Identification of tree crowns requires very high-density LiDAR coverage of 10 or more pulses per square meter of area or LiDAR in combination with very high-resolution aerial photographs. If the tree canopy is not dense enough to reflect a sufficient portion of the LiDAR pulse, then the canopy height will be underestimated. Because tree species vary in the structure of their canopies, errors in individual tree height estimation will vary (Hyyppä et al., 2008).

The other component to individual tree identification is an algorithm that can identify and cluster points associated with a single tree crown, even if there is overlap with other trees located lower in the canopy. Methods for tree-crown clustering using LiDAR point clouds have been described by Morsdorf et al. (2004) and summarized by Hyyppä et al. (2008). Using LiDAR point data at densities of 20–30 points per square meter of area, Morsdorf et al. (2004) used a modified k-means clustering analysis of the LiDAR 3D points to identify the elliptical crowns of boreal pine species. The highest pixels from the digital canopy model (DCM) were used as seed points for the k-means clustering. The results showed a common difficulty in segmenting individual trees. Even with high-density LiDAR, groups of trees located within 1 m of each other were clustered as a single tree crown.

Very high-resolution aerial photographs have been added to LiDAR data to increase the accuracy of tree-crown detection, particularly when low–point density LiDAR data are used in the analysis. LiDAR can be used to remove areas from the analysis that are not tree crowns and also to identify the treetops. The multispectral images can aid in tree identification through the spectral signatures from the tree canopy. Various segmentation methods have been described, ranging from customized tree-crown extraction algorithms to commercially implemented image segmentation programs such as eCognition® (http://www.definiens-imaging.com).

In addition to identification of individual trees, the spatial neighborhood information contained in LiDAR point clouds are used to describe the spatial arrangement of individual trees in the forest through point pattern analysis or a similar method (McElhinny et al., 2005). Identified individual key tree species and spatial structures can be used to obtain estimates of timber volume, biomass, carbon cycling, and fuel volumes for input to fire models. Several laser-derived variables have been used to estimate biophysical properties of forest stands, such as individual tree height, the mean of individual tree heights, or canopy cover density, calculated as the fraction of total laser pulses emitted that strike the canopy. Georeferenced training plots are systematically established to build models relating LiDAR survey to forest parameters. Regression models have been developed to predict mean tree height, mean diameter, stand volume, and basal area from laser-derived canopy metrics that can cover areas of 50–1,000 km². The response variable in the multiple regression model is a field variable—height or basal area, for example. The predictor variables are the LiDAR metrics, which are often expressed as their natural logs. Many LiDAR metrics have been described for characterizing forests, but most are correlated, and only five or six contribute independent information. The use of LiDAR allows the survey of much larger areas than ground-based forest surveys; however, the precision and accuracy of the resulting models are influenced by the laser sampling density (Næsset, Bollandsås, and Gobakken, 2005).

In other applications, single-story and multistory tree vertical structure classes have been characterized using laser-derived tree height variances (Zimble et al., 2003). Computer models are used to predict the behavior of forest fires for both management and fire fighting. In a Douglas fir–western hemlock forest, LiDAR data were collected at a point density of 3.5 pulses per square meter of area with four returns measured. LiDAR metrics were used to predict canopy height, canopy base height, canopy bulk density, and total canopy fuel weight for wildfire simulation models.

2.5.2 Patch analysis of vegetation

Hierarchical patch dynamics often is used to model the characteristics of forests and analyze the structure of ecological systems (Wu and Loucks, 1995). The fundamental unit of this method is the patch, a spatial unit that can be distinguished from surrounding areas and characterized by composition and spatial configuration. The ecological systems are composed of hierarchical groups of patches that differ in size, shape, and successional stages at particular scales. Patches at lower scales in the hierarchy are nested within patches at higher levels. LiDAR measurements are useful to detect patches of forests by providing information on spatial changes based on heights and densities of vegetation in an area.

Gaps are one type of patch in forests and are important for the recruitment of young trees, but they pose special problems for optical remote sensing because the illumination is transient and affected by the canopy and understory reflectances and shadows. LiDAR data combined with airborne imagery (see Figure 2.17) were found to be highly effective for gap feature extraction, overcoming the limitations of spectral data analysis alone (Koukoulas and Balckburn, 2004). Zhang (2008) developed a method to detect gaps caused by lightning strikes and hurricanes using DCM derived from LiDAR measurements.

In this method, a filtered canopy surface, approximating the midpoint of a fluctuating top canopy surface, was generated first by applying an alternative sequential filter (ASF) to the DCM (see Figure 2.18). Then, the elements of a gap pixel set (G) in a DCM were selected by comparing the height of a pixel with that of the filtered canopy surface at the same location:

$$G = \{z[i, j] \mid B[i, j] > \lambda\, ASF_N[i, j] \ \& \ ASF_N[i, j] > h_{min}\}$$
$$B[i, j] = ASF_N[i, j] - z[i, j], \tag{2.13}$$

where $z[i, j]$ is the height of DCM pixel at row i and column j, which are determined by horizontal coordinates of a pixel. $ASF_N[i, j]$ is the filtered

(a)   (b)

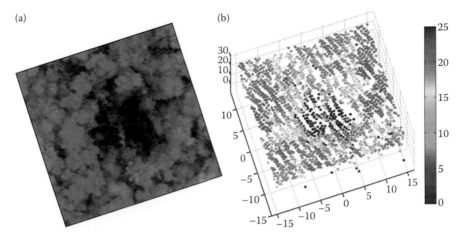

Figure 2.17 **(See color insert.)** (a) Color-infrared aerial photograph and (b) first return
LiDAR measurements for a gap and surrounding trees. The gap area is obvious
in both the aerial photograph and LiDAR measurements because of differences in spectral reflectance and elevation. (After Zhang, K., "Identification
of Gaps in Mangrove Forests with Airborne LIDAR," *Remote Sensing of
Environment* 112: 2309–2325, 2008. With permission from Elsevier.)

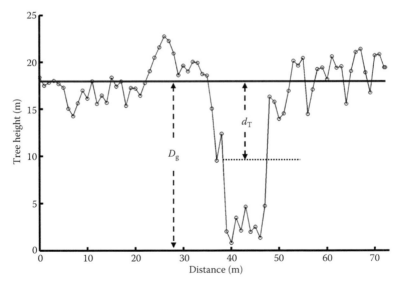

Figure 2.18 A tree height profile (line with circle) extracted from the DCM with a pixel
size of 1 m. The filtered surface (solid line) was derived using the ASF with
11 sequential operations and a window size increment of 2. Parameter d_T
is the fixed depth threshold and D_g is the height of the gap. The parameter
$\lambda = d_T/D_g$ determines the relative location below a canopy top to delineate the gap boundary. (Modified from Zhang, K., "Identification of Gaps
in Mangrove Forests with Airborne LIDAR," *Remote Sensing of Environment*
112: 2309–2325, 2008. With permission from Elsevier.)

canopy surface height at i and j, N is the number of windows used by ASF, λ is a predefined constant with a value less than one, and h_{min} is the minimum height of the canopy surface. The parameter λ determines the relative location below the filtered canopy surface to delineate the gap boundary, and h_{min} was introduced to prevent open spaces in grass or brush areas of low canopy heights from being detected as gaps. The application of this method to a LiDAR data set for the mangrove forest in South Florida shows that the method is able to extract lightning and hurricane gap polygons from LiDAR points (see Figure 2.19). A number of metrics for gaps can be established based on extracted polygons to measure sizes, shapes, and connectivity of gaps in various development stages to aid in understanding forest dynamics (Zhang et al., 2008).

Another application of LiDAR to patch level analysis is to classify vegetation using the distribution of neighboring heights from the DCMs

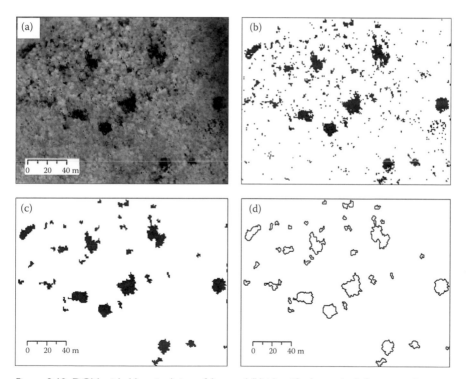

Figure 2.19 DCM with (a) a pixel size of 1 m and (b) identified gap pixels for a sample area. The DCM is represented by a gray-scale image with black color indicating small height values and white color indicating large height values. There are many tiny gap patches in (b). (c) Gap areas after tiny patches are removed. (d) Gap areas after small inside holes are filled. (After Zhang, K., "Identification of Gaps in Mangrove Forests with Airborne LIDAR," *Remote Sensing of Environment* 112: 2309–2325, 2008. With permission from Elsevier.)

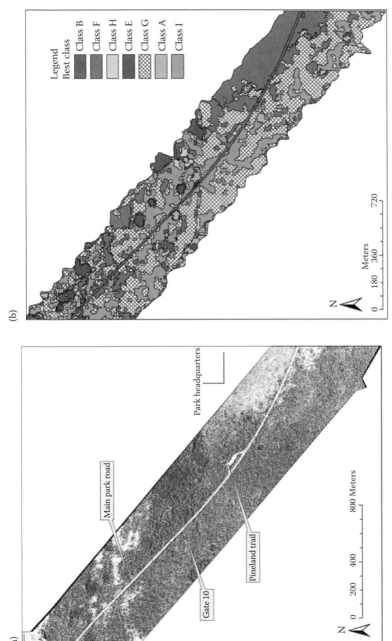

Figure 2.20 **(See color insert.)** (a) Aerial photograph for an area in Everglades National Park. (b) Map showing the classification of the vegetation communities based on the height distributions of LiDAR measurements in a moving window. The classified vegetation classes include A: Hammock edge, B: Transition, E: Mature hammock center, F: Open ground with herbaceous plants, G: Pine trees with herbaceous understory, H: Pine forest edge, and I: Pine trees with shrub understory.

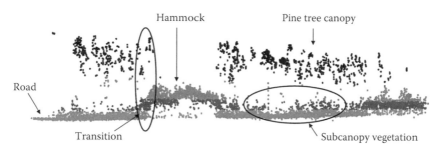

Figure 2.21 **(See color insert.)** 3D display of raw LiDAR points and vegetation classes. Colors indicate elevation ranges in meters (NAVD88): gray, 0.2–3.1 m; red, 3.2–5.4 m; green, 5.5–9.4 m; light blue, 9.5–14 m; dark blue, 14–18.8 m.

(Houle, 2006; Zhang et al., 2008). For example, to classify pine and hammock forests in Everglades National Park (see Figure 2.20), a study area was first divided into an array of square cells with a size of 3 × 3 m. Second, a 10 × 10 m window was generated for each cell. Third, the frequencies (as percentages) of LiDAR points occurring in user-defined vertical intervals within the window were computed based on LiDAR elevations. The intervals were determined empirically based on the published vegetation structure for the study area to distinguish ground with herbaceous growth, low shrubs, tall shrubs, low height trees, and tall trees (see Figure 2.21). Nine height intervals were chosen and labeled Layers 1–9 as shown in Table 2.1.

The nine-layer image was imported into the eCognition® software for patch object identification using hierarchical segmentations. Finally, segmented patches are clustered based on their properties to produce the vegetation classes (see Figure 2.20b). Although it is difficult to quantitatively estimate the accuracy of the boundaries between these classes, comparison of the results with high-resolution aerial photographs and ground truth surveys demonstrated that the method classified the vegetation communities reasonably well.

Table 2.1 Intervals for LiDAR height frequency analysis

Label	Height interval (m)	Description
Layer 1	0–0.5	Open ground and herbaceous plants
Layer 2	0.5–1	Forbs and low shrubs
Layer 3	1–2	Low shrubs
Layer 4	2–3	Shrubs
Layer 5	3–5	Shrubs and low trees
Layer 6	5–7	Medium broadleaf trees
Layer 7	7–9	Tall broadleaf trees
Layer 8	9–12	Low pine trees
Layer 9	>12	Tall pine trees

2.6 CONCLUSION

Airborne LiDAR as an active remote-sensing technology provides highly accurate elevation measurements of objects on the Earth's surface, adding a new dimension to the traditional optical remote-sensing imagery. The LiDAR technology has been extensively used in mapping topography, vegetation, buildings, and other objects on the Earth's surface in the past two decades. The large volumes of 3D point data from LiDAR remote sensing demand new algorithms to process the data and extract useful features, including topography, buildings, and vegetation for various applications. Great progress has been made in the development of algorithms for filtering LiDAR data, construction of building models, and extraction of biophysical variables as demonstrated in this chapter. However, the development of adaptive filters and fusion of LiDAR data with aerial photographs and high-resolution satellite images and the application of this fused imagery remain active research fields.

ACKNOWLEDGMENTS

This work has been supported by grants from National Science Foundation, National Oceanic and Atmospheric Administration, and the Florida Center of Excellence. We thank Mr. Dan McGillicuddy for drawing the airplane in Figure 2.1.

REFERENCES

Ackermann, F. 1999. Airborne laser scanning–present status and future expectations. *ISPRS Journal of Photogrammetry & Remote Sensing* 54: 64–67.

Axelsson, P. 2000. DEM generation from laser scanner data using adaptive TIN models. *International Archives of Photogrammetry and Remote Sensing* 33(Part B3): 85–92.

Carter, W. E., R. L. Shrestha, and K. C. Slatton. 2007. Geodetic laser scanning. *Physics Today* (December): 41–47.

Douglas, D. H., and T. K. Peucker. 1973. Algorithms for the reduction of the number of points required to represent a digitized line or its caricature. *The Canadian Cartographer* 10: 112–122.

Flood, M. 2001. LiDAR actitivities and research priorities in the commercial sector. *International Archives of Photogrammetry and Remote Sensing* 34(Part 3/W4): 3–7.

Haugerud, R. A., and D. J. Harding. 2001. Some algorithms for virtual deforestation (VDF) of LiDAR topographic survey data. *International Archives of Photogrammetry and Remote Sensing* 34(Part 3/W4): 211–218.

Houle, P., and Department of Earth and Environment, Florida International University. 2006. Use of Airborne Laser Mapping (LiDAR) for the assessment of landscape structure in the pine forest of Everglades National Park. In *Geoscience and Remote Sensing Symposium IEEE International Conference, July 31–August 4, Miami, FL,* 1960–1963.

Hu, Y., and C. V. Tao. 2005. Hierarchical recovery of digital terrain models from single and multiple return lidar data. *Photogrammetric Engineering and Remote Sensing* 71(4): 425–433.

Hyyppä, J., H. Hyyppä, X. Yu, H. Kaartinen, A. Kukko, and M. Holopainen. 2008. Forest inventory using small-footprint airborne Lidar. In *Topographic Laser Ranging and Scanning: Principles and Processing,* edited by J. Shan and C.K. Toth. Boca Raton, FL: CRC Press.

Jensen, J. R. 2000. *Remote Sensing of the Environment,* 2nd ed. Upper Saddle River, NJ: Prentice-Hall.

Kampa, K., and K. C. Slatton. 2004. An adaptive multiscale filter for segmenting vegetation in ALSM data. In *2004 IEEE International Geoscience and Remote Sensing Symposium Proceedings, Anchorage, AK.*

Koukoulas, S., and G. A. Balckburn. 2004. Quantifying the spatial properties of forest canopy gaps using lidar imagery and GIS. *International Journal of Remote Sensing* 25(15): 3049–3071.

Kraus, K., and N. Pfeifer. 1998. Determination of terrain models in wood areas with airborne laser scanner data. *ISPRS Journal of Photogrammetry & Remote Sensing* 53: 193–203.

Liu, X. 2008. Airborne LiDAR for DEM generation: some critical issues. *Progress in Physical Geography* 32(1): 31–49.

McElhinny, C., P. Gibbons, C. Brack, and J. Bauhus. 2005. Forest and woodland stand structural complexity: Its definition and measurement. *Forest Ecology and Management* 218: 1–24.

Meng, X., N. Currit, and K. Zhao. 2010. Ground filtering algorithms for airborne LiDAR data: a review of critical issues. *Remote Sensing of Environment* 2(3): 833–860.

Meng, X., L. Wang, and N. Currit. 2009. Morphology-based building detection from airborne LiDAR data. *Photogrammetric Engineering & Remote Sensing* 75(4): 427–442.

Meng, X., L. Wang, J. L. Silván, and N. Currit. 2009. A multi-directional ground filtering algorithm for airborne LiDAR. *ISPRS Journal of Photogrammetry and Remote Sensing* 64(1): 117–124.

Morsdorf, F., E. Meier, B. Kötz, K. Itten, M. Dobbertin, and B. Allgöwer. 2004. LiDAR-based geometric reconstruction of boreal type forest stands at single tree level for forest and wildland fire management. *Remote Sensing of Environment* 92: 353–362.

Næsset, E., O. M. Bollandsås, and T. Gobakken. 2005. Comparing regression methods in estimation of biophysical properties of forest stands from two different inventories using laser scanner data. *Remote Sensing of Environment* 94: 541–553.

Omasa, K., F. Hosoi, and A. Konishi. 2007. 3D lidar imaging for detecting and understanding plant responses and canopy structure. *Journal of Experimental Botany* 58(4): 881–898.

Petrie, G., and C. K. Toth. 2008. Introduction to laser ranging, profiling, and scanning. In *Topographic Laser Ranging and Scanning: Principles and Processing,* edited by J. Shan and C. K. Toth. Boca Raton, FL: CRC Press.

Pfeifer, N., P. Stadler, and C. Briese. 2001. Derivation of digital terrian models in the SCOP++ environment. Paper read at OEEPE Workshop on Airborne Laserscanning and Interferometric SAR for Digital Elevation Models, Stockholm.

Roering, J. J., L. L. Stimely, B. H. Mackey, and D. A. Schmidt. 2009. Using DInSAR, airborne LiDAR, and archival air photos to quantify landsliding and sediment transport. *Geophysical Research Letters* 36: 5.

Schwalbe, E., H. G. Maas, and F. Seidel. 2005. 3-D building model generation from airborne laser scanner data using 2-D data and orthogonal point cloud projections. Paper read at ISPRS Workshop Laser Scanning, September 12–14, Enschede, the Netherlands.

Shan, J., and C. K. Toth, eds. 2008. *Topographic Laser Ranging and Scanning: Principles and Processing.* Boca Raton, FL: CRC Press.

Silván, J. M., and L. Wang. 2006. A Multi-resolution approach for filtering LiDAR altimetry data. *ISPRS Journal of Photogrammetry and Remote Sensing* 61(1): 11–22.

Sithole, G., and G. Vosselman. 2004. Experimental comparison of filter algorithms for bare-Earth extraction from airborne laser scanning point clouds. *ISPRS Journal of Photogrammetry & Remote Sensing* 59: 85–101.

Slatton, K. C., W. E. Carter, R. L. Shrestha, and W. Dietrich. 2007. Airborne laser swath mapping: achieving the resolution and accuracy required for geosurficial research. *Geophysical Research Letters* 34: 5.

Vierling, K. T., L. A. Vierling, W. A. Gould, S. Martinuzzi, and R. M. Clawges. 2008. Lidar: shedding new light on habitat characterization and modeling. *Frontiers in Ecology and the Environment* 6(2): 90–98.

Vosselman, G. 2000. Slope based filtering of laser altimetry data. Paper read at 19th ISPRS Congress, Commission 4, Amsterdam, the Netherlands.

Vosselman, G. 2008. Analysis of planimetric accuracy of airborne laser scanning surveys. *International Archives of Photogrammetry and Remote Sensing* XXI: 217–222.

Wehr, A. 2008. LiDAR systems and calibration. In *Topographic Laser Ranging and Scanning: Principles and Processing*, edited by J. Shan and C.K. Toth. Boca Raton, FL: CRC Press.

Whitman, D., K. Zhang, S. P. Leatherman, and W. Robertson. 2003. Airborne laser topographic mapping: application to hurricane storm surge hazards. In *Earth Sciences in the City: A Reader*, edited by G. Heiken, R. Fakundivny, and J. Sutter, 363–376. Washington, DC: American Geophysical Union.

Wu, J., and O. L. Loucks. 1995. From balance of nature to hierarchical patch dynamics: a paradigm shift in ecology. *The Quarterly Review of Biology* 70(4): 439–466.

Yan, J., K. Zhang, C. Zhang, S. C. Chen, and G. Narasimhan. 2007. A graph reduction method for 2D snake problems. Paper read at IEEE Computer Society Conference on Computer Vision and Pattern Recognition, Minneapolis, MN.

Zhang, K. 2008. Identification of gaps in mangrove forests with airborne LiDAR. *Remote Sensing of Environment* 112: 2309–2325.

Zhang, K., S. C. Chen, P. Singh, K. Saleem, and N. Zhao. 2006. A 3D visualization system for hurricane storm surge flooding. *IEEE Computer Graphics and Applications* 26: 18–25.

Zhang, K., S. C. Chen, D. Whitman, M. L. Shyu, J. Yan, and C. Zhang. 2003. A progressive morphological filter for removing non-ground measurements from airborne LiDAR data. *IEEE Transactions on Geoscience and Remote Sensing* 41: 872–882.

Zhang, K., and Z. Cui. 2007. Airborne LiDAR data processing and analysis tools: ALDPAT 1.0. Miami: Florida International University.

Zhang, K., M. Simard, M. Ross, V. H. Rivera-Monroy, P. Houle, P. Ruiz, R. R. Twilley, and K. Whelam. 2008. Airborne laser scanning quantification of disturbances from hurricanes and lightning strikes to mangrove forests in Everglades National Park. *Sensor* 8: 2262–2292.

Zhang, K., and D. Whitman. 2005. Comparison of three algorithms for filtering airborne LiDAR data. *Photogrammetric Engineering and Remote Sensing* 71: 313–324.

Zhang, K., D. Whitman, S. P. Leatherman, and W. Robertson. 2005. Quantification of the changes caused by Hurricane Floyd along Florida's Atlantic Coast using airborne LiDAR survey. *Journal of Coastal Research* 21: 123–134.

Zhang, K., J. Yan, and S. C. Chen. 2006. Automatic construction of building footprints from airborne LiDAR data. *IEEE Transactions on Geoscience and Remote Sensing* 44(9): 2523–2533.

Zhang, K., J Yan, and S. C. Chen. 2008. A Framework for automated construction of building models from airborne LiDAR measurements. In *Topographic Laser Ranging and Scanning: Principles and Processing*, edited by J. Shan and C. K. Toth. Boca Raton, FL: CRC Press.

Zhao, K., S. Popescu, X. Meng, Y. Pang, and M. Agca. 2011. Counting lidar points to characterize forest canopies and machine learning. *Remote Sensing of Environment* 115(8): 1978–1996.

Zimble, D. A., D. L. Evans, G. C. Carlson, R. C. Parker, S. C. Grado, and P. D. Gerard. 2003. Characterizing vertical forest structure using small-footprint airborne lidar. *Remote Sensing of Environment* 87: 171–82.

Chapter 3

Advanced algorithms for land use and cover classification

Mahesh Pal

CONTENTS

Support vector machines (SVMs) and random forest (RF) represent promising developments in machine learning research and have been used widely in the remote-sensing community within past decades. In spite of their popularity for land use and cover classification, SVMs still have some problems. This chapter discusses various issue related to the design of SVMs and compare its performance with RF classifiers in terms of classification accuracy, computation cost, various user-defined parameters, and feature selection. Two data sets—one multispectral and other hyperspectral—are used to compare the performance of SVMs and RF classifiers. Results suggest the usefulness of RF classifiers in comparison to SVMs in terms of classification accuracy, computation cost, and feature selection. Results also suggest that such issues as the requirement of a set of user-defined parameters, choice of a suitable multiclass classification approach, and kernel function need attention while using SVMs for land cover classification.

3.1 INTRODUCTION

Remote sensing includes the interpretation of measured electromagnetic energy reflected from or emitted from a target. Sensors mounted on aircraft or satellite platforms record this electromagnetic radiation in digital form, which is then processed by a computer. Computer-processing applications range from the calibration of the data for the effects of such factors as the changing response of sensors over time to the identification of patterns in multi- and hyperspectral data that relate to ground features. Classification of satellite images is one of the most commonly applied techniques used to process remotely sensed data. Image classification is the process of creating a meaningful digital thematic map from an image data set. The classes shown on the map are derived either from known cover types or by algorithms that search the data for similar pixels. Once data values are known for the distinct cover types in the image, a computer algorithm can be used to divide or segment the image into regions that correspond to each cover type or class. The classified image can be converted to a land use map if the use of each area of land is known. Image classification can be done using a single image data set, multiple images acquired at different times, or image data with additional information such as elevation measurements, soil information, or expert knowledge about the area.

Classification is a method by which labels are attached to pixels in view of their character (Richards, 1993). This character is generally their response in different spectral ranges. Labeling is implemented through pattern classification procedures. The term *pattern* refers to the set of radiance measurements obtained in the various wavebands for a given pixel, and *spectral pattern classification* refers to the family of classification procedures that utilizes this pixel-by-pixel spectral information as the basis for land cover classification. In contrast, spatial pattern recognition involves the classification of image pixels on the basis of their spatial relationship with surrounding pixels. Temporal pattern recognition uses change in spectral reflectance over time as the basis of object identification. A number of techniques exist in the literature for classification of remotely sensed data (Swain and Davis, 1978; Richards, 1993; Schowengerdt, 1997; Mather and Koch, 2011).

The classification process has two main stages. In the first stage, the number and nature of the categories are determined, whereas in the second stage every unknown or unseen element is assigned to one of the categories according to its level of resemblance (or similarity) to the basic pattern. These stages are often called classification and identification, respectively. In the context of remote sensing, the categories could be land cover features or cloud types, and the assignment to one of the categories is carried out by allocating numerical labels, corresponding to the classes, to individual pixels. Hence, in remote sensing, classification basically means determining the class membership of each pixel in an image by comparing the characteristics of that pixel to those of categories known a priori.

The methodology of pattern recognition applied to a particular problem depends on the data, the data model, and the information that one is expecting to find within the data (Bezdek, 1981). A number of methodologies have been developed and employed for image classification using remotely sensed data within the past 30 years. Statistical image classification techniques are ideally suited for data in which the distribution of the data within each of the classes can be assumed to follow a normal distribution in multispectral space (Swain and Davis, 1978). The most commonly used pixel-based statistical classifier known as the maximum likelihood approach is found to have some limitations in resolving interclass confusion if the data used are not normally distributed. As a result, in recent years, and following advances in computer technology, alternative classification strategies have been proposed.

Artificial intelligence–based approaches have been widely used in image classification in the past two decades. The major contribution of the artificial intelligence to pattern analysis has been the study of how domain-specific and heuristic knowledge can be represented and used to control the process of extracting meaningful descriptors and objects from images. Research in this field led to the adoption of artificial neural networks (ANNs), which have been extensively tested for different applications by the remote-sensing community over the past 20 years (Key et al., 1989; Benediktsson et al., 1990; Hepner et al., 1990; Heermann and Khazenie, 1992; Civco, 1993; Paola and Schowengerdt, 1995; Schaale and Furrer, 1995; Tso and Mather, 2001; Mas and Flores, 2008). The popularity of ANN-based classifiers may be attributed to their ability to learn and generalize well with test data. In particular, ANNs make no prior assumptions about the statistics of input data and can construct complex decision boundaries (Lee and Landgrebe, 1997). This property makes ANNs an attractive solution for classification of remotely sensed data. The multilayer feed-forward network is one of the most widely used neural network architecture in remote sensing (Mas and Flores, 2008). The learning process of the feed-forward neural network can be viewed as minimizing a cost function that depends on synaptic weights and biases of the network (Bishop, 1995). The most commonly adopted error function is the mean square error. Among the various learning algorithms, the back-propagation algorithm is one of the most widely used algorithms in remote sensing (Paola and Schowengerdt, 1995; Mas and Flores, 2008). A number of studies have reported that the use of back-propagation ANN classifiers have problems in setting various parameters during training (Foody and Arora, 1997; Wilkinson, 1997; Kavzoglu and Mather, 2003; Mas and Flores, 2008). The choice of network architecture (i.e., number of hidden layers and nodes in each layer, learning rate as well as momentum), weight initialization, and number of iterations required for training are some of the important parameters that affects the learning performance of ANN classifiers. The other shortcoming of the conventional back-propagation learning algorithm is its slow convergence rate, and it even may produce a local minimum.

To deal with the problems of designing an ANN, Huang et al. (2006) proposed a modified neural network approach based on the use of a single hidden layer. This approach requires random initialization of the hidden layer parameters and uses a least-mean square method based on Moore–Penrose's generalized inverse (Rao and Mitra, 1972) to compute the weights of the output layer. The computational requirements of this approach were found to be much lower than that of a classical learning algorithm, such as a back-propagation neural network based on gradient-descent methods. Another advantage of this neural network approach over back-propagation algorithm includes the requirement of one user-defined parameter (i.e., number of hidden nodes in hidden layer) and better generalization performance with remote-sensing data (Pal, 2009).

Classification techniques such as support vector machines (SVM) (Vapnik, 1995) and random forest (RF) classifier have been used for land use and cover classifications within the past decade and found to be working well in terms of classification accuracy in comparison to ANN classifier. Keeping their superior performance for land use and cover classifications in mind, this chapter presents a summary of the results of these two classification algorithms, namely SVM and RF, with multi- and hyperspectral data. Section 3.2 describes both classification algorithms in brief and Section 3.3 discusses the results and usefulness of both classification algorithms in terms of classification accuracy, computation cost, and various issues related to the design of these two classifiers. Section 3.4 discusses their effectiveness in feature selection, and Section 3.5 discusses further developments in the use of both classification algorithms and a summary of this chapter.

3.2 CLASSIFICATION ALGORITHMS

This chapter discusses two classification algorithms: SVM, a kernel-based classification algorithm; and RF, a decision tree–based classification algorithm. A summary of both classification algorithms is provided in following sections.

3.2.1 SVM

SVM is based on statistical learning theory and has the aim of determining the location of decision boundaries that produce the optimal separation of classes (Vapnik, 1995). SVM has been used extensively for land use and cover classifications within the past decades (Gualtieri and Cromp, 1998; Huang et al., 2002; Zhu and Blumberg, 2002; Foody and Mathur, 2004a;

Melgani and Bruzzone, 2004; Pal and Mather, 2004, 2005, 2006; Camps-Valls and Bruzzone, 2005; Mazzoni et al., 2007; Waske and Benediktsson, 2007; Dixon and Candade, 2008; Lizarazo, 2008; Mathur and Foody, 2008a, 2008b; Pal, 2008; Cao et al., 2009; Kavzoglu and Colkesen, 2009; Lardeux et al., 2009; Tuia et al., 2009; Otukei and Blaschke, 2010; Pal and Foody, 2010; Sesnie et al., 2010; Mountrakis, Im, and Ogole, 2011; Yang, 2011). In case of a two-class classification problem in which the classes are linearly separable, the SVM selects from among the infinite number of linear decision boundaries the one that minimizes the generalization error. Thus, the selected decision boundary will be the one that leaves the greatest margin between the two classes, in which the margin is defined as the sum of the distances to the hyperplane from the closest points of the two classes (Vapnik, 1995). The problem of maximizing the margin can be solved using a standard quadratic programming optimization technique. The data points closest to the hyperplane are used to measure the margin and these data points are termed *support vectors*. Only the support vectors are needed to form the classification decision boundaries, and these typically represent a very small proportion of the total training set (Foody and Mathur, 2004b). If regions likely to furnish support vectors can be predicted, then only a small training set, comprising the support vectors, may be acquired for the classification (Foody and Mathur, 2004b; Pal and Foody, forthcoming).

In this chapter a training set of n pixels, represented by $\{x_i, y_i\}$, $i = 1, ..., n$, where $x = [x_1, x_2, ..., x_f]^T \in R^f$ is input vector with f number of input features (wavebands) and $y = [y_1, y_2, ... y_q]^T \in R^q$ is the class vector with q classes used for all classification algorithms, is considered.

For a two-class classification problem (i.e., $y \in \{-1, +1\}$), these training pixels are linearly separable if there exists a weight vector **g** (determining the orientation of a discriminating plane) and a scalar b (determining the offset of the discriminating plane from the origin), such that $y_i (g \cdot x_i + b) - 1 \geq 0$ and the hypothesis space can be defined by the set of functions given by

$$f_{g,b} = sign\,(g \cdot x + b). \tag{3.1}$$

The SVM finds the separating hyperplanes for which the distance between the classes, measured along a line perpendicular to the hyperplane, is maximized. This can be achieved by solving the following constrained optimization problem:

$$\min_{g,b} \frac{1}{2} \|g\|^2. \tag{3.2}$$

If the two classes are not linearly separable, the SVM tries to find the hyperplane that maximizes the margin, while at the same time, minimizing a quantity proportional to the number of misclassification errors. The restriction that all training pixels of a given class lie on the same side of the optimal hyperplane can be relaxed by the introduction of a *slack variable* $\xi_i \geq 0$, and the trade-off between margin and misclassification error can be controlled by a positive user-defined constant C such that $\infty > C > 0$ (Cortes and Vapnik, 1995). Thus, for nonseparable data, Equation 3.2 can be written as follows:

$$\min_{g,\, b,\, \xi_1,\, \cdots\, \xi_k} \left[\frac{1}{2} \|g\|^2 + C \sum_{i=1}^{n} \xi_i \right]. \tag{3.3}$$

SVM also can be extended to handle nonlinear decision surfaces. Boser et al. (1992) proposed a method of projecting the input data onto a high-dimensional feature space through some nonlinear mapping and formulating a linear classification problem in that feature space. A kernel function, which satisfies Mercer's theorem, is used to reduce the computational cost of dealing with high-dimensional feature space (Vapnik, 1995). A kernel function is defined as $K(x_i, x_j) = \Phi(x_i) \cdot \Phi(x_j)$, and with the use of a kernel function, Equation 3.1 becomes:

$$f(x) = sign \left(\sum_i \lambda_i y_i \, K(x_i, x_j) + b \right), \tag{3.4}$$

where λ_i is a Lagrange multiplier.

Further and more detailed discussion on SVM can be found in Vapnik (1995), Cristianini and Shawe-Taylor (2000), and Camps-Valls and Bruzzone (2009).

3.2.2 RF classifier

RF classifier has a combination of tree classifiers in which each classifier is generated using a random vector sampled independently from the input vector, and each tree casts a unit vote for the most popular class to classify an input vector (Breiman, 1999). The RF classifier discussed in this chapter uses randomly selected features or a combination of features at each node to grow a tree. Bagging (or bootstrap sampling), a method to generate a training data set by randomly drawing with replacement n samples (i.e., pixels in the case of land use and cover classification), where n is the size of the original training set (Breiman, 1996), was used for each feature or feature combination selected in the design of RF. Each bootstrap training set consists about 67% of samples from the original training set, and thus about one-third of the samples are left out from every tree grown. Left-out samples are called out-of-bag (out of the bootstrap sample). These out-of-bag samples can be used as a test set for the tree grown on the non-out-of-bag samples. Any sample from the test or out-of-bag data set is classified by taking the most popular voted class from all of the tree predictors in the forest (Breiman, 1999). In general, the design of a decision tree requires the use of an attribute selection measure and a pruning method. There are many approaches to select the attributes used for decision tree induction, and most of these approaches assign a quality measure directly to the attribute. The most frequently used attribute selection measures in the design of decision tree classifier are information gain ratio criterion (Quinlan, 1993) and Gini index (Brieman et. al., 1984). RF classifier uses the Gini index as an attribute selection measure, which measures the impurity of an attribute with respect to the classes. For a given training set T, by selecting one sample at random and saying that it belongs to some class C_i, the Gini index can be written as follows:

$$\sum_{j \neq i} \sum (f(C_i, T) / |T|) (f(C_j, T) / |T|), \tag{3.5}$$

where $f(C_i, T)/|T|$ is the probability that selected pixel belongs to class C_i.

The design of RF consists of growing a tree to the maximum depth on new training data using a combination of features, and these full-grown trees are not pruned. This is one of the major advantages of RF classifier over other decision tree methods like the one proposed by Quinlan (1993). This RF classifier is advantageous because the choice of the pruning methods, and not the attribute selection measures, affect the performance of tree-based classifiers (Mingers, 1989; Pal and Mather, 2003). Breiman (1999) suggested that the generalization error always converges without pruning the tree in the design of a random forest and overfitting is not a problem because of the strong law of large numbers (Feller, 1968).

3.3 EXPERIMENTAL RESULTS

To discuss the function of SVM and RF algorithms for land use and cover classifications, results with two remotely sensed data are reported: (1) a multispectral image acquired on June 19, 2000, by an Enhanced Thematic Mapper Plus (ETM+) sensor over an area located near the town of Littleport in eastern England and (2) a digital airborne imaging spectrometer (DAIS) hyperspectral image acquired on June 29, 2000 over the region of La-Mancha Alta in central Spain (Strobl et al., 1996; Pal and Mather, 2003). The image over eastern England covers an agricultural area that has a classification problem with seven land cover types: wheat, sugar beet, potato, onion, peas, lettuce, and beans. For the multispectral data set, a subimage consisting of six wavebands and covering the area of 307 pixel (columns) by 330 pixel (rows) was used with both classification

Figure 3.1 **(See color insert.)** ETM+ image of the study area Near Littleport, England.

algorithms (see Figure 3.1). For the La-Mancha Alta area, the data set consists of 65 wavebands with an area of 512 pixels by 512 pixels and the classification problem involves the identification of eight classes: wheat, water body, salt lake, hydrophytic vegetation, vineyards, bare soil, pasture lands, and built-up area (see Figure 3.2).

Figure 3.2 DAIS image of the study area over La-Mancha, Spain. (From Pal, M., *International Journal of Remote Sensing* 27, 2877–2894, 2006; Pal, M. and Mather, P. M., Some issue in classification of DAIS hyperspectral data, *International Journal of Remote Sensing* 27, 2895–2916, 2006. With permission.)

Table 3.1 Number of training and test pixels used with multispectral data set

Class name	Number of training pixels	Number of test pixels
Wheat	400	300
Sugar beat	400	300
Potato	400	300
Onion	400	300
Peas	400	300
Lettuce	400	300
Beans	300	237

Reference images used to collect the training and test pixels of both study areas were prepared after field visits. A total of 2,700 and 800 pixels for training and 2,037 and 3,800 pixels for testing acquired by random sampling were used with multispectral data and hyperspectral data set, respectively (Pal and Mather, 2003) (see Tables 3.1 and 3.2).

3.4 CLASSIFICATION WITH SVM

Several user-defined parameters are required to achieve optimal performance by SVM classifier for land use and cover classification. When dealing with multiclass land cover classification problems, selecting a suitable multiclass method, suitable value of regularization parameter (C), type of kernel, and kernel specific parameters need to be selected for the effective implementation of SVM.

3.4.1 Multiclass approaches

SVMs were initially designed to solve binary (two-class) classification problems. Several methods are proposed in the literature to create multi-class classifiers using two-class methods (Allwein et al., 2001; Hsu and Lin,

Table 3.2 Number of training and test pixels used with hyperspectral data set

Class name	Number of training pixels	Number of test pixels
Water	100	500
Dry salt lake	100	500
Hydrophytic vegetation	100	500
Wheat	100	500
Vineyards	100	500
Bare soil	100	500
Pasture lands	100	300
Built-up area	100	500

Table 3.3 Comparison of carious multiclass approaches with ETM+ data

Multiclass approach	Classification accuracy (%)	Training time (sec.)
One against one	87.87	6.40
One against rest	86.55	30.37
Directed acyclic graph	87.63	6.50
Bound-constrained approach	87.29	79.60
Crammer and Singer approach	87.43	20,838.00
ECOC (exhaustive approach)	89.00	48,396.00

2002) and used for remote-sensing data (Pal, 2005a; Mathur and Foody, 2008a). The most common multiclass approaches used with remote-sensing data include one vs. one, one vs. rest, directed acyclic graph (DAG), and error corrected output coding (ECOC). These approaches create many binary classifiers during the training process and combine their output to determine the class label of a test pixel. Westin and Watkins (1998) proposed a multiclass approach by modifying the binary class objective function of SVM that allows for the simultaneous computation of multi-class classification by solving a single optimization problem. A study by Mathur and Foody (2008a) suggested the suitability of this approach for land use and cover classification in comparison with the *one vs. one* or *one vs. rest* approaches for their data set. Most of the studies for land use and cover classification suggest the usefulness of a *one against one* multiclass approach (Foody and Mathur, 2004a; Melgani and Bruzzone, 2004; Pal and Mather, 2004, 2005; Kavzoglu and Colkesen, 2009). Pal (2005a) discussed the results of six multiclass approaches and suggested the suit-ability of the *one against one* approach for the ETM+ data set in terms of classification accuracy and computational cost (see Table 3.3).

3.4.2 Choice of kernel function

The use of several kernels is discussed for classification of various remotely sensed data sets using SVM (Huang et al., 2002; Pal, 2002; Fauvel et al., 2006a; Watanachaturaporn et al., 2008). Results from these studies suggest that it is difficult to use any one kernel function that provides the best gener-alization with different remotely sensed data sets. Studies by Pal (2002) and Watanachaturaporn et al. (2004) have suggested improved performance by radial basis kernel function for land cover classification using hyperspectral and multispectral data sets. Another study by Watanachaturaporn et al. (2008) with a multispectral data set found that a linear kernel function achieves a comparable accuracy to that of a radial basis kernel function and requires less computational cost, thus suggesting the need for detailed analysis with different kernel functions to suit a given data set. Recently, several new kernel functions considering spatial and spectral characteristics of remotely sensed data have been proposed for land use and cover classification (Camps-Valls et al., 2006, 2008; Fauvel et al., 2012). An improved performance was achieved by these kernels in comparison to radial basis function kernel when used with high-resolution remotely sensed data sets. Mercier and Lennon (2003), Sap and Kohram (2008), and Fauvel et al. (2010) proposed new kernel functions for the classification of hyperspectral data and found them to perform well in comparison to other kernel functions.

With a suitable choice of kernel function and multiclass approach, SVM is found to outperform the ANN classifier in terms of classification accuracy and computation cost (Foody and Mathur, 2004a; Pal and Mather, 2005; Watanachaturaporn et al., 2008), but it still requires the selection of two more user-defined parameters that include regularization parameter (C) and a kernel-specific parameter. In most of the studies, a grid search using cross-validation or trial-and-error method was employed to find the optimal value of both parameters (Pal and Mather, 2004; Bazi and Melgani, 2006). Zhuo et al. (2008) proposed the use of a genetic algorithm-based parameter selection approach for the classification of hyperspectral data with SVM and found it to be computationally efficient. A grid search–based cross-validation method becomes computationally demanding with the increasing size of the training samples and the decreasing step size. Anther problem with the grid search approach is the use of training pixels only to find the suitable values of user-defined parameters, which may result in inferior performance with the testing pixels. Results with ETM+ data (Pal and Mather, 2005) suggest the suitability of the SVM classifier in comparison with the ANN classifier (see Table 3.4).

3.4.3 Ensemble of SVM

Within the past decade, several researches proposing the combination of multiple classifiers (called ensemble classifier) have been reported for land

Table 3.4 Classification accuracy and computational cost by SVM and RF with ETM+

Classifier	Classification accuracy (%)	Computational cost (sec.)
ANN	85.10	3480.00
SVM	87.87	12.98
SVM with stump kernel with *coefficient* = 0	86.11	3.88
SVM with exponential kernel with $\gamma = 2$	89.45	3.20
SVM with perceptron kernel with *coefficient* = 0	89.74	1.63
Laplacian kernel with $\gamma = 2$	87.23	3.60
SVM with boosting	85.90	515.14
SVM with bagging	89.45	115.47
Random forest	87.90	18.00

use and cover classifications (Benediktsson et al., 1997; Giacinto and Roli, 1997; Roli et al., 1997; Chan et al., 2001; Briem et al., 2002; Pal and Mather, 2003; Fauvel et al., 2006b; Lee and Ersoy, 2007; Pal, 2007). The resulting classifier is generally found to be more accurate than any of the individual classifiers making up the ensemble. In a recent study, Lin and Li (2008) proposed an ensemble approach with SVM classifier by introducing new kernel functions (stump, Laplacian, exponential, perceptron). This approach combines infinite hypothesis into SVM through the use of these kernel functions. The infinite ensemble approach combines the infinite number of base classification models, which is not possible with traditional ensemble learning approaches, such as bagging (Breiman, 1996) or boosting (Freund and Schapire, 1996). Pal (2008) compared three ensemble approaches using SVM as a base classifier. Results from this study suggest that the infinite ensemble of SVM provides a significant increase in classification accuracy in comparison with the radial basis kernel function–based SVM (see Table 3.4). In comparison with infinite ensemble approach, bagging achieves comparable performance, whereas boosting decreases the performance of SVM with the ETM+ data set.

3.5 CLASSIFICATION WITH RF

RF is found to provide excellent classification performance with remotely sensed data (Gislason et al., 2006; Lawrence et al., 2006; Chan and Paelinckx, 2008; Watts et al., 2009; Stumpf and Kerle, 2011) and most often comparable or better than SVM (Pal, 2005b; Sesnie et al., 2010). Although RF is not as widely used in land use or cover classification as SVM, it has several other advantages over SVM: (a) it can produce a ranked list of all wavebands, which can later be used for feature selection; (b) it can be used to remove outliers; and (c) it can handle missing data as well as categorical and continuous variables simultaneously.

The design of RF classifier requires the selection of two user-defined parameters, namely the number of features (wavebands) used at each node to generate a tree and the total number of trees to be grown. Only selected features are searched through for the best split at each node of the tree, thus, the RF classifier consists of t trees, where t is a user-defined parameter. To classify test pixels, each pixel of the data set is passed down to each of the t trees and assigned a class having the most of the t votes. Studies using multispectral remote-sensing data suggest that changing the values of both user-defined parameters has no significant effect on classification accuracy except increasing computational cost (Pal, 2005b). Selection of both user-defined parameters is much easier in comparison with the parameters required by SVM. Table 3.4 provides classification accuracy and computational cost required by SVM and RF classifiers with multispectral data set.

3.6 FEATURE SELECTION USING SVM AND RF

The availability of increased spectral resolution of airborne and spaceborne sensors has allowed for the collection of hyperspectral images in hundreds of contiguous spectral wavebands. Images acquired by hyperspectral sensors provide detailed information about the spectral signatures of different objects at a fine spectral resolution, thus providing increased capabilities for land cover classification (Chang, 2007). Theoretically, a large number of features should result in better classification, but practical experience suggests that this is not always the case (Benediktsson and Sveinsson, 1997; Tadjudin and Landgrebe, 1999; Pal, 2006). To achieve an acceptable classification accuracy with hyperspectral data, large numbers of training pixels are required, which may be quite difficult and expensive to acquire (Foody and Mathur, 2004b; Chi and Bruzzone, 2005). When small numbers of fixed training pixels are used to classify hyperspectral data, it was found that the addition of more features leads to a reduction in classifier performance (Hughes, 1968). This behavior is known as the Hughes phenomenon. Thus, problems in acquiring sufficient training pixels requires some way to mitigate the problem of redundancy of features (Zhong et al., 2008; Pal, 2009; Pal and Foody, 2010).

Feature selection is one of the ways to alleviate the Hughes phenomenon in the classification of hyperspectral data with limited training pixels. Feature subset selection is the process of identifying and removing as much irrelevant and redundant information from the original data as possible (Liu, 2005). The design of efficient and robust feature selection algorithms has been an important issue addressed by the remote-sensing community and several feature selection algorithms have been proposed and used in past (Jain and Zongker, 1997; Serpico and Bruzzone, 2001; Kavzoglu and Mather, 2002; Korycinski et al., 2003; Bajcsy and Groves, 2004; Su et al., 2008). Studies by Pal (2006), Bazi and Melgani (2006), Zhang and Ma (2009), and Pal and Foody (2010) have suggested that SVM can be an effective tool for feature selection. SVM works by determining the importance of different features using the weight value calculated during the training stage as the ranking criterion (SVM-RFE; Guyon et al., 2002). On the other hand, RF produces a list of ranked features during the classification process itself (Pal, 2006). Classification results with a smaller subset of selected features suggest the suitability of SVM-RFE and RF-based feature selection procedures using hyperspectral data sets (Pal, 2006; Pal and Foody, 2010) (see Table 3.5).

3.7 CONCLUSION AND FUTURE DIRECTIONS

This chapter briefly discusses the use of SVM- and RF-based classification approaches for land use and cover classification. In spite of the extensive

Table 3.5 Results of feature selection with SVM-RFE and RF algorithm

	Classification accuracy (%)	
Feature selection approach	65 Features	20 Top-ranked features
SVM-RFE	91.76	92.55
RF	91.76	91.42

use and popularity of classification and feature selection using remotely sensed data in past decades, the SVM classifier is not problem free. Issues such as the requirement of a set of user-defined parameters, choice of a suitable multiclass classification approach, and kernel function (Pal, 2002; Mountrakis et al., 2011) need to be carefully discussed and identified in advance for land use and cover classification problems. Other problems associated with using SVM include (a) kernel functions have to satisfy Mercer's condition and (b) the nonavailability of probabilistic output. In case of both classification and feature selection, the RF classification algorithm may be an attractive alternative to the SVM because of their ease in use and smaller computation cost, especially with regard to the aforementioned concerns with SVM-based classification. One major advantage in the working of SVM in relation to RF includes the use of a smaller number of training pixels (i.e., support vectors) to create the model. Studies by Foody and Mather (2004b) and Mathur and Foody (2008b) suggested ways to identify these training pixels to reduce the cost of training data collection.

Recently, a Bayesian extension of the SVM, called the relevance vector machine (RVM; Tipping, 2001) received significant attention as an alternative to the SVM for land use and cover classification. Studies by Demir and Ertürk (2007), Foody (2008), and Mianji and Zhang (2011) suggested that RVM can be used effectively for land use and cover classifications and that it has the ability to use non-Mercer kernels, provide probabilistic output, and require fewer number of training pixels to create a model (i.e., relevance vectors) for a given data set, and it does not need to define the parameter C. Some of the issues related to the design of RVM and how to identify training pixels used as relevance vectors are discussed in Pal and Foody (forthcoming).

REFERENCES

Allwein, E. L., R. E. Schapire, and Y. Singer. 2001. Reducing multiclass to binary: a unifying approach for margin classifiers. *Journal of Machine Learning Research* 1: 113–141.

Bajcsy, P., and P. Groves. 2004. Methodology for hyperspectral band selection. *Photogrammetric Engineering and Remote Sensing* 70: 793–802.

Bazi, Y., and F. Melgani. 2006. Toward an optimal SVM classification system for hyperspectral remote sensing images. *IEEE Transactions on Geoscience and Remote Sensing* 44: 3374–3385.

Benediktsson, J. A., and J. R. Sveinsson. 1997. Feature extraction for multisource data classification with artificial neural networks. *International Journal of Remote Sensing* 18: 727–740.

Benediktsson, J. A., J. R. Sveinsson, O. K. Ersoy, and P. H. Swain. 1997. Parallel consensual neural networks. *IEEE Transactions on Neural Networks* 8: 54–65.

Benediktsson, J. A., P. H. Swain, and O. K. Erase. 1990. Neural network approaches versus statistical methods in classification of multisource remote sensing data. *IEEE Transactions on Geoscience and Remote Sensing* 28. 540–551.

Bezdek, J. 1981. *Pattern Recognition with Fuzzy Objective Function Algorithms.* New York: Plenum Press.

Bishop, C. M. 1995. *Neural Networks for Pattern Recognition.* Oxford, UK: Oxford University Press.

Boser, B., I. Guyon, and V. N. Vapnik. 1992. A training algorithm for optimal margin classifiers. *Proceedings of 5th Annual Workshop on Computer Learning Theory*, 144–152. Pittsburgh, PA: Association for Computing Machinery.

Breiman, L. 1996. Bagging predictors. *Machine Learning* 26: 123–140.

Breiman, L. 1999. Random forests: random features. *Technical Report 567.* Berkeley: Statistics Department, University of California–Berkeley. ftp://ftp.stat.berkeley. edu/pub /users/breiman (accessed June 20, 2004).

Breiman, L., J. H. Friedman, R. A. Olshen, and C. J. Stone. 1984. *Classification and Regression Trees.* Monterey, CA: Wadsworth.

Briem, G. J., J. A. Benediktsson, and J. R. Sveinsson. 2002. Multiple classifiers applied to multisource remote sensing data. *IEEE Transactions on Geoscience and Remote Sensing* 40: 2291–2299.

Camps-Valls, G., and L. Bruzzone. 2005. Kernel-based methods for hyperspectral image classification. *IEEE Transactions on Geoscience and Remote Sensing* 43: 1351–1362.

Camps-Valls, G., and L. Bruzzone. 2009. *Kernel methods for Remote Sensing Data Analysis.* Chichester, UK: Wiley & Sons.

Camps-Valls, G., L. Gómez-Chova, J. Muñoz-Marí, J. L. Rojo-Álvarez, and M. Martínez-Ramón. 2008. Kernel-based framework for multitemporal and multisource remote sensing data classification and change detection. *IEEE Transactions on Geoscience and Remote Sensing* 46: 822–1834.

Camps-Valls, G., L. Gomez-Chova, J. Muñoz-Marí, J. Vila-Francés, and J. Calpe-Maravilla. 2006. Composite kernels for hyperspectral image classification. *IEEE Geoscience and Remote Sensing Letters* 3: 93–97.

Cao, X., J. Chen, B. Matsushita, H. Imura, and L. Wang. 2009. An automatic method for burn scar mapping using support vector machines. *International Journal of Remote Sensing* 30: 577–594.

Chan, J. C.-W., and D. Paelinckx. 2008. Evaluation of random forest and Adaboost tree-based ensemble classification and spectral band selection for ecotope mapping using airborne hyperspectral imagery. *Remote Sensing of Environment* 112: 2999–3011.

Chan, J. C.-W., C. Huang, and R. Defries. 2001. Enhanced algorithm performance for land cover classification from remotely sensed data using bagging and boosting. *IEEE Transactions on Geoscience and Remote Sensing* 39: 693–695.

Chang, C.-I. 2007. *Hyperspectral Data Exploitation: Theory and Applications.* Hoboken, NJ: Wiley & Sons.

Chi, M., and L. Bruzzone. 2005. A semilabeled-sample-driven bagging technique for ill-posed classification problems. *IEEE Geosciences and Remote Sensing Letters* 2: 69–73.

Civco, D. L. 1993. Artificial neural networks for land-cover classification and mapping. *International Journal of Geographical Information Systems* 7: 173–186.

Cortes, C., and V. N. Vapnik. 1995. Support vector networks. *Machine Learning* 20: 273–297.

Cristianini, N., and J. Shawe-Taylor. 2000. *An Introduction to Support Vector Machines.* Cambridge: Cambridge University Press.

Demir, B., and S. Erturk. 2007. Hyperspectral image classification using relevance vector machines. *IEEE Geoscience and Remote Sensing Letters* 4: 586–590.

Dixon, B., and N. Candade. 2008. Multispectral landuse classification using neural networks and support vector machines: one or the other, or both? *International Journal of Remote Sensing* 29: 1185–1206.

Fauvel, M., J. Chanussot, and J. A. Benediktsson. 2006a. Evaluation of kernels for multiclass classification of hyperspectral remote sensing data. In *Proceedings of IEEE International Conference on Acoustics, Speech and Signal Processing, May 14–19, Toulouse, France.*

Fauvel, M., J. Chanussot, and J. A. Benediktsson. 2006b. Decision fusion for the classification of urban remote sensing images. *IEEE Transactions on Geoscience and Remote Sensing* 44: 2828–2838.

Fauvel, M., J. Chanussot, and J. A. Benediktsson. 2012. A spatial-spectral kernel based approach for the classification of remote sensing images. *Pattern Recognition* 45: 381–392.

Fauvel, M., A. Villa, J. Chanussot, and J. A. Benediktsson. 2010. Mahalanobis kernel for the classification of hyperspectral images. Presented at the IEEE International Geoscience and Remote Sensing Symposium, July, 25–30, Honolulu, HI.

Feller, W. 1968. *An Introduction to Probability Theory and Its Application,* 3rd ed. New York: Wiley & Sons.

Foody, G. M. 2008. RVM-based multi-class classification of remotely sensed data. *International Journal of Remote Sensing* 29: 1817–1823.

Foody, G. M., and M. K. Arora. 1997. An evaluation of some factors affecting the accuracy of classification by an artificial neural network. *International Journal of Remote Sensing* 18: 799–810.

Foody, G. M., and A. Mathur. 2004a. A relative evaluation of multiclass image classification by support vector machines. *IEEE Transactions on Geoscience and Remote Sensing* 42: 1335–1343.

Foody, G. M., and A. Mathur. 2004b. Toward intelligent training of supervised image classifications: directing training data acquisition for SVM classification. *Remote Sensing of Environment* 93: 107–117.

Freund, Y., and R. Schapire. 1996. Experiments with a new boosting algorithm. In *Machine Learning: Proceedings of the Thirteenth International Conference,* 148–156. San Mateo, CA: Morgan Kaufmann.

Giacinto, G., and F. Roli. 1997. Ensembles of neural networks for soft classification of remote sensing images. In *Proceedings of the European Symposium on Intelligent Techniques*, 166–170. Bari, Italy: European Network for Fuzzy Logic and Uncertainty Modelling in Information Technology.

Gislason, P. O., J. A. Benediktsson, and J. R. Sveinsson. 2006. Random forests for land cover classification. *Pattern Recognition Letters* 27: 294–300.

Gualtieri, J. A., and R. F. Cromp. 1998. Support vector machines for hyperspectral remote sensing classification. In *Proceedings of the 27th AIPR Workshop: Advances in Computer Assisted Recognition, Washington, DC,* 221–232. Bellingham, WA: Society of Photo-Optical Instrumentation Engineers.

Guyon, I., J. Weston, S. Barnhill, and V. Vapnik. 2002. Gene selection for cancer classification using support vector machines. *Machine Learning* 46: 389–422.

Heerman, P. D., and N. Khazenie. 1992. Classification of multispectral remote sensing data using a back propagation neural network. *IEEE Transactions on Geoscience and Remote Sensing* 30: 81–88.

Hepner, G. F., T. Logan, N. Ritter, and N. Bryant. 1990. Artificial neural network classification using a minimal training set: comparison to conventional supervised classification. *Photogrammetric Engineering and Remote Sensing* 56: 469–473.

Hsu, C.-W., and C.-J. Lin. 2002. A comparison of methods for multi-class Support Vector Machines. *IEEE Transaction on Neural Networks* 13: 415–425.

Huang, G.-B., Q.-Y. Zhu, and C.-K. Siew. 2006. Extreme learning machine: theory and applications. *Neurocomputing* 70: 489–501.

Huang, C., L. S. Davis, and J. R. G. Townshend. 2002. An assessment of support vector machines for land cover classification. *International Journal of Remote Sensing* 23: 725–749.

Hughes, G. F. 1968. On the mean accuracy of statistical pattern recognizers. *IEEE Transactions on Information Theory* IT-14: 55–63.

Jain, A., and D. Zongker. 1997. Feature selection: evaluation, application, and small sample performance. *IEEE Transactions on Pattern Analysis and Machine Intelligence* 19: 153–158.

Kavzoglu, T., and I. Colkesen. 2009. A kernel functions analysis for support vector machines for land cover classification. *International Journal of Applied Earth Observation and Geoinformation* 11: 352–359.

Kavzoglu, T., and P. M. Mather. 2002. The role of feature selection in artificial neural network applications. *International Journal of Remote Sensing* 23: 2787–2803.

Kavzoglu, T., and P. M. Mather. 2003. The use of backpropagating artificial neural networks in land cover classification. *International Journal of Remote Sensing* 24: 4907–4938.

Key, J., A. Maslanic, and A. J. Schweiger. 1989. Classification of merged AVHRR and SMMR arctic data with neural network. *Photogrammetric Engineering and Remote Sensing* 55: 1331–1338.

Korycinski, D., M. M. Crawford, J. W. Barnes, and J. Ghosh. 2003. Adaptive feature selection for hyperspectral data analysis using a binary hierarchical classifier and Tabu search. In *Proceedings of the 2003 International Geoscience and Remote Sensing Symposium*, 21–25 July, Toulouse, France, 297–299.

Lardeux, C., P.-L. Frison, C. Tison, J.-C. Souyris, B. Stoll, B. Fruneau, and J.-P. Rudant. 2009. Support vector machine for multifrequency SAR polarimetric data classification. *IEEE Transactions on Geoscience and Remote Sensing* 47: 4143–4152.

Lawrence, R. L., S. D. Wood, and R. L. Sheley. 2006. Mapping invasive plants using hyperspectral imagery and Breiman Cutler classifications (RandomForest). *Remote Sensing of Environment* 100: 356–362.

Lee, C., and D. A. Landgrebe. 1997. Decision boundary feature extraction for neural networks. *IEEE Transactions on Neural Networks* 8: 75–83.

Lee, J., and O. K. Ersoy. 2007. Consensual and hierarchical classification of remotely sensed multispectral images. *IEEE Transactions on Geoscience and Remote Sensing* 45: 2953–2963.

Lin H.-T., and L. Li. 2008. Support vector machinery for infinite ensemble learning. *Journal of Machine Learning Research* 9: 285–312.

Liu, H. 2005. Evolving feature selection. *IEEE Intelligent Systems* 20: 64–76.

Lizarazo, I. 2008. SVM based segmentation and classification of remotely sensed data. *International Journal of Remote Sensing* 29: 7277–7283.

Mas, J. F., and J. J. Flores. 2008. The application of artificial neural networks to the analysis of remotely sensed data. *International Journal of Remote Sensing* 29: 617–663.

Mathur, A., and G. M. Foody. 2008a. Multiclass and binary SVM classification: implications for training and classification users. *IEEE Geoscience and Remote Sensing Letters* 5: 241–245.

Mathur, A., and G. M. Foody. 2008b. Crop classification by support vector machine with intelligently selected training data for an operational application. *International Journal of Remote Sensing* 29: 2227–2240.

Mather, P. M., and M. Koch. 2011. *Computer Processing of Remotely-Sensed Images: An Introduction*, 4th ed. Chichester, UK: Wiley & Sons.

Mazzoni, D., M. J. Garay, R. Davies, and D. Nelson. 2007. An operational MISR pixel classifier using support vector machines. *Remote Sensing of Environment* 107: 149–158.

Melgani, F., and L. Bruzzone. 2004. Classification of hyperspectral remote sensing images with support vector machines. *IEEE Transactions on Geoscience and Remote Sensing* 42: 1778–1790.

Mercier, G., and M. Lennon. 2003. Support vector machines for hyperspectral image classification with spectral-based kernels. In *Proceedings of IEEE Geoscience and Remote Sensing Symposium, Vancouver, BC, Canada* 1: 288–290. Piscataway, NJ: IEEE International.

Mianji, F.A., and Y. Zhang. 2011. Robust hyperspectral classification using relevance vector machine. *IEEE Transactions on Geoscience and Remote Sensing* 49: 2100–2112.

Mingers, J. 1989. An empirical comparison of pruning methods for decision tree induction. *Machine Learning* 4: 227–243.

Mountrakis, G., J. Im, and C. Ogole. 2011. Support vector machines in remote sensing: a review. *ISPRS Journal of Photogrammetry and Remote Sensing* 66: 247–259.

Otukei, J. R., and T. Blaschke. 2010. Land cover change assessment using decision trees, support vector machines and maximum likelihood classification algorithms. *International Journal of Applied Earth Observation and Geoinformation* 12: 627–631.

Pal, M. 2002. *Factors influencing the accuracy of remote sensing classifications: a comparative study*. http://etheses.nottingham.ac.uk/archive/00000314/01/full_phd.pdf.

Pal, M. 2005a. Multiclass approaches for support vector machine based land cover classification. In *8th Annual International Conference, Map India 2005.* http://www.mapindia.org/2005/papers/pdf/54.pdf (accessed December 9, 2011).

Pal, M. 2005b. Random forest classifier for remote sensing classifications. *International Journal of Remote sensing* 26: 217–222.

Pal, M. 2006. Support vector machine-based feature selection for land cover classification: a case study with DAIS hyperspectral data. *International Journal of Remote Sensing* 27: 2877–2894.

Pal, M. 2007. Ensemble learning with decision tree for remote sensing classification. In *Proceedings of World Academy of Science, Engineering and Technology, December 14–46, Bangkok* 36: 258–260. http://www.waset.org/journals/waset/v36/v36-47.pdf.

Pal, M. 2008. Ensemble of support vector machines for land cover classification. *International Journal of Remote Sensing* 29: 3043–3049.

Pal, M. 2009. Margin based feature selection for hyperspectral data. *International Journal of Applied Earth Observations and Geoinformation* 11: 212–220.

Pal, M. 2009. Extreme-learning-machine-based land cover classification. *International Journal of Remote Sensing* 30: 3835–3841.

Pal, M., and G. M. Foody. 2010. Feature selection for classification of hyperspectral data by SVM. *IEEE Transactions on Geoscience and Remote Sensing* 48: 2297–2307.

Pal, M., and G. M. Foody. Forthcoming. Evaluation of SVM, RVM and SMLR for accurate image classification with limited ground data. *IEEE Journal of Selected Topics in Applied Earth Observations and Remote Sensing.*

Pal, M., and P. M. Mather. 2003. An assessment of the effectiveness of decision tree methods for land cover classification. *Remote Sensing of Environment* 86: 554–565.

Pal, M., and P. M. Mather. 2004. Assessment of the effectiveness of support vector machines for hyperspectral data. *Future Generation Computer Systems* 20: 1215–1225.

Pal, M., and P. M. Mather. 2005. Support vector machines for classification in remote sensing. *International Journal of Remote Sensing* 26: 1007–1011.

Pal, M., and P. M. Mather. 2006. Some issue in classification of DAIS hyperspectral data. *International Journal of Remote Sensing* 27: 2895–2916.

Paola, J. D., and R. A. Schowengerdt. 1995. A review and analysis of backpropagation neural networks for classification of remotely sensed multispectral imagery. *International Journal of Remote Sensing* 16: 3033–3058.

Quinlan, J. R. 1993. *C4.5: Programs for Machine Learning.* San Mateo, CA: Morgan Kaufmann.

Rao, C. R., and S. K. Mitra. 1972. *Generalized Inverse of Matrices and Its Applications.* New York: Wiley & Sons.

Richards, J. A. 1993. *Remote Sensing Digital Image Analysis: An Introduction.* Berlin: Springer-Verlag.

Roli, F., G. Giacinto, and G. Vernazza. 1997. Comparison and combination of statistical and neural networks algorithms for remote-sensing image classification. In *Neurocomputation in Remote Sensing Data Analysis,* edited by J. Austin, I. Kanellopoulos, F. Roli, and G. Wilkinson, 117–124. Berlin: Springer-Verlag.

Sap, M. N. M., and M. Kohram. 2008. Spectral angle based kernels for the classification of hyperspectral images using support vector machines. In *Second Asia International Conference on Modelling & Simulation, May 13–15, Kuala Lumpur, Malaysia, 559–563.* Los Alamitos, CA: IEEE Computer Society.

Schaale, M., and R. Furrer. 1995. Land surface classification by neural networks. *International Journal of Remote Sensing* 16: 3003–3031.

Schowengerdt, R. A. 1997. *Remote Sensing Models and Methods for Image Processing.* New York: Academic Press.

Serpico S. B., and Bruzzone, L. 2001. A new search algorithm for feature selection in hyperspectral remote sensing images. *IEEE Transactions on Geoscience and Remote Sensing* 39: 1360–1367.

Sesnie, S. E., B. Finegan, P. E. Gessler, S. Thessler, Z. R. Bendana, and A. M. S. Smith. 2010. The multispectral separability of Costa Rican rainforest types with support vector machines and Random Forest decision trees. *International Journal of Remote Sensing* 31: 2885–2909.

Strobl, P., R. Richter, F. Lehmann, A. Mueller, B. Zhukov, and D. Oertel. 1996. Preprocessing for the airborne imaging spectrometer DAIS 7915. *SPIE Proceedings* 2758: 375–382.

Stumpf, A., and N. Kerle. 2011. Object-oriented mapping of landslides using Random Forests. *Remote Sensing of Environment* 115: 2564–2577.

Su, H., Y. Sheng, and P. Du. 2008. A new band selection algorithm for hyperspectral data based on fractal dimension. *The International Archives of the Photogrammetry, Remote Sensing and Spatial Information Sciences* 37(Beijing): 279–284.

Swain, P., and S. Davis. 1978. *Remote Sensing: The Quantitative Approach.* New York: McGraw-Hill.

Tadjudin, S., and D. A. Landgrebe. 1999. Covariance estimation with limited training samples. *IEEE Transactions on Geoscience and Remote Sensing* 37: 2113–2118.

Tipping, M. E. 2001. Sparse Bayesian learning and the relevance vector machine. *Journal of Machine Learning Research* 1: 211–244.

Tso, B. C. K., and P. M. Mather. 2001. *Classification Methods for Remotely sensed Data.* London: Taylor and Francis.

Tuia, D., F. Pacifici, M. Kanevski, and W. J. Emery. 2009. Classification of very high spatial resolution imagery using mathematical morphology and support vector machines. *IEEE Transactions on Geoscience and Remote Sensing* 47: 3866–3879.

Vapnik, V. N. 1995. *The Nature of Statistical Learning Theory.* New York: Springer-Verlag.

Waske, B., and J. A. Benediktsson. 2007. Fusion of support vector machines for classification of multisensor data. *IEEE Transactions on Geoscience and Remote Sensing* 45: 3858–3866.

Watanachaturaporn, P., M. K. Arora, and P. K. Varshney. 2004. Evaluation of factors affecting support vector machines for hyperspectral classification. Presented at the *American Society for Photogrammetry & Remote Sensing (ASPRS) 2004 Annual Conference,* Denver, CO.

Watanachaturaporn, P., M. K. Arora, and P. K. Varshney. 2008. Multisource classification using support vector machines: an empirical comparison with decision tree and neural network classifiers. *Photogrammetric Engineering & Remote Sensing* 74: 239–246.

Watts, J. D., R. L. Lawrence, P. R. Miller, and C. Montagne. 2009. Monitoring of cropland practices for carbon sequestration purposes in north central Montana by Landsat remote sensing. *Remote Sensing of Environment* 113: 1843–1852.

Westin, J., and C. Watkins. 1998. *Multi-class Support Vector Machines*. Technical Report CSD-TR-98-04. University of London: Royal Holloway.

Wilkinson, G. G. 1997. Open questions in neurocomputing for Earth observation. In *Neuro-Computational in Remote Sensing Data Analysis*, 3–13. New York: Springer-Verlag.

Yan, G., K. Lishan, L. Fujiang, and M. Linlu. 2007. Evolutionary support vector machine and its application in remote sensing imagery classification. In *Geoinformatics 2007: Remotely Sensed Data and Information, Proceedings of the SPIE*, edited by W. Ju and S. Zhao, 6752, 67523D. Bellingham, WA. The International Society for Optical Engineering.

Yang, X. 2011. Parameterizing support vector machines for land cover classification. *Photogrammetric Engineering and Remote Sensing* 77: 27–37.

Zhang, R., and J. Ma. 2009. Feature selection for hyperspectral data based on recursive support vector machines. *International Journal of Remote Sensing* 30: 3669–3677.

Zhong, P., P. Zhang, and R. Wang. 2008. Dynamic learning of SMLR for feature selection and classification of hyperspectral data. *IEEE Geoscience and Remote Sensing Letters* 5: 280–284.

Zhu, G., and D. G. Blumberg. 2002. Classification using ASTER data and SVM algorithms: the case study of Beer Sheva, Israel. *Remote Sensing of Environment* 80: 233–240.

Zhuo, L., J. Zheng, F. Wang, X. Li, B. Ai, and J. Qian. 2008. A genetic algorithm based wrapper feature selection method for classification of hyperspectral images using support vector machine. *The International Archives of the Photogrammetry, Remote Sensing and Spatial Information Sciences* 37: 397–402.

Chapter 4

Global digital elevation model development from satellite remote-sensing data

Dean B. Gesch

CONTENTS

Elevation, as a measurement of topography (one of Earth's most fundamental geophysical properties), is often a key variable in Earth science studies. Because of its importance in the Earth sciences, topography has been widely represented by regularly spaced measurements across the land surface in the form of digital elevation models (DEMs). Elevation data are collected with a number of methods, including surveys conducted at ground level, as well as with remote-sensing techniques from both airborne and spaceborne platforms. Remote-sensing techniques used to collect topographic information include stereo-optical imagery, interferometric synthetic aperture radar (InSAR), radar altimetry, and laser altimetry (or light detection and ranging [LiDAR]). Elevation models derived from these approaches often are used together to develop global DEMs for Earth science studies, and the important differences (and similarities) among elevation measurements from the differing remote-sensing methods must be well understood. Satellite remote sensing is an ideal method for collecting data suitable for the development of global elevation models at a range of resolutions over broad areas. Spaceborne systems have the advantage of acquiring consistent quality data over the globe. Future prospects are excellent for a better understanding of existing data and for the collection of new data suitable for improving global DEMs.

4.1 INTRODUCTION

Topography, or the shape of the land surface, is one of Earth's most fundamental geophysical properties. Topography exerts a primary influence on many physical processes active on the land surface and in the atmosphere. Surface topography is also an expression of subsurface processes in the Earth's crust and upper mantle. Elevation, as a measurement of topography, is often a key variable in Earth science studies. Related topographic parameters, such as slope gradient, aspect, and curvature, are widely used in hydrologic, ecologic, and climatic studies. Because of its importance in the Earth sciences, topography has been widely represented by regularly spaced measurements across the land surface in the form of digital elevation models (DEMs). Because of the widespread use of DEMs in studies of fundamental Earth surface and near-surface processes (Tarolli et al., 2009; Wilson, 2012), as well as in radiometric and geometric processing of remote-sensing imagery, the development and improvement of global DEMs have become important research topics.

Elevation data are collected with a number of methods, including surveys conducted at ground level as well as with remote-sensing techniques from both airborne and spaceborne platforms (Nelson et al., 2009). Consequently, the resulting data vary widely in terms of spatial resolution, accuracy, precision, and overall information content and quality (Gao,

2007). Remote-sensing techniques used to collect topographic information include stereo-optical imagery, interferometric synthetic aperture radar (InSAR), radar altimetry, and laser altimetry (or LiDAR). Each of these methods measures elevation in a fundamentally different way, and the postprocessing required to generate DEMs is distinct for each. Elevation models derived from these approaches, however, often are used together in developing global DEMs for Earth science studies, and the important differences (and similarities) among elevation measurements from the differing remote-sensing methods must be well understood.

4.2 HISTORY OF GLOBAL DEMs

The requirements for global digital topographic data increased markedly in the 1980s with the advent of Earth system science and an increasing research focus on global change. In 1988, the Topographic Science Working Group, convened by the U.S. National Aeronautics and Space Administration (NASA), documented the diverse scientific applications of elevation data and recommended producing a global elevation data set (Topographic Science Working Group, 1988). NASA's Earth Observing System (EOS), which was developed and became operational in the 1990s, further added to the requirements for high-quality global topographic data for both satellite data product generation and interdisciplinary science investigations of physical processes (Gesch, 1994).

4.2.1 ETOPO5

Early large-area (regional, continental, and global) DEMs were mostly recompilations of existing cartographic sources and were not directly based on remote-sensing data. ETOPO5 (National Geophysical Data Center, 1988) was the first widely used global elevation model. ETOPO5 included land and sea floor elevations on a 5-arc-minute latitude-longitude grid. The 5-arc-minute grid spacing equates to about 10 kilometers (km) at the equator. Although it was used widely for studies and applications requiring global elevation data, ETOPO5's limited horizontal resolution and vertical accuracy did not meet the topographic data requirements levied by EOS for emerging global change research (Gesch, 1994).

4.2.2 GTOPO30

To meet the needs of EOS and other global change research programs, in the late 1990s, the U.S. Geological Survey (USGS) developed GTOPO30 (Gesch et al., 1999), a global 1 km DEM. With elevations spaced every 30 arc-seconds on a latitude-longitude grid, GTOPO30 represented a vast improvement in resolution and quality over ETOPO5; consequently, GTOPO30 has been widely used for geometric, radiometric, and atmospheric correction of topographic effects on satellite image products as well as in investigations of land surface and atmospheric processes. Typical of global DEMs, GTOPO30 was compiled from eight different sources of topographic information because no consistent source data with global coverage were available.

The primary sources for GTOPO30, accounting for nearly 80% of the land surface, were Digital Terrain Elevation Data (DTED) and the digital chart of the world (DCW), both products of the U.S. National Geospatial-Intelligence Agency (NGA). The DTED, which has elevations spaced every 3 arc-seconds, was aggregated to the 30-arc-second grid spacing of GTOPO30. DTED was produced with both cartographic and photogrammetric methods. The DCW, a cartographic database derived from 1:1,000,000-scale topographic maps, supplied the source hypsography (contours and spot heights) and hydrography features that were input to the contour-to-grid interpolation procedures to produce elevation grid values (Gesch and Larson, 1998).

Since its release, GTOPO30 has been the global DEM of choice for many large-area applications, as evidenced by numerous citations in the scientific literature. It also has been the primary source data set for other global elevation products, including the Global Land One-km Base Elevation (GLOBE) DEM (Hastings and Dunbar, 1998), and ETOPO2, a 2-arc-minute (approximately 4 km) compilation that included both land and sea floor elevations (Smith and Sandwell, 1997). Despite the prevalent use of GTOPO30, its relatively coarse 1 km resolution did not meet the needs for all large-area applications of elevation data. These unmet requirements for publicly available high-quality higher resolution topographic data led to the ensuing development of improved elevation measurements derived from remote-sensing sources.

4.3 SATELLITE REMOTE-SENSING SOURCES OF ELEVATION DATA

The requirements for global DEMs have been well documented (Topographic Science Working Group, 1988; Gesch, 1994), especially for medium to coarse resolution data over broad areas. Even for applications that require high spatial resolution elevation data, however, the locations where such data are needed can vary widely. Thus, satellite remote sensing is an ideal method for collecting data suitable for the development of global elevation models at a range of resolutions over broad areas. Data collected by airborne platforms are also useful for producing high-quality elevation models; however, spaceborne systems have the advantage of acquiring

Table 4.1 Selected satellite remote-sensing sources of elevation data

Remote-sensing technique	Representative missions/systems	Typical grid spacing of derived DEMs	Coverage
Stereo-optical imagery	SPOT, ASTER PRISM	1-arc-sec. (30 m) 5 m	Near global
Synthetic aperture radar	SRTM TanDEM-X	1-arc-sec. (30 m) 12 m	Near global
Radar altimetry	ERS-1/2, Envisat	30-arc-sec. (1 km)	Near global
Laser altimetry	ICESat/GLAS	Variable	Global (profiles)

consistent quality data over the globe, and, in most cases, the derived DEMs are nonproprietary data sets available to all classes of users with few or no restrictions. The system descriptions in Sections 4.3.1 through 4.3.4 do not include all satellite remote-sensing sources of topographic data, but they do include the primary sources that have been successfully used to produce broad-area coverage of high-quality elevation data. Table 4.1 lists some of the characteristics of these systems.

4.3.1 Stereo-optical imagery

Stereo-optical images are perhaps the most common remote-sensing source used to produce elevation data. The approach is a mature technology and has a long-term well-established heritage. Stereo-aerial photography collected with specialized mapping cameras on airborne platforms has long been used to generate elevation information via photogrammetry for topographic maps. More recently, digital stereo-optical imagery has become a primary data source for production of global DEMs.

4.3.1.1 SPOT

Beginning in 1986 when SPOT stereo images became available, the promise of routinely generating elevation models from satellite digital data became a reality. The first SPOT satellites were not true topographic mapping systems, as the stereo imagery was not collected simultaneously on the same pass; instead, the stereo parallax was obtained by across-track pointing of the sensor on different passes along the same orbital path. The repeat-pass configuration of the early SPOT systems led to limitations imposed by variable solar illumination conditions and base-to-height (B/H) ratios and by cloud cover, as both images of the stereo pair had to be cloud free to derive reliable terrain measurements for the entire satellite scene. Despite these limitations, numerous investigators successfully accomplished DEM generation from repeat-pass across-track SPOT data (Day and Muller, 1988; Welch, 1989; Theodossiou and Dowman, 1990; Sasowsky et al., 1992). With the launch of the SPOT 5 platform in 2002, high-resolution (5 meter [m]) panchromatic stereo images were collected along-track (or inline), thereby adding the advantage of a fixed B/H ratio. The inline single-pass imaging configuration of SPOT 5 also reduces the limitations of cloud cover associated with the multitemporal repeat-pass imaging configuration of the previous SPOT systems. SPOT 5 stereo data now are routinely used to produce DEMs for large areas under the product name Reference3D (Bouillon et al., 2006).

4.3.1.2 ASTER

Japan has developed and launched a number of spaceborne stereo-optical imaging systems, including two prominent systems: ASTER and PRISM. ASTER (Yamaguchi et al., 1998), which began collecting data in 2000, has acquired well over 1 million stereo pairs. ASTER operates aboard NASA's Terra platform and collects 15 m resolution simultaneous stereo images with an inline (nadir and backward-looking) imaging configuration. These 15 m images are routinely processed to derive DEMs with 30 m postings (Hirano et al., 2003; Fujisada et al., 2005; Toutin, 2008). One of the significant advances in DEM generation from ASTER stereo data is the elimination of the requirement for ancillary control points. Initially, external geodetic control points with precise horizontal and vertical coordinates had to be supplied to produce an absolute (georeferenced) DEM; otherwise, the resulting product was a relative DEM. Because of improvements to the locational information from the satellite ephemeris and the sensor model for the ASTER instrument, ASTER DEMs are now adequately georeferenced (with absolute coordinates) without the need for external control points. Such an enhancement facilitates processing of large volumes of data efficiently for the production of large-area DEM coverage.

Even though ASTER is an on-demand system with an average 8% duty cycle, because of its long on-orbit service time, it has acquired a sufficient number of stereo images to create a near-global DEM (Abrams et al., 2010). The ASTER Global Digital Elevation Model (GDEM) was generated from more than 1.2 million ASTER scenes. Released in 2009, the ASTER GDEM covers the global land surface from 83 degrees north to 83 degrees south latitude with 1-arc-second (30 m) elevation data. Because of its broad spatial (near-global) coverage and its grid spacing, the ASTER GDEM represents a significant addition to the collection of global DEMs derived from satellite remote-sensing data. Although some problems have been identified with data artifacts and anomalies, with effective spatial resolution, and with the vertical accuracy not meeting specifications for some areas (Reuter et al., 2009; Guth, 2010; Hirt et al., 2010; Zhao et al., 2010; Chrysoulakis et al., 2011; Slater et al., 2011), the ASTER GDEM has been used in a number of topographic applications (Hugenholtz and Barchyn, 2010; Ni et al., 2010; Nuth and Kaab, 2011; Zouzias et al., 2011; Wang et al., 2012).

4.3.1.3 PRISM

PRISM, launched in 2006 aboard the Japanese ALOS platform, collects inline stereo 2.5 m resolution images with a backward, nadir, and forward-looking camera configuration. PRISM data have been used successfully to produce high-accuracy DEMs (Takaku and Tadono, 2009b) that have been used in a variety of photogrammetric applications (Falala et al., 2009; Krauss et al., 2009; Gonçalves, 2010; Nikolakopoulos et al., 2010). There is an ongoing global data-collection strategy for PRISM, and the data are a valuable source for the development of large-area topographic data coverage (Takaku and Tadono, 2009a).

4.3.1.4 Characteristics common to stereo-optical systems

Stereo-optical imagery acquired from spaceborne platforms will continue to be a primary source of elevation information for Earth science applications. In addition to the systems highlighted thus far, several other high-resolution (2.5 m or better) sensors have been developed, launched, and operated by national governments (Ahmed et al., 2007) and commercial companies (Deilami and Hashim, 2011).

All stereo-optical satellite-imaging systems share common characteristics that affect DEM production and the resulting products. All systems are "first-return" in nature, meaning that the image-derived elevations represent the top surface of the imaged features. In open terrain, these elevations represent ground level, but in areas of dense vegetation or built structures, the elevations represent the height at or near the top of the canopy or the top surface of the structures. An elevation model that includes elevations of vegetation and manmade features is often referred to as a digital surface model (DSM) instead of a DEM, which denotes "bare Earth" elevations (Maune, 2007). Thus, elevation data sets derived from stereo-optical images processed with an image correlation method to measure parallax are more accurately termed DSMs. DSMs can be processed into bare Earth DEMs, but such filtering can be computationally and labor intensive and may require ancillary data.

All stereo-optical images require sufficient contrast for conjugate features to be matched, so some areas will fail to correlate between a stereo image pair. No parallax can be measured in areas of failed correlation, so reliable elevations cannot be derived directly. Areas of water, ice, snow, and homogenous surfaces (such as sand or mud flats) often fail to correlate in image-matching techniques. Elevations for those features must be derived from nearby valid elevations or from ancillary data sources—or else those areas must be left as voids. Areas obscured by shadows caused by low–sun angle illumination conditions are also problematic when deriving elevations from stereo-optical images.

Cloud cover is perhaps the most consistent constraint on the use of stereo-optical imagery for DEM production. The problem with ground-obscuring clouds is greatly reduced with single-pass satellite systems as any clouds present will be at the same location on each half of a stereo pair, but perpetually cloudy locations of the Earth can require many passes to collect suitable data for DEM generation. Thus, for global DEM generation purposes, there are clear advantages in a long-term acquisition strategy and a DEM production approach that uses multiple stereo pairs over a given area, as was done for the ASTER GDEM.

4.3.2 Synthetic aperture radar

Synthetic aperture radar (SAR), or imaging radar, is an active remote-sensing technique that has been extensively employed to produce digital elevation data. Although similar in principle to stereo-optical imagery in that data collected from slightly different perspectives are used to measure positions of conjugate features, the processing involved to derive Earth surface elevations is substantially different. A number of different methods exist for deriving elevations from radar imagery (Toutin and Gray, 2000), but the predominant ones, radargrammetry and interferometry, are highlighted in Sections 4.3.2.1 and 4.3.2.2.

4.3.2.1 Radargrammetry

Radargrammetry grew out of the well-established methods of stereo photogrammetry. In radargrammetry, the magnitude (or intensity) of the radar return forms an image, and the images of corresponding features in a pair are used to derive elevations, much like what is done with stereo photos. With SAR imagery, the parallax is contained within the difference in range from a feature to the two receiving antennas. As with stereo photogrammetry, precise knowledge of the imaging geometry (platform and sensor position and attitude) is exploited to derive elevations of the imaged features on the ground. Because of its heritage from aerial photogrammetry, radargrammetric processing was used on airborne radar data beginning in earnest in the 1970s (Leberl, 1976). Satellite radargrammetry became feasible in the 1980s with the availability of NASA's Shuttle Imaging Radar (SIR) data (Leberl et al., 1985; Leberl et al., 1986) and continued in the 1990s with European Remote-Sensing Satellite (ERS-1) and Radarsat data (Marinelli et al., 1997; Toutin, 1999). Radargrammetry progressed further in the 2000s as new satellite radar sources came online (Li et al., 2006; d'Ozouville et al., 2008; Raggam et al., 2010; Toutin, 2010), and it continues to be an active research topic (Renga and Moccia, 2009).

4.3.2.2 SAR interferometry

SAR interferometry is a technique that takes advantage of the phase (or timing) information in a radar return signal. Interferometry uses two SAR antennas with a known baseline separation to collect stereo radar images. The difference in phase between the radar returns is measured for each pixel. This phase difference is proportional to the difference in path length from each antenna to the surface and is thus influenced by elevation. A phase-difference "image" is generated by registering the two SAR images and then performing cross-correlation. Contours in the phase-difference image correspond to contours of topographic height. Incorporating platform position and attitude information or control points allows the phase differences to be converted to terrain heights to create a DEM. In the literature, interferometric SAR often is referred to as IFSAR, or sometimes as InSAR.

4.3.2.2.1 European Remote-Sensing Satellites

With the availability of ERS-1 satellite data, interferometric SAR for topographic data generation became an active research area in the early 1990s (Corr, 1993; Coulson, 1993). Interferometric processing of ERS-1 data suffered from some of the same constraints as early stereo-optical systems, namely that it was not a true topographic mapping system because the image pairs required for elevation extraction via interferometry were acquired with a temporal delay on repeat passes. Variation in the ERS-1 orbits affected the usability of the images. For successful elevation extraction, the proper baseline and coherence between the images had to exist. Changes in the scattering properties of a target between image acquisitions resulted in degraded coherence, which affected correlation, so measured phase difference became a less reliable source of terrain height information. Despite these limitations, numerous investigators successfully demonstrated DEM generation from ERS-1 interferometry, and their results have been well documented in the scientific literature. With the launch of ERS-2 in 1995 (4 years after the launch of ERS-1) and its subsequent operation in a tandem mode with ERS-1 (1 day repeat-pass orbits), some of the constraints on multitemporal data acquisitions affecting phase coherence were minimized. The tandem configuration of ERS-1/2 also led to numerous successful demonstrations of DEM generation (Rocca et al., 1997; Sansosti et al., 1999).

4.3.2.2.2 Shuttle Radar Topography Mission

NASA's SIR-C/X-SAR instrument flown aboard the space shuttle also facilitated important research and technique development on repeat-pass interferometry (Moreira et al., 1995; Coltelli et al., 1996). This progress was crucial for setting the stage for subsequent operational global DEM production from SAR interferometry.

The Shuttle Radar Topography Mission (SRTM) (Farr et al., 2007), a joint project of NASA and NGA, was the first true satellite remote-sensing mission designed specifically to collect data for the goal of producing a global DEM. SRTM flew aboard the space shuttle *Endeavour* in February 2000 on an 11-day mission and collected interferometric data for 80% of the world's landmass (all land between 60 degrees north and 56 degrees south latitude). SRTM used imaging radar hardware that previously flew on the shuttle in 1994, and that hardware was supplemented by a 60 m mast that included the required second antenna to form the space-based, single-pass interferometer. Kobrick (2006) provided a detailed account of how the mission came to be formulated and the significance of the engineering innovations accomplished in building and successfully operating the system. Those engineering accomplishments alone are sufficient to recognize the significance of SRTM, but the resulting near-global DEM is a noteworthy achievement for the Earth science community. For the first time, the vast majority of the Earth's land surface had been mapped at high resolution with a consistent source in a short period of time.

About 4 years after SRTM flew, the final edited (Slater et al., 2006) SRTM data sets became available to the user community. Since that time, SRTM data have been validated and used extensively in many different applications (Gesch et al., 2006), and the scientific literature is rife with documentation of the successful use of SRTM data products. In SRTM, the many advantages of a global DEM have been and are continuing to be realized, including a consistent high-quality source of publicly available data, and the data set has facilitated advances in the scientific use of elevation models in numerous Earth science fields (Yang et al., 2011).

SRTM data, which were produced for public distribution at a 1-arc-second (30 m) grid spacing over the United States (and its territories) and at a 3-arc-second (90 m) spacing over non-U.S. territories, have been incorporated into several coarse-level global DEMs. SRTM30 (Showstack, 2003) is a 30-arc-second (1 km) DEM that mimics the spatial framework of GTOPO30 but with averaged SRTM data covering all land areas between 60 degrees north and 56 degrees south latitude. In areas of SRTM voids, and in areas beyond the SRTM coverage mask, GTOPO30 elevation values were used as backfill to provide complete global coverage for SRTM30. ETOPO1 (Amante and Eakins, 2009) includes SRTM30 data along with other sources, including ocean bathymetry, to provide global elevation values on a 1-arc-minute (2 km) grid.

4.3.2.2.3 TanDEM-X

The next stage in global DEM generation from SAR interferometry is being accomplished with the TanDEM-X mission (Huber et al., 2010) launched by Germany in 2010. The TanDEM-X mission configuration includes a companion satellite to the TerraSAR-X platform, flying in close formation to provide a single-pass interferometric system. The interferometric baseline, controlled by the physical separation of the satellites, is variable and is being optimized throughout the data acquisition mission phase to collect data that meet strict accuracy specifications over varying terrain conditions. The goal of the mission is to produce global elevation data at a 12 m grid spacing.

As with stereo-optical systems, satellite interferometric SAR systems progressed from research to operational status. The most significant advance in moving to an operational system for global mapping was to fly a single-pass spaceborne configuration, first accomplished by SRTM and now continued by TanDEM-X. Single-pass, across-track radar image collection, along with fixed or controllable baselines optimized for effective interferometry, overcame the problems of degraded coherence associated with repeat-pass multitemporal acquisitions. Also similar to stereo-optical systems, the operational spaceborne IFSAR systems, SRTM and TanDEM-X, are first-return systems. SRTM collected C-band radar data to produce its DEMs, and TanDEM-X uses X-band data, which are both shorter radar wavelengths that do not effectively penetrate foliage to reach the ground in most conditions. Thus, in areas of vegetation canopy, as well as areas with built structures, SRTM and TanDEM-X will contain non-ground-level elevations. A distinct advantage of SAR systems over stereo-optical instruments is that clouds do not hinder SAR imaging of the land surface.

4.3.3 Radar altimetry

Satellite radar altimeters, originally designed to measure sea surface heights and ice sheet topography, have been used to produce gridded elevation data over land areas (Frey and Brenner, 1990). Radar altimeter instruments on the ERS and Envisat platforms have collected data that have facilitated progress in using radar altimetry for large-area topographic applications (Berry, 2000; Dowson and Berry, 2000; Bennett et al., 2005). The derived elevation data typically have moderate to low spatial resolution but have high vertical accuracy. As such, the data are useful for validating or checking the quality of other large-area elevation data sets (Berry et al., 2000a; Hilton et al., 2003; Berry et al., 2007). A known limitation of radar altimetry that had to be overcome for its use for land topography measurement is the errors induced by higher relief (and thus higher slope) terrain (Brenner et al., 1983). Correction of these slope-induced errors has allowed radar altimeter data over land to be incorporated into the ACE (altimeter corrected elevations) (Berry et al., 2000b) global 30-arc-second DEM, and its follow-on ACE2 (Berry et al., 2010), which has data available at 3, 9, and 30 arc-seconds and at 5 arc-minutes.

Radar altimeter data have been used effectively to produce updated high-quality DEMs over Greenland (Bamber et al., 2001) and Antarctica (Bamber et al., 2009; Griggs and Bamber, 2009). In addition to land topography, radar altimeter data have been successfully processed to map ocean floor bathymetry (Smith and Sandwell, 1997; Smith and Sandwell, 2004), and these bathymetry data have been merged with land data from SRTM30 to produce a global elevation data set that covers both land surface and ocean floor (Becker et al., 2009).

4.3.4 Laser altimetry

Laser altimetry, commonly known as LiDAR, is a useful technique for collecting elevation data with centimeter-level vertical accuracy. Laser altimetry is an active remote-sensing technique, somewhat analogous to imaging radar, in which laser energy is emitted toward a target, and the range to the target is determined from the timing of the return energy. Given precise knowledge of the position and attitude of the laser instrument, that range can be used to calculate a highly accurate elevation of the target in real-world coordinates.

Airborne LiDAR is a standard operational technology for the collection of high-resolution, high-accuracy elevation data over local and regional areas. For global elevation applications, satellite laser altimetry is playing an increasingly important role. Spaceborne laser altimetry was successfully demonstrated aboard the space shuttle. The Shuttle Laser Altimeter (SLA) (Garvin et al., 1998) flew twice in the latter half of the 1990s and collected data that proved useful for numerous topographic applications, including validation of existing global DEMs (Gesch, 1998; Harding et al., 1999).

The concepts of satellite laser altimetry for topography successfully explored with the SLA provided the impetus for NASA to develop a dedicated altimetry mission. The ICESat platform carrying the Geoscience Laser Altimeter System (GLAS) (Zwally et al., 2002; Schutz et al., 2005) was launched in 2003 and collected large-footprint LiDAR data (approximately 70 m footprints) over most of the global land surface (between 86 degrees north and south latitude) for more than 6 years. Because of the high vertical accuracy afforded by laser altimetry, a primary mission objective of GLAS was to collect repeat topographic measurements for monitoring of the polar ice sheets, and this was accomplished successfully. Because GLAS is a profiling instrument, it is not possible to have contiguous coverage of its data as a base for a full global DEM; however, because of the convergence of ICESat's orbital paths near the poles, the data have nearly continuous coverage in much of the polar regions. Thus, ICESat laser altimetry has contributed significantly to the development of a new high-quality Antarctic DEM (Bamber et al., 2009; Griggs and Bamber, 2009).

In addition to its primary mission objective of polar monitoring, ICESat has proven to be useful for collecting data for numerous other topographic applications. Because of its high vertical accuracy, the data are extremely useful as reference control for validation of other global DEM source data, particularly SRTM (Bhang et al., 2007; Braun and Fotopoulos, 2007) and

ASTER GDEM (Zhao et al., 2010). Another advantage of the ICESat altimeter is that it collects full waveform return data, which subsequently can be processed to derive elevations for multiple surfaces within a laser footprint, such as canopy top and ground level. This capability has been successfully exploited to characterize the first-return nature of SRTM elevation data in vegetated areas (Carabajal and Harding, 2006).

4.4 QUALITY AND ACCURACY OF SATELLITE REMOTE-SENSING-DERIVED DEMs

As the quality of the remote-sensing sources of elevation information has improved, the overall quality of the derived global DEMs has increased correspondingly. It is important, however, for users to have a quantitative understanding of the accuracy and characteristics of the global DEMs that are available for use in Earth science applications. Fortunately, the primary global DEMs, SRTM and ASTER GDEM, have been extensively validated and the accuracy information has been well documented.

4.4.1 Vertical accuracy

Absolute vertical accuracy is the most common measure of the quality of DEMs and is often expressed as a root mean square error (RMSE) as measured against independent reference ground control. When errors follow a normal distribution with a mean of zero, the RMSE is equivalent to the statistical standard deviation of the measured errors. Absolute vertical accuracy is also commonly expressed as linear error at 90% confidence (LE90) or linear error at 95% confidence (LE95). There are simple conversion factors to convert among the three common error metrics of RMSE, LE90, and LE95 (Maune et al., 2007).

4.4.1.1 SRTM accuracy

The mission specification for SRTM data for absolute vertical accuracy was 16 m (LE90), which equates to an RMSE of 9.73 m. Rodriguez et al., (2006) thoroughly documented the extensive accuracy testing performed for SRTM, which showed that for all continents the absolute vertical accuracy specification was met and exceeded by a significant margin (the values ranged from 5.6 to 9 m LE90). Other investigators have tested SRTM and reported accuracy results that show the data generally meet vertical accuracy specifications (Falorni et al., 2005; Gorokhovich and Voustianiouk, 2006; Shortridge, 2006; Shortridge and Messina, 2011), and it was consistently noted that errors are strongly correlated with relief, slope, some aspect conditions, and presence of forest cover. An accuracy

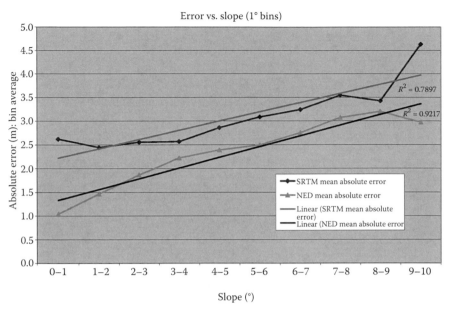

Figure 4.1 **(See color insert.)** Increasing elevation error with increasing slope for SRTM and the USGS National Elevation Dataset, as measured versus more than 13,000 geodetic control points in the conterminous United States.

assessment of SRTM against more than 13,000 high-accuracy geodetic control points in the conterminous United States and comparison with the USGS National Elevation Dataset (Gesch, 2007) exhibited similar findings that error increases as overall terrain slope increases (see Figure 4.1).

4.4.1.2 ASTER GDEM accuracy

The production goal for the ASTER GDEM for absolute vertical accuracy was 20 m (LE95), which equates to an RMSE of 10.2 m. Before public release of the data set, extensive validation of the GDEM was carried out for Japan, the conterminous United States, and a sample of selected other locations around the globe (ASTER GDEM Validation Team, 2009). Results of the validation show that where absolute vertical accuracy testing was performed against reference control points, the GDEM meets or exceeds the accuracy goal. In other areas that were tested, the accuracy goal was not met, a finding documented by other investigators after the 2009 public release of the GDEM (Zhao et al., 2010; Chrysoulakis et al., 2011; Slater et al., 2011). The validation reports for the GDEM also noted the presence of numerous data artifacts and anomalies, which may not have a noticeable effect on overall accuracy statistics, but certainly can be a barrier for the effective use of the GDEM in some applications (Abrams et al., 2010).

Several of the GDEM assessments found an overall negative bias of ASTER GDEM with respect to reference elevation points (see Figure 4.2). The results illustrated in Figure 4.2 are for the accuracy assessment of GDEM against more than 13,000 geodetic control points in the conterminous United States. The overall RMSE was 9.34 m, and when segmented by land cover class, the RMSE ranged from 6.41 to 11.28 m. Ideally, a DEM data set will exhibit a mean error of zero, indicating that there is no systematic bias in the elevations. As can be seen in Figure 4.2, the mean error (grouped for all land cover types) is –3.69 m, which indicates an overall negative bias in GDEM elevation values. The land cover classes on the right side of the figure (pasture, developed open space, cropland, grassland, and shrubland) would have minimal or no vegetation significantly above ground level that would affect elevation measurement by the first-return nature of ASTER's stereo-optical imagery system. Thus, it is likely that the mean errors seen in these classes are more indicative of the magnitude of the true bias present in the GDEM, a number on the order of –5 to –6 m. Some of the land cover classes on the left side of the figure (woody wetlands, forest, and high-intensity developed land) would include significant amounts of trees or buildings that would register elevations above ground level by the ASTER first-return system. The observed overall bias of –3.69 m is likely a reflection of a dampening of the true –5 to –6 m bias because of measured elevations of above-ground-level objects in some areas. This overall negative elevation bias in the GDEM was observed in other assessments (Zhao et al., 2010; Slater et al., 2011).

4.4.2 Horizontal resolution

The grid spacing for both SRTM and ASTER GDEM is 1 arc-second (approximately 30 m). Numerous evaluations of these data sets, however, have shown that the true ground resolution, or topographic information content, for most areas is coarser than the DEM post spacing. Part of the SRTM production process included adaptive smoothing (Farr et al., 2007), which had the effect of decreasing the effective ground resolution in some areas, with more smoothing done in flatter, low-relief areas. Although different methods were used to measure the true resolution of SRTM data, several investigators have come to the same conclusion that the data do not reflect a 30 m resolution even though the elevation values are supplied at that density (Smith and Sandwell, 2003; Guth, 2006; Pierce et al., 2006). Similar reports show that the true resolution of GDEM is much less than its 1-arc-second post spacing implies (ASTER GDEM Validation Team, 2009; Guth, 2010).

4.4.3 Effects of data characteristics on applications

Both SRTM and ASTER GDEM have characteristics that can affect the use of the global DEMs, so users should be aware of these properties to judge

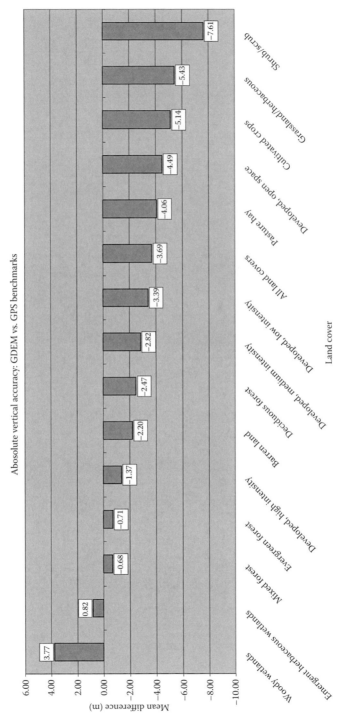

Figure 4.2 Mean error for ASTER GDEM (version 1, 2009 release) as measured versus more than 13,000 geodetic control points in the conterminous United States and segmented by land cover type. Note that most land cover types exhibit a negative elevation bias.

whether they will hamper or preclude effective use in a specific application. Data voids are perhaps the most obvious characteristic that may affect DEM usage. For SRTM, voids were caused by the radar signal interacting with steep terrain (thus causing shadowing, foreshortening, or layover) and by smooth surfaces such as water or sand not reflecting enough energy (Hall et al., 2005). Because many DEM applications require continuous elevation surfaces, filling SRTM voids has been an active development area (Grohman et al., 2006; Crippen et al., 2007; Luedeling et al., 2007; Reuter et al., 2007). For ASTER GDEM, void areas occur where no reliable elevation information could be derived from the available ASTER stereo pairs, most often because of perpetual cloud cover (Abrams et al., 2010).

Another prominent characteristic of SRTM and ASTER GDEM that has been discussed is the first-return nature of the data sets, where non-ground-level elevations are included for areas that contain trees or built structures. Although this might be viewed as an impediment for some DEM uses, especially in low-relief areas (LaLonde et al., 2010), it has been exploited to successfully characterize forest canopy conditions (Kellndorfer et al., 2004; Walker et al., 2007a; Walker et al., 2007b; Yu et al., 2010).

Global DEMs have been used extensively for hydrological applications. The components of DEM quality (vertical accuracy, horizontal resolution, data voids, and non-ground-level elevations) all can have significant effects on the results of such applications. Researchers have documented well the relative advantages and limitations of using SRTM and ASTER DEMs for hydrological modeling in various physical settings and conditions (Lehner et al., 2008; Datta and Schack-Kirchner, 2010; Deng et al., 2011; Karlsson and Arnberg, 2011; Wang et al., 2012).

4.5 A NEW MULTISCALE REMOTE-SENSING-DERIVED GLOBAL DEM

The ready availability of near-global coverage of high-resolution elevation data from a consistent satellite remote-sensing source, namely SRTM, has facilitated the recent development (Danielson and Gesch, 2008) of a notably enhanced global DEM. The new model, called Global Multiresolution Terrain Elevation Data 2010 (GMTED2010) (Danielson and Gesch, 2011), is a multiresolution, multiparameter data set that provides a significant upgrade over GTOPO30 for large-area (regional, continental, global) applications. In contrast to GTOPO30, which provides a "representative" elevation on a 30-arc-second grid, GMTED2010 provides data at three grid spacings: 30 arc-seconds (1 km), 15 arc-seconds (500 m), and 7.5 arc-seconds (250 m) (see Figure 4.3). In addition, GMTED2010 includes seven different products at each resolution generated with the following

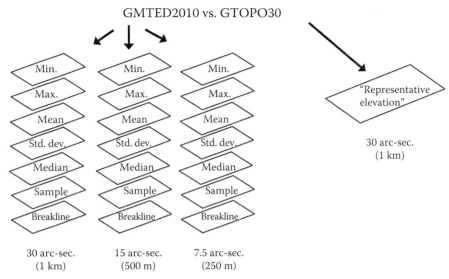

Figure 4.3 Multiple elevation representations provided in the new GMTED2010 global DEM, as compared with the single elevation representation in GTOPO30, the previous best available global DEM.

aggregation methods: minimum elevation, maximum elevation, mean elevation, median elevation, elevation standard deviation, systematic subsampling, and breakline emphasis. In each case, the primary source data (void-filled 1-arc-second SRTM data) were processed within the aggregation window defined by the target output resolution. For areas outside the SRTM coverage mask, other high-quality regional DEMs were processed and merged with the generalized SRTM data to complete global coverage. Details on the source data and processing methods are documented in Danielson and Gesch (2011).

Because GMTED2010 was generated primarily from SRTM data, it retains all of the qualities afforded by the satellite remote-sensing-based data set, namely a consistent source collected over a short time frame with well-documented characteristics. Because it is a multisource global DEM, GMTED2010 has accompanying "spatially referenced metadata," which is a polygonal data set that includes footprints of all the source data. This data set allows any individual grid value to be traced back to its source, and all of the source data characteristics are stored as polygon attributes. GMTED2010 has been subject to a thorough accuracy assessment (Danielson and Gesch, 2011). At the 30-arc-second level, the GMTED2010 mean elevation product exhibits an absolute vertical accuracy of 26.72 m RMSE, a notable improvement over an accuracy of 66.09 m RMSE for GTOPO30 as measured against the same reference control points. ICESat laser altimeter data have been used to assess the Australian portion of

GMTED2010, with results showing even better accuracy than the global evaluation versus control points (Carabajal et al., 2011).

4.6 CONCLUSION AND FUTURE PROSPECTS

Satellite remote sensing of the Earth's topography has been an active science topic for the past several decades, with activities spanning the research to operations continuum. Satellite sources of terrain information are the reason numerous global DEMs exist in the 21st century. These DEMs have now been improved with ancillary data layers that include information important to the user, such as height error, source data, and processing parameters. Future prospects are excellent for a better understanding of existing data and for collection of new data. Even though SRTM flew more than 10 years ago, many studies of its accuracies and characteristics are still being published. Evaluations and applications of the ASTER GDEM are also an active research area. Newer investigations are making synergistic use of several types of remote-sensing-derived elevation to extract more information than could be attained from any one of the sources alone (Crippen, 2010; Smith and Berry, 2011). New global elevation data sets include a second version of the ASTER GDEM (released in October 2011) and, on the horizon, the TanDEM-X data products. This progress will continue to advance topographic science, a critical component of Earth system science.

NOTE

Any use of trade, product, or firm names is for descriptive purposes only and does not imply endorsement by the U.S. government.

REFERENCES

Abrams, M., B. Bailey, H. Tsu, and M. Hato. 2010. The ASTER global DEM. *Photogrammetric Engineering and Remote Sensing* 76(4): 344–348.

Ahmed, N., A. Mahtab, R. Agrawal, P. Jayaprasad, S. K. Pathan, Ajai, D. K. Singh, and A. K. Singh. 2007. Extraction and validation of Cartosat-1 DEM. *Journal of the Indian Society of Remote Sensing* 35(2): 121–127.

Amante, C., and B. W. Eakins. 2009. ETOPO1 1 arc-minute global relief model: procedures, data sources and analysis. *NOAA Technical Memorandum NESDIS NGDC-24*, 19.

ASTER GDEM Validation Team. 2009. ASTER global DEM validation summary report. https://lpdaac.usgs.gov/lpdaac/products/aster_products_table/routine/global_digital_elevation_model/v1/astgtm (accessed July 6, 2011).

Bamber, J. L., S. Ekholm, and W. B. Krabill. 2001. A new, high-resolution digital elevation model of Greenland fully validated with airborne laser altimeter data. *Journal of Geophysical Research B: Solid Earth* 106(B4): 6733–6745.

Bamber, J. L., J. L. Gomez-Dans, and J. A. Griggs. 2009. A new 1 km digital elevation model of the Antarctic derived from combined satellite radar and laser data-part 1: data and methods. *Cryosphere* 3(1): 101–111.

Becker, J. J., D. T. Sandwell, W. H. F. Smith, J. Braud, B. Binder, J. Depner, D. Fabre, et al. 2009. Global bathymetry and elevation data at 30 arc seconds resolution: SRTM30_PLUS. *Marine Geodesy* 32(4): 355–371.

Bennett, J., J. D. Garlick, and P. A. M. Berry. 2005. Contribution of Envisat RA-2 to global SRTM evaluation with altimeter data. *European Space Agency (Special Publication)* ESA SP(572): 987–990.

Berry, P. A. M. 2000. Topography from land radar altimeter data: possibilities and restrictions. *Physics and Chemistry of the Earth, Part A: Solid Earth and Geodesy* 25(1): 81–88.

Berry, P. A. M., J. D. Garlick, and R. G. Smith. 2007. Near-global validation of the SRTM DEM using satellite radar altimetry. *Remote Sensing of Environment* 106(1): 17–27.

Berry, P. A. M., J. E. Hoogerboord, and R. A. Pinnock. 2000a. Identification of common error signatures in global digital elevation models based on satellite altimeter reference data. *Physics and Chemistry of the Earth, Part A: Solid Earth and Geodesy* 25(1): 95–99.

Berry, P. A. M., R. A. Pinnock, R. D. Hilton, and C. P. D. Johnson. 2000b. ACE: A new global digital elevation model incorporating satellite altimeter derived heights. *European Space Agency, (Special Publication)* ESA SP(461): 783–791.

Berry, P. A. M., R. G. Smith, and J. Benveniste. 2010. ACE2: the new global digital elevation model. *Gravity, Geoid and Earth Observation: International Association of Geodesy Symposia, 2010* 135(3): 231–238.

Bhang, K. J., F. W. Schwartz, and A. Braun. 2007. Verification of the vertical error in C-band SRTM DEM using ICESat and Landsat-7, Otter Tail County, MN. *IEEE Transactions on Geoscience and Remote Sensing* 45(1): 36–44.

Bouillon, A., M. Bernard, P. Gigord, A. Orsoni, V. Rudowski, and A. Baudoin. 2006. SPOT 5 HRS geometric performances: using block adjustment as a key issue to improve quality of DEM generation. *ISPRS Journal of Photogrammetry and Remote Sensing* 60(3): 134–146.

Braun, A., and G. Fotopoulos. 2007. Assessment of SRTM, ICESat, and survey control monument elevations in Canada. *Photogrammetric Engineering and Remote Sensing* 73(12): 1333–1342.

Brenner, A. C., R. A. Bindschadler, R. H. Thomas, and H. J. Zwally. 1983. Slope-induced errors in radar altimetry over continental ice sheets. *Journal of Geophysical Research* 88(C3): 1617–1623.

Carabajal, C. C., and D. J. Harding. 2006. SRTM C-band and ICESat laser altimetry elevation comparisons as a function of tree cover and relief. *Photogrammetric Engineering and Remote Sensing* 72(3): 287–298.

Carabajal, C. C., D. J. Harding, J.-P. Boy, J. J. Danielson, D. B. Gesch, and V. P. Suchdeo. 2011. Evaluation of the Global Multi-resolution terrain elevation data 2010 (GMTED2010) using ICESat geodetic control. In *International Symposium on Lidar and Radar Mapping: Technologies & Applications, May 26–29, 2011,*

Nanjing, China, Proceedings of SPIE Vol. 8286, edited by J. Li, paper number 82861Y. Bellingham, WA: Society of Photo-Optical Instrumentation Engineers.

Chrysoulakis, N., M. Abrams, Y. Kamarianakis, and M. Stanislawski. 2011. Validation of ASTER GDEM for the area of Greece. Photogrammetric Engineering and Remote Sensing 77(2): 157–165.

Coltelli, M., G. Fornaro, G. Franceschetti, R. Lanari, M. Migliaccio, J. R. Moreira, K. P. Papathanassiou, G. Puglisi, D. Riccio, and M. Schwäbisch. 1996. SIR-C/X-SAR multifrequency multipass interferometry: a new tool for geological interpretation. Journal of Geophysical Research E: Planets 101(E10): 23127–23148.

Corr, D. 1993. Production of DEMs from ERS-1 radar data. Mapping Awareness 7(10): 18–22.

Coulson, S. N. 1993. SAR interferometry with ERS-1. Earth Observation Quarterly 40: 20–23.

Crippen, R. E. 2010. Global topographical exploration and analysis with the SRTM and ASTER elevation models. In Elevation Models for Geoscience, edited by C. Fleming, S. H. Marsh, and J. R. A. Giles, 5–15. London: Geological Society, Special Publications 345.

Crippen, R. E., S. J. Hook, and E. J. Fielding. 2007. Nighttime ASTER thermal imagery as an elevation surrogate for filling SRTM DEM voids. Geophysical Research Letters 34(1): article no. L01302.

Danielson, J. J., and D. B. Gesch. 2008. An enhanced global elevation model generalized from multiple higher resolution source datasets. The International Archives of the Photogrammetry, Remote Sensing, and Spatial Information Sciences XXXVII (Part B4, Beijing 2008): 1857–1863.

Danielson, J. J., and D. B. Gesch. 2011. Global multi-resolution terrain elevation data 2010 (GMTED2010). U.S. Geological Survey Open-File Report 2011-1073, 26.

Datta, P., and H. Schack-Kirchner. 2010. Erosion relevant topographical parameters derived from different DEMs: a comparative study from the Indian Lesser Himalayas. Remote Sensing 2(8): 1941–1961.

Day, T., and J. P. Muller. 1988. Quality assessment of digital elevation models produced by automatic stereo matchers from SPOT image pairs. International Archives of Photogrammetry and Remote Sensing 27: 148–159.

Deilami, K., and M. Hashim. 2011. Very high resolution optical satellites for DEM generation: a review. European Journal of Scientific Research 49(4): 542–554.

Deng, L., Y. Liang, and C. Zhang. 2011. Research on acquisition methods of high-precision DEM for distributed hydrological model. IFIP Advances in Information and Communication Technology 346 AICT(PART 3): 390–402.

Dowson, M., and P. A. M. Berry. 2000. Envisat RA-2: potential for land topographic mapping. European Space Agency, (Special Publication) ESA SP(461): 957–965.

d'Ozouville, N., B. Deffontaines, J. Benveniste, U. Wegmüller, S. Violette, and G. de Marsily. 2008. DEM generation using ASAR (ENVISAT) for addressing the lack of freshwater ecosystems management, Santa Cruz Island, Galapagos. Remote Sensing of Environment 112(11): 4131–4147.

Falala, L., P. Favé, and P. Gigord. 2009. Analysis of ALOS-PRISM image geometry for advanced photogrammetric applications. International Geoscience and Remote Sensing Symposium (IGARSS) 5: V188–V191.

Falorni, G., V. Teles, E. R. Vivoni, R. L. Bras, and K. S. Amaratunga. 2005. Analysis and characterization of the vertical accuracy of digital elevation models from the Shuttle Radar Topography Mission. *Journal of Geophysical Research F: Earth Surface* 110(2): article no. F02005.

Farr, T. G., P. A. Rosen, E. Caro, R. Crippen, R. Duren, S. Hensley, M. Kobrick, et al. 2007. The Shuttle Radar Topography Mission. *Reviews of Geophysics* 45(RG2004). doi:10.1029/2005RG000183.

Frey, H., and A. C. Brenner. 1990. Australian topography from SEASAT overland altimetry. *Geophysical Research Letters* 17(10): 1533–1536.

Fujisada, H., G. B. Bailey, G. G. Kelly, S. Hara, and M. J. Abrams. 2005. ASTER DEM performance. *IEEE Transactions on Geoscience and Remote Sensing* 43(12): 2707–2714.

Gao, J. 2007. Towards accurate determination of surface height using modern geoinformatic methods: possibilities and limitations. *Progress in Physical Geography* 31(6): 591–605.

Garvin, J., J. Bufton, J. Blair, D. Harding, S. Luthcke, J. Frawley, and D. Rowlands. 1998. Observations of the earth's topography from the Shuttle Laser Altimeter (SLA): laser-pulse echo-recovery measurements of terrestrial surfaces. *Physics and Chemistry of the Earth* 23(9–10): 1053–1068.

Gesch, D. B. 1994. Topographic data requirements for EOS global change research. *U.S. Geological Survey Open-File Report 94-626*, 60.

Gesch, D. B. 1998. Accuracy assessment of a global elevation model using Shuttle Laser Altimeter data. In *Proceedings, 1998 IEEE International Geoscience and Remote Sensing Symposium, Seattle, Washington, July 6–10, 1998* (CD-ROM). Piscataway, NJ: Institute of Electrical and Electronics Engineers, Inc.

Gesch, D. B. 2007. The National Elevation Dataset. In *Digital Elevation Model Technologies and Applications: The DEM Users Manual*, 2nd ed., edited by D. Maune, 99–118. Bethesda, MD: American Society for Photogrammetry and Remote Sensing.

Gesch, D. B., and K. S. Larson. 1998. Techniques for development of global 1-kilometer digital elevation models. In *Proceedings, Pecora Thirteen, Human Interactions with the Environment—Perspectives from Space, Sioux Falls, South Dakota, August 20–22, 1996*, 568–572 (CD-ROM). Bethesda, MD: American Society of Photogrammetry and Remote Sensing.

Gesch, D. B., J. P. Muller, and T. G. Farr. 2006. Special issue foreword: the Shuttle Radar Topography Mission: data validation and applications. *Photogrammetric Engineering and Remote Sensing* 72(3): 233–235.

Gesch, D. B., K. L. Verdin, and S. K. Greenlee. 1999. New land surface digital elevation model covers the earth. *Eos, Transactions, American Geophysical Union* 80(6): 69–70.

Gonçalves, J. A. 2010. Automatic image orientation and DSM extraction from ALOS-PRISM triplet images. *International Geoscience and Remote Sensing Symposium (IGARSS)*: 2295–2298.

Gorokhovich, Y., and A. Voustianiouk. 2006. Accuracy assessment of the processed SRTM-based elevation data by CGIAR using field data from USA and Thailand and its relation to the terrain characteristics. *Remote Sensing of Environment* 104(4): 409–415.

Griggs, J. A., and J. L. Bamber. 2009. A new 1 km digital elevation model of Antarctica derived from combined radar and laser data part 2: validation and error estimates. *Cryosphere* 3(1): 113–123.

Grohman, G., G. Kroenung, and J. Strebeck. 2006. Filling SRTM voids: the delta surface fill method. *Photogrammetric Engineering and Remote Sensing* 72(3): 213–216.

Guth, P. L. 2006. Geomorphometry from SRTM: comparison to NED. *Photogrammetric Engineering and Remote Sensing* 72(3): 269–277.

Guth, P. L. 2010. Geomorphometric comparison of ASTER GDEM and SRTM. *Conference Proceedings Special Joint Symposium of ISPRS Technical Commission IV and AutoCarto 2010 in conjunction with ASPRS/CaGIS 2010 Specialty Conference, Orlando, Florida, November 15–19, 2010*, CD-ROM. Bethesda, MD: American Society for Photogrammetry and Remote Sensing.

Hall, O., G. Falorni, and R. L. Bras. 2005. Characterization and quantification of data voids in the Shuttle Radar Topography Mission data. *IEEE Geoscience and Remote Sensing Letters* 2(2): 177–181.

Harding, D. J., D. B. Gesch, C. C. Carabajal, and S. B. Luthcke. 1999. Application of the Shuttle Laser Altimeter in an accuracy assessment of GTOPO30, a global 1-kilometer digital elevation model. *International Archives of Photogrammetry and Remote Sensing* 32: 81–85.

Hastings, D. A., and P. K. Dunbar. 1998. Development and assessment of the Global Land One-km Base Elevation digital elevation model (GLOBE). *International Archives of Photogrammetry and Remote Sensing* 32(4): 218–221.

Hilton, R. D., W. E. Featherstone, P. A. M. Berry, C. P. D. Johnson, and J. F. Kirby. 2003. Comparison of digital elevation models over Australia and external validation using ERS-1 satellite radar altimetry. *Australian Journal of Earth Sciences* 50(2): 157–168.

Hirano, A., R. Welch, and H. Lang, 2003. Mapping from ASTER stereo image data: DEM validation and accuracy assessment. *ISPRS Journal of Photogrammetry and Remote Sensing* 57(5-6): 356–370.

Hirt, C., M. S. Filmer, and W. E. Featherstone. 2010. Comparison and validation of the recent freely available ASTER-GDEM ver1, SRTM ver4.1 and GEODATA DEM-9s ver3 digital elevation models over Australia. *Australian Journal of Earth Sciences* 57(3): 337–347.

Huber, S., M. Younis, and G. Krieger. 2010. The TanDEM-X mission: overview and interferometric performance. *International Journal of Microwave and Wireless Technologies* 2(3–4): 379–389.

Hugenholtz, C. H., and T. E. Barchyn. 2010. Spatial analysis of sand dunes with a new global topographic dataset: new approaches and opportunities. *Earth Surface Processes and Landforms* 35(8): 986–992.

Karlsson, J. M., and W. Arnberg. 2011. Quality analysis of SRTM and HYDRO1K: a case study of flood inundation in Mozambique. *International Journal of Remote Sensing* 32(1): 267–285.

Kellndorfer, J., W. Walker, L. Pierce, C. Dobson, J. A. Fites, C. Hunsaker, J. Vona, and M. Clutter. 2004. Vegetation height estimation from Shuttle Radar Topography Mission and National Elevation Datasets. *Remote Sensing of Environment* 93(3): 339–358.

Kobrick, M. 2006. On the toes of giants: how SRTM was born. *Photogrammetric Engineering and Remote Sensing* 72(3): 206–210.

Krauss, T., M. Schneider, and P. Reinartz, P. 2009. Orthorectification and DSM generation with ALOS-PRISM data in urban areas. *International Geoscience and Remote Sensing Symposium (IGARSS)* 5: V33–V36.

LaLonde, T., A. Shortridge, and J. Messina. 2010. The influence of land cover on Shuttle Radar Topography Mission (SRTM) elevations in low-relief areas. *Transactions in GIS* 14(4): 461–479.

Leberl, F. 1976. Imaging radar applications to mapping and charting. *Photogrammetria* 32(3): 75–100.

Leberl, F. W., G. Domik, and M. Kobrick. 1985. Mapping with aircraft and satellite radar images. *Photogrammetric Record* 11(66): 647–665.

Leberl, F., G. Domik, J. Raggam, J. Cimino, and M. Kobrick. 1986. Multiple incidence angle SIR-B experiment over Argentina: stereo-radargrammetric analysis. *IEEE Transactions on Geoscience and Remote Sensing* GE-24(4): 482–491.

Lehner, B., K. Verdin, and J. Jarvis. 2008. New global hydrography derived from spaceborne elevation data. *Eos, Transactions, American Geophysical Union* 89(10): 93–94.

Li, Z., G. Liu, and X. Ding. 2006. Exploring the generation of digital elevation models from same-side ERS SAR images: topographic and temporal effects. *Photogrammetric Record* 21(114): 124–140.

Luedeling, E., S. Siebert, and A. Buerkert. 2007. Filling the voids in the SRTM elevation model – a TIN-based delta surface approach. *ISPRS Journal of Photogrammetry and Remote Sensing* 62(4): 283–294.

Marinelli, L., T. Toutin, and I. Dowman. 1997. DTM Generation by radargrammetry: status and prospects. *Bulletin—Societe Francaise de Photogrammetrie et de Teledetection* 148: 89–95.

Maune, D., ed. 2007. *Digital Elevation Model Technologies and Applications: The DEM Users Manual*, 2nd ed. Bethesda, MD: American Society for Photogrammetry and Remote Sensing.

Maune, D. F., J. B. Maitra, and E. J. McKay. 2007. Accuracy standards & guidelines. In *Digital Elevation Model Technologies and Applications: The DEM Users Manual*, 2nd ed., edited by D. Maune, 65–97. Bethesda, MD: American Society for Photogrammetry and Remote Sensing.

Moreira, J., M. Schwaebisch, G. Fornaro, R. Lanari, R. Bamler, D. Just, U. Steinbrecher, H. Breit, M. Eineder, G. Franceschetti, D. Geudtner, and H. Rinkel. 1995. X-SAR interferometry: first results. *IEEE Transactions on Geoscience and Remote Sensing* 33(4): 950–956.

National Geophysical Data Center. 1988. *Data announcement 88-MGG-02, Digital relief of the surface of the Earth*. Boulder, CO: National Oceanic and Atmospheric Administration.

Nelson, A., H. I. Reuter, and P. Gessler. 2009. DEM production methods and sources. In *Geomorphometry: Concepts, Software, Applications*, edited by T. Hengl, and H. I. Reuter, 65–85. Amsterdam: Elsevier.

Ni, W., Z. Guo, G. Sun, and H. Chi. 2010. Investigation of forest height retrieval using SRTM-DEM and ASTER-GDEM. *International Geoscience and Remote Sensing Symposium (IGARSS)*: 2111–2114.

Nikolakopoulos, K.G., A. D. Vaiopoulos, and P. I. Tsombos. 2010. DSM from ALOS data, the case of Andritsena, Greece. *Proceedings of SPIE – The International Society for Optical Engineering* 7831: article no. 78310K.

Nuth, C., and A. Kaab. 2011. Co-registration and bias corrections of satellite elevation data sets for quantifying glacier thickness change. *Cryosphere* 5(1): 271–290.

Pierce, L., J. Kellndorfer, W. Walker, and O. Barros. 2006. Evaluation of the horizontal resolution of SRTM elevation data. *Photogrammetric Engineering and Remote Sensing* 72(11): 1235–1244.

Raggam, H., K. Gutjahr, R. Perko, and M. Schardt. 2010. Assessment of the stereo-radargrammetric mapping potential of TerraSAR-X multibeam spotlight data. *IEEE Transactions on Geoscience and Remote Sensing* 48(2): 971–977.

Renga, A., and A. Moccia. 2009. Performance of stereoradargrammetric methods applied to spaceborne monostatic-bistatic synthetic aperture radar. *IEEE Transactions on Geoscience and Remote Sensing* 47(2): 544–560.

Reuter, H.I., A. Nelson, and A. Jarvis. 2007. An evaluation of void-filling interpolation methods for SRTM data. *International Journal of Geographical Information Science* 21(9): 983–1008.

Reuter, H.I., A. Nelson, P. Strobl, W. Mehl, and A. Jarvis. 2009. A first assessment of ASTER GDEM tiles for absolute accuracy, relative accuracy and terrain parameters. *International Geoscience and Remote Sensing Symposium (IGARSS)* 5: V240–V243.

Rocca, F., C. Prati, and A. Ferretti. 1997. An overview of ERS-SAR interferometry. *European Space Agency, (Special Publication)* ESA SP (414 Part 1): xxvii–xxxvi.

Rodríguez, E., C. S. Morris, and J. E. Belz. 2006. A global assessment of the SRTM performance. *Photogrammetric Engineering and Remote Sensing* 72(3): 249–260.

Sansosti, E., R. Lanari, G. Fornaro, G. Franceschetti, M. Tesauro, G. Puglisi, and M. Coltelli. 1999. Digital elevation model generation using ascending and descending ERS-1/ERS-2 tandem data. *International Journal of Remote Sensing* 20(8): 1527–1547.

Sasowsky, K. C., G. W. Petersen, and B. M. Evans. 1992. Accuracy of SPOT digital elevation model and derivatives: utility for Alaska's North Slope. *Photogrammetric Engineering and Remote Sensing* 58: 815–824.

Schutz, B. E., H. J. Zwally, C. A. Shuman, D. Hancock, and J. P. DiMarzio. 2005. Overview of the ICESat mission. *Geophysical Research Letters* 32(21): 1–4.

Shortridge, A. 2006. Shuttle Radar Topography Mission elevation data error and its relationship to land cover. *Cartography and Geographic Information Science* 33(1): 65–75.

Shortridge, A., and J. Messina. 2011. Spatial structure and landscape associations of SRTM error. *Remote Sensing of Environment* 115(6): 1576–1587.

Showstack, R. 2003. Digital elevation maps produce sharper image of Earth's topography. *Eos, Transactions, American Geophysical Union* 84(37): 363.

Slater, J. A., G. Garvey, C. Johnston, J. Haase, B. Heady, G. Kroenung, and J. Little. 2006. The SRTM data "finishing" process and products. *Photogrammetric Engineering and Remote Sensing* 72(3): 237–247.

Slater, J. A., B. Heady, G. Kroenung, W. Curtis, J. Haase, D. Hoegemann, C. Shockley, and K. Tracy. 2011. Global assessment of the new ASTER global digital elevation model. *Photogrammetric Engineering and Remote Sensing* 77(4): 335–349.

Smith, B., and D. Sandwell. 2003. Accuracy and resolution of Shuttle Radar Topography Mission data. *Geophysical Research Letters* 30(9): 20-1–20-4.

Smith, R. G., and P. A. M. Berry. 2011. Evaluation of the differences between the SRTM and satellite radar altimetry height measurements and the approach taken for the ACE2 GDEM in areas of large disagreement. *Journal of Environmental Monitoring* 13: 1646–1652.

Smith, W. H. F., and D. T. Sandwell. 1997. Global sea floor topography from satellite altimetry and ship depth soundings. *Science* 277(5334): 1956–1962.

Smith, W. H. F., and D. T. Sandwell. 2004. Conventional bathymetry, bathymetry from space, and geodetic altimetry. *Oceanography* 17(1): 8–23.

Takaku, J., and T. Tadono. 2009a. High resolution DSM generation from ALOS PRISM - Status updates on over three year operations. *International Geoscience and Remote Sensing Symposium (IGARSS)* 3: III769–III772.

Takaku, J., and T. Tadono. 2009b. PRISM on-orbit geometric calibration and DSM performance. *IEEE Transactions on Geoscience and Remote Sensing* 47(12): 4060–4073.

Tarolli, P., J. R. Arrowsmith, and E. R. Vivoni. 2009. Understanding earth surface processes from remotely sensed digital terrain models. *Geomorphology* 113(1–2): 1–3.

Theodossiou, E., and I. Dowman. 1990. Heighting accuracy of SPOT. *Photogrammetric Engineering and Remote Sensing* 56: 1643–1649.

Topographic Science Working Group. 1988. *Topographic Science Working Group Report to the Land Processes Branch, Earth Science and Applications Division, NASA Headquarters*, 64. Houston, TX: Lunar and Planetary Institute.

Toutin, T. 1999. Error tracking of radargrammetric DEM from RADARSAT images. *IEEE Transactions on Geoscience and Remote Sensing* 37(5): 2227–2238.

Toutin, T. 2008. ASTER DEMs for geomatic and geoscientific applications: a review. *International Journal of Remote Sensing* 29(7): 1855–1875.

Toutin, T. 2010. Impact of Radarsat-2 SAR ultrafine-mode parameters on stereo-radargrammetric DEMs. *IEEE Transactions on Geoscience and Remote Sensing* 48(10): 3816–3823.

Toutin, T., and L. Gray. 2000. State-of-the-art of elevation extraction from satellite SAR data. *ISPRS Journal of Photogrammetry and Remote Sensing* 55(1): 13–33.

Walker, W. S., J. M. Kellndorfer, E. LaPoint, M. Hoppus, and J. Westfall. 2007a. An empirical InSAR-optical fusion approach to mapping vegetation canopy height. *Remote Sensing of Environment* 109(4): 482–499.

Walker, W. S., J. M. Kellndorfer, and L. E. Pierce. 2007b. Quality assessment of SRTM C- and X-band interferometric data: implications for the retrieval of vegetation canopy height. *Remote Sensing of Environment* 106(4): 428–448.

Wang, W., X. Yang, and T. Yao. 2012. Evaluation of ASTER GDEM and SRTM and their suitability in hydraulic modelling of a glacial lake outburst flood in southeast Tibet. *Hydrological Processes* 26(2): 213–225.

Welch, R. 1989. Desktop mapping with personal computers. *Photogrammetric Engineering and Remote Sensing* 55: 1651–1662.

Wilson, J. P. 2012. Digital terrain modeling. *Geomorphology* 137: 107–121.

Yamaguchi, Y., A. B. Kahle, H. Tsu, T. Kawakami, and M. Pniel. 1998. Overview of Advanced Spaceborne Thermal Emission and Reflection Radiometer (ASTER). *IEEE Transactions on Geoscience and Remote Sensing* 36(4): 1062–1071.

Yang, L., X. Meng, and X. Zhang. 2011. SRTM DEM and its application advances. *International Journal of Remote Sensing* 32(14): 3875–3896.

Yu, Y., S. Saatchi, L.S. Heath, E. Lapoint, R. Myneni, and Y. Knyazikhin. 2010. Regional distribution of forest height and biomass from multisensor data fusion. *Journal of Geophysical Research: Biogeosciences* 115(3): article no. G00E12.

Zhao, G., H. Xue, and F. Ling. 2010. Assessment of ASTER GDEM performance by comparing with SRTM and ICESat/GLAS data in central China. *18th International Conference on Geoinformatics, Geoinformatics*: article no. 5567970.

Zouzias, D., G. C. Miliaresis, and K. S. Seymour. 2011. Interpretation of Nisyros volcanic terrain using land surface parameters generated from the ASTER Global Digital Elevation Model. *Journal of Volcanology and Geothermal Research* 200(3–4): 159–170.

Zwally, H. J., B. Schutz, W. Abdalati, J. Abshire, C. Bentley, A. Brenner, J. Bufton, et al. 2002. ICESat's laser measurements of polar ice, atmosphere, ocean, and land. *Journal of Geodynamics* 34(3–4): 405–445.

Chapter 5

Digital elevation model generation from satellite interferometric synthetic aperture radar

Zhong Lu, Hyung-Sup Jung, Lei Zhang,
Wonjin Lee, Chang-Wook Lee, and Daniel Dzurisin

CONTENTS

An accurate digital elevation model (DEM) is a critical data set for characterizing the natural landscape, monitoring natural hazards, and georeferencing satellite imagery. The ideal interferometric synthetic aperture radar (InSAR) configuration for DEM production is a single-pass two-antenna system. Repeat-pass single-antenna satellite InSAR imagery, however, also can be used to produce useful DEMs. DEM generation from InSAR is advantageous in remote areas where the photogrammetric approach to DEM generation is hindered by inclement weather conditions. There are many sources of errors in DEM generation from repeat-pass InSAR imagery, for example, inaccurate determination of the InSAR baseline, atmospheric delay anomalies, and possible surface deformation because of tectonic, volcanic, or other sources during the time interval spanned by the images. This chapter presents practical solutions to identify and remove various artifacts in repeat-pass satellite InSAR images to generate a high-quality DEM.

5.1 INTRODUCTION

An accurate digital elevation model (DEM) is a critical data set for the studies of hydrology, glaciology, forestry, geology, oceanography, and land environment, and it has been used extensively as a base layer in hazards-related geographic information systems (GIS) to model natural hazards (Maune, 2001). In volcano monitoring, DEMs calculated before, during, and after an eruption can be used to understand the eruption progress, estimate the thickness and volume of lava flows or ash deposits, and simulate potential mudflows. In glacier monitoring, DEMs can be used to determine the magnitude and direction of the gravitational force that drives ice flow and ice dynamics. Timely DEMs can be important for characterizing hazards associated with earthquakes, landslides, flooding, snow avalanches, and other processes. In addition, DEMs are indispensible in geocoding satellite images during geometric processing so that the georeferenced satellite images can be used as GIS data sets for hazard monitoring and resource management.

Interferometric synthetic aperture radar (InSAR) utilizes two or more SAR images of the same area to extract the landscape topography and patterns of surface change (Massonnet and Feigl, 1998; Rosen et al., 2000; Lu et al., 2007). An InSAR image, or interferogram, can be produced by combining the phase components of two coregistered SAR images of the same area acquired from similar vantage points. An interferogram formed in this way depicts range changes between the radar and the ground resolution elements and can be used to derive both the landscape topography and subtle changes in surface elevation if the SAR images are acquired at different times.

For InSAR purposes, the spatial separation between two SAR antennas, or between two vantage points of the same SAR antenna, is called the baseline. Two antennas can be mounted on a single platform for simultaneous, single-pass InSAR. This is the usual implementation for airborne and shuttle systems, such as topographic SAR (TOPSAR) (Zebker et al., 1992) and shuttle radar topography mission (SRTM) (Farr et al., 2007). Single-pass two-antenna InSAR is the ideal configuration for generating high-resolution, precise DEMs over large regions. For satellite systems, however, SAR images used for InSAR mapping can be acquired only by using a single antenna in nearly identical repeating orbits. In this case, known as repeat-pass InSAR, even though successive observations of the target area are separated in time, the SAR observations will be highly correlated if the backscattering properties of the surface have not changed in the interim. This is the typical implementation for past and present satellite SAR sensors, including European Space Agency (ESA) European Remote-sensing Satellite 1 (ERS-1) (operated 1991–2000, C-band, wavelength $\lambda = 5.66$ centimeter [cm]), Japan Aerospace Exploration Agency (JAXA) Japanese

Earth Resources Satellite 1 (JERS-1) (1992–1998, L-band, λ = 23.5 cm), ESA European Remote-sensing Satellite 2 (ERS-2) (1995–2011, C-band, λ = 5.66 cm), Canadian Space Agency (CSA) Canadian Radar Satellite 1 (RADARSAT-1) (1995–present, C-band, λ = 5.66 cm), ESA European Environmental Satellite (Envisat) (2002–present, C-band, λ = 5.63 cm), JAXA Japanese Advanced Land Observing Satellite (ALOS) (2006–2011, L-band, λ = 23.6 cm), CSA RADARSAT-2 (2007–present, C-band, λ = 5.55 cm), German Aerospace Agency (DLR) TerraSAR-X (2007–present, X-band, λ = 3.1 cm), Italian COSMO-SkyMed satellite constellation (2007–present, X-band, λ = 3.1 cm), and DLR TerraSAR-X Add-on for Digital Elevation Measurements (TanDEM-X) (2010 present, X-band, λ = 3.1 cm).

Repeat-pass InSAR has proven capable of mapping ground-surface deformation with subcentimeter accuracy for X-band and C-band sensors (λ =2–8 cm), or few-centimeter accuracy for L-band sensors (λ = 15–30 cm) at spatial resolutions of tens-of-meters over image swaths tens to hundreds of kilometers wide (Massonnet and Feigl, 1998; Lu, 2007). Because spatial baselines of repeat-pass satellite radars are not zero, however, the derived interferograms contain information about the landscape topography. Therefore, repeat-pass InSAR can be used to generate DEMs as well (e.g., Small et al., 1995; Zebker et al., 1995; Ruffino et al., 1998; Sansosti et al., 1999; Hensley et al., 2001; Lu et al., 2003).

This chapter discusses techniques and issues related to DEM generation from repeat-pass satellite InSAR images. The chapter also addresses the two main reasons to use this method. First, past, current, and near-future radar satellites all are equipped with a single antenna, providing a large volume of archived imagery that can be exploited to generate DEMs. Second, InSAR is probably the most practical means of constructing DEMs of areas where the photogrammetric approach to DEM generation is hindered by inclement weather or difficult logistical factors (Maune, 2001).

5.2 DEM GENERATION FROM REPEAT-PASS InSAR

5.2.1 Technical issues on DEM generation from repeat-pass InSAR

InSAR processing involves the combination of two or more complex SAR images of the same terrain, and typically includes steps such as precise registration of an interferometric SAR image pair, interferogram generation, removal of curved Earth phase trend, adaptive filtering, phase unwrapping, precision estimate of interferometric baseline, and generation of a surface deformation image or DEM map (e.g., Lu, 2007).

Generally speaking, the phase of an interferogram from two repeat-pass SAR images is a superposition of phases resulting from several processes:

$$\Delta\phi(x) = -\frac{4\pi}{\lambda}\left[\Delta d(x) + \frac{B_\perp}{R\sin\theta}\,h(x)\right] + \Delta\phi_{atmo}(x) + \Delta\phi_{orbit}(x) + \Delta\phi_n(x), \quad (5.1)$$

where $\Delta\phi$ is the InSAR phase measurement, x is the pixel location index, λ the radar wavelength, R is the slant range distance, $B_\perp$ the perpendicular baseline, θ the SAR look angle, Δd the surface displacement in the radar look direction, h the topographic height, $\Delta\phi_{atmo}$ the atmospheric phase delay artifact, $\Delta\phi_{orbit}$ the baseline error, and $\Delta\phi_n$ the phase noise resulting from temporal decorrelation and other noise sources. Because our goal is to derive an accurate DEM (h) based on the observed InSAR phase ($\Delta\phi$), the other phase components (Δd, $\Delta\phi_{atmo}$, $\Delta\phi_{orbit}$, and $\Delta\phi_n$) are treated as error sources.

Before we discuss the error sources that affect DEM accuracy, let us briefly look at the sensitivity of the InSAR phase measurement ($\Delta\phi$) to the surface deformation (Δd) and topographic height (h). If we assume that the error terms related to atmosphere, orbit, and noise are zero, the interferogram phase [after removing the contribution from the effect of ellipsoid Earth (Lu, 2007)] can be simplified as follows:

$$\Delta\phi \approx -\frac{4\pi}{\lambda}\Delta d - \frac{4\pi}{\lambda}\frac{B_\perp}{H\tan\theta}h, \quad (5.2)$$

where H is the satellite altitude above a reference Earth surface. If Δd is negligible in Equation 5.2, the interferogram phase value in Equation 5.2 can be used to calculate height h. This is the principle of how InSAR phase measurements can be used to produce a DEM. For the ESA ERS-1/-2 satellites, H is about 800 kilometers (km), θ is about 23 degrees, λ is 5.66 cm, and $B_\perp$ should be less than the critical baseline, beyond which an interferogram loses coherence (Massonnet and Feigl, 1998). Therefore, Equation 5.2 can be approximated as follows:

$$\Delta\phi \approx -\frac{4\pi}{\lambda}\Delta d - \frac{2\pi}{9,600}B_\perp h. \quad (5.3)$$

For an interferogram with $B_\perp$ of 100 meters (m), which is within the critical baseline of 1,100 m for ERS-1/-2 SAR, 1 m of topographic relief produces a phase value of about 4 degrees. Producing the same phase value, however, requires only 0.3 millimeters (mm) of surface deformation. Therefore, it is evident that the interferogram phase value can be much more sensitive to changes in topography (i.e., the surface deformation Δd) than to the topography itself (i.e., h). That explains why repeat-pass InSAR is capable of detecting surface deformation at theoretical subcentimeter accuracy (Lu, 2007). The accuracy of a DEM derived from repeat-pass InSAR, however, can only reach to meter level if artifacts in InSAR images are negligible. Several specific error sources can affect DEM accuracy.

First, a major error source in repeat-pass InSAR DEM generation is the baseline uncertainty due to inaccurate determination of SAR antenna positions. Errors in this value propagate into very large systematic errors of terrain height. If precision satellite orbit data are available, they should be used for InSAR processing. The interferogram baseline should always be refined using control points or areas with known elevations (or from an existing low-resolution DEM) via a least squares approach (Rosen et al., 1996). In this approach, areas (or pixels) of the interferogram that are used to refine the baseline should have negligible deformation, or known deformation, obtained from an independent source.

Second, because the phase of the radar signal is used to calculate elevation, errors in phase measurement can contribute to the topographic inaccuracy. Random phase errors generally are caused by thermal noise in the SAR system and by decorrelation or incoherence that is in turn caused by volume scattering and environmental change of the imaged surface. Surface changes tend to accumulate with time, so a longer time interval between passes of an InSAR pair can result in poorer coherence (Lu and Freymueller, 1998). The elevation error from a given phase error is inversely proportional to the perpendicular component of baseline length (see Equation 5.2). The topography effect does not appear if the baseline is zero and the sensitivity of interferometric phase value to the topography increases with lengthening baseline. Therefore longer baselines are necessary for accurate DEMs. Unless the volume scattering is negligible and an ideal spectrum filtering is applied (Gatelli et al., 1994), longer baselines can cause geometric decorrelation. This results in an increase in the phase error and, consequently, the elevation error. Therefore, for DEM generation, we should choose interferograms with the largest available baseline within the limit of correlation.

A third critical error source in deriving DEMs from repeat-pass InSAR is atmospheric delay anomalies caused by small variations in the index of refraction along the line of propagation (Goldstein, 1997; Gray et al., 2000). Changes in the total electron content of the ionosphere as well as changes in water vapor content of the troposphere will compromise the quality of the observed interferogram by varying the phase signals. Height errors due to atmospheric anomalies are typically not as large as those resulting from baseline errors, but they are less systematic and harder to detect. Ionosphere artifacts are more severe on longer wavelength (e.g., L-band) and methods of reducing ionospheric artifacts are still an active research topic. The methods to mitigate effects of atmospheric delays in InSAR data can be grouped as follows: (a) integration of InSAR observations with data from dense GPS networks (e.g., Emardson et al., 2003); (b) integration with multispectrum atmospheric water vapor observations (e.g., precipitable water vapor products from moderate resolution imaging spectroradiometer [MODIS] or medium resolution imaging spectrometer [MERIS]; Li et al., 2003); (c) time-series InSAR techniques (e.g., Ferretti et al., 2001; Berardino et al., 2002); (d) integration with short-term predictions from operational weather models (e.g., Foster et al., 2006; Gong et al., 2011); and (e) some combination of methods a through d. In remote and cloud-prone areas, complex environmental conditions typically limit the feasibility of methods a and b. Method c exploits the statistical properties of atmospheric phase components in time-series SAR observations. Its limitation is that it requires a large number of SAR acquisitions and prior knowledge of radar backscattering properties of the study area to properly set parameters during data processing. Method d, utilizing numerical weather models, is not limited by the aforementioned facts, but it requires sufficient initial boundary data and is also computation intensive (Gong et al., 2011). To minimize atmospheric artifacts in repeat-pass InSAR DEM generation, the most practical approach is to choose interferograms with relatively long baselines, as the effect of atmospheric anomalies on DEM heights is inversely proportional to baseline length and to average multiple interferograms to reduce the atmospheric effects.

Finally, we must take into account any possible surface deformation resulting from tectonic, volcanic, or other loading sources over the time interval spanned by repeat-pass interferograms (Lu et al., 2003). Even for areas without tectonic motions, ground surface deformation resulting from variations of groundwater tables (e.g., Lu and Danskin, 2001), permafrost (Rykhus and Lu, 2008), and changes in soil moisture (Gabriel et al., 1989) should be considered. Interferograms with shorter temporal separations are preferred for generating DEMs. The ESA ERS-1/ERS-2 Tandem data meet this requirement in most cases. During the ERS-1/ERS-2 Tandem mission in 1995–1996, interferometric pairs were acquired by ERS-1 and ERS-2 satellites, each of which repeat every 35 days, to follow one another by 1 day in the same orbital geometry. Thus, a point on the surface was imaged by one satellite (ERS-1) on a given day and by the other satellite (ERS-2) on the following day. In cases in which tandem data are not available or not appropriate for DEM generation, deformation rates should be estimated independently and removed from the interferograms used for DEM production.

5.2.2 DEM generation from repeat-pass InSAR: A case study

In this case study, we demonstrate how repeat-pass InSAR images were used to derive DEMs of Okmok volcano, Umnak Island, Alaska, and how

InSAR-derived DEMs of Okmok both before and after its 1997 eruption were used to estimate lava flow thickness.

Okmok volcano, a broad shield topped with a 10 km-wide caldera, occupies most of the northeastern end of Umnak Island, Alaska (see Figure 5.1). The caldera was formed by eruptions about between 8,000 and 2,000 years ago (Miller et al., 1988). A dozen eruptions have occurred in the 20th century, including one in 1997. These eruptions all originated from Cone A, a cinder cone located on the southern edge of the caldera floor. Abundant ash emissions and mafic lava flows originating from Cone A have spread across the caldera floor. The 1997 eruption of Okmok volcano began in early February and ended in late April. The eruption was a moderate Hawaiian to Strombolian type with an ash plume reaching to 10,000 m above sea level, and erupted basaltic a'a' lava flows traveled a few kilometers from cone A. ERS-1/-2 InSAR data were used to map the preeruptive, coeruptive, and posteruptive deformation (Lu et al., 1998, 2000, 2005, 2010a). The authors measured about 140 cm of subsidence associated with the 1997 eruption of Okmok volcano. This subsidence occurred during an interval beginning 16 months before the eruption and ending 5 months after the

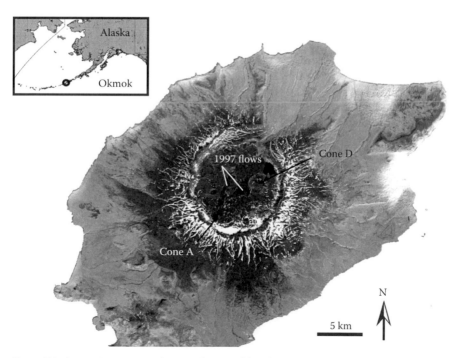

Figure 5.1 A terrain-corrected, georeferenced Landsat-7 Enhanced Thematic Mapper+ image (Band 8) of Okmok volcano, Umnak Island, Alaska. The image was acquired on August 18, 2000. The location of Okmok volcano relative to the rest of Alaska and the Aleutian arc is shown in the inset.

eruption. This subsidence was preceded by about 18 cm of uplift between 1992 and 1995, centered in the same location as the coeruptive subsidence source, and was followed by progressive inflation afterward (Lu et al., 2010a).

Because SRTM was acquired in February 2000 (Farr et al., 2007), we can use the SRTM DEM (see Figure 5.2) to represent Okmok's topography after the 1997 eruption. Here, utilizing multitemporal repeat-pass InSAR images from ERS-1/ERS-2, we constructed a DEM that represents the topography of Okmok volcano before the 1997 eruption. The difference of the preeruption and posteruption DEMs can render a three-dimensional (3D) distribution of the lava flows erupted in 1997.

As discussed, to produce accurate DEMs using repeat-pass ERS-1 and ERS-2 SAR images, atmospheric anomalies need to be carefully considered, because images used for InSAR processing are acquired at different times. Also, a compromise between baseline and interferometric coherence has to be made to select InSAR pairs suitable for DEM generation. Finally, for tectonically or volcanically active regions, any deformation signal must be removed from the interferograms used for DEM generation.

On the basis of the available ERS-1/-2 SAR images acquired before the 1997 eruption at Okmok volcano, we generated seven interferograms. Three interferograms were used to estimate ground surface deformation

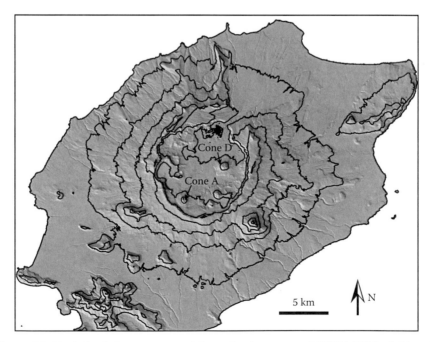

Figure 5.2 Shaded relief image created from the 1-arc-second SRTM DEM of Okmok volcano, Alaska. The contour interval is 200 m.

(a)   93.06.14-93.08.23   (b)   93.09.11-93.10.16   (c)   95.05.22-95.09.04

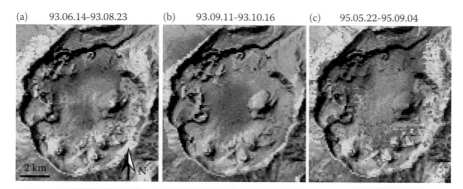

Figure 5.3 **(See color insert.)** Deformation interferograms of Okmok volcano during three different time periods: (a) June 14 to August 23, 1993, with the perpendicular component of baseline, $B_\perp$, equal to 32 m; (b) September 11 to October 16, 1993, with $B_\perp = 25$ m; and (c) May 22 and September 4, 1995, with $B_\perp = 22$ m. The inflation was estimated and removed from those interferograms used for DEM generation (see Table 5.1). A full cycle of colors represents 28 mm surface deformation along the satellite look direction. Areas of coherence loss are uncolored.

(see Figure 5.3) and the other four InSAR images (see Table 5.1) were used for DEM generation. In general, interferometric coherence is maintained reasonably well within the caldera, and it is lost around the caldera rim where terrain is rugged and persistent snow patches are present. This is sufficient as we only needed to determine a preeruption DEM within the caldera floor, part of which is covered by lava flows from the 1997 eruption. The caldera floor subsided about 1.4 m during the April 1997 eruption and inflated at about 10 cm/year from 1997 to 2000 (Lu et al., 1998, 2000, 2005, 2010a). Therefore, over areas that are not covered by 1997 lava flows, the topographic change between 1993 (or 1995) (i.e., preeruption DEM) and 2000 (i.e., posteruption DEM) is about 1.1 m at maximum.

We produced three InSAR images with small baselines (see Figure 5.3), which were used to estimate and remove the volcanic inflation during the time periods of the interferograms used for DEM generation (see Table 5.1). Because these interferograms (see Figure 5.3) have very small baselines, they are insensitive to DEM errors. Therefore, we can use either the posteruption

Table 5.1 Interferometric data acquisition parameters for DEM generation over Okmok volcano

Orbit 1	Orbit 2	Date 1	Date 2	$B_\perp$ (m)
ERS1_22376	ERS2_02703	October 25, 1995	October 26, 1995	83
ERS1_10781	ERS1_11282	August 7, 1993	September 11, 1993	403
ERS1_11783	ERS1_12284	October 16, 1993	November 20, 1993	395
ERS1_11010	ERS1_11511	August 23, 1993	September 27, 1993	690

SRTM DEM (see Figure 5.2) or the preexisting low-resolution DEM (Lu et al., 1998) to remove topographic effects for deformation analysis. Because these interferograms have shorter time separation and are temporally close to the interferograms used for preeruption DEM generation, they better portray the deformation that occurred in the interferograms used for DEM generation (see Table 5.1).

To remove the deformation signal from the DEM interferograms (see Table 5.1), we estimated the location and magnitude of the inflation source responsible for the surface deformation using a point source model embedded in an elastic homogeneous half-space (Mogi, 1958). We interpreted this source to represent a magma chamber at depth. Deformation predicted by the best-fitting model was then removed from the DEM interferograms. We concluded that the magma body was located over the center of the caldera at about 3 km deep and that the maximum inflation was about 18 mm and 4 mm per 35 days during the summers of 1993 and 1995, respectively.

Errors in deformation estimates will transfer into errors in the created DEM. If we neglect atmosphere, baseline, and noise terms in Equation 5.1, we can obtain the following:

$$\Delta h = -\frac{\lambda}{4\pi} \frac{H \tan \theta}{B_\perp} \Delta\phi, \tag{5.4}$$

where Δh is the height error due to an error in interferogram phase ($\Delta\phi$) resulting from the inaccurate estimate of deformation phase. We calculated the uncertainty of the estimated deformation during the summers of 1993 and 1995 to be less than 4 mm per 35 days, which corresponds to an interferometric phase value of 0.9 radians. This error will propagate into an error of less than 3 m in the preeruption DEM elevation with a low-frequency spatial characteristic.

The baseline vectors for all the interferograms were calculated using precision vectors (Massmann, 1995). The baseline vectors were further refined using the posteruption DEM from the SRTM data based on the approach described by Rosen et al. (1996) and using ground points with known elevation from SRTM. About 100 points were selected, all of them lying within the caldera but distant from the 1997 lava flows. We used a least squares approach to estimate the baseline components, and all ground points were weighted equally.

An unwrapped interferometric phase image together with the precision baseline vectors and imaging geometry were needed to derive the topographic heights. The following hierarchical approach, similar to the one proposed by Lanari et al. (1996), was used to facilitate the phase unwrapping procedure (Goldstein et al., 1988; Costantini, 1998). We started with the interferogram having the smallest baseline (i.e., the tandem pair acquired on October 25 and 26, 1995) (see Table 5.1). We first subtracted the topographic phase from the interferogram using the SRTM DEM. The residual fringes were unwrapped (see Figure 5.4a), and the topographic phase was added back to this result. A DEM based on this tandem interferogram was then produced. Next, we unwrapped the August–September 1993 interferogram with $B_\perp = 403$ m (see Table 5.1), because the coherence for this interferogram was better than the October–November 1993 pair (with $B_\perp = 395$ m) (see Table 5.1). The simulated topographic phase based on the SRTM DEM was removed from the interferogram. The resulting residual interferogram

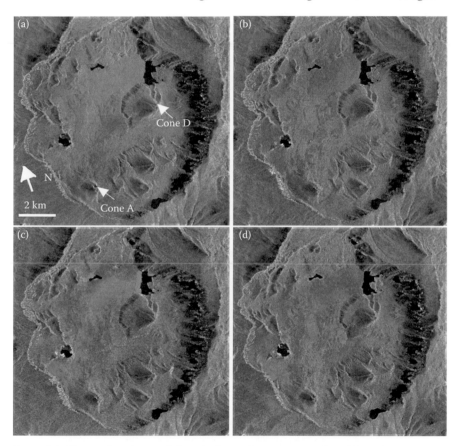

Figure 5.4 **(See color insert.)** Residual interferograms produced by subtracting the topographic phase from the original interferograms (see Table 5.1). (a) The tandem interferogram with $B_\perp = 83$ m and the SRTM DEM was used to remove the topographic phase. (b) The interferogram with $B_\perp = 403$ m and the SRTM DEM was used to remove the topographic phase. (c) The interferogram with $B_\perp = 395$ m and the DEM produced from the interferogram with $B_\perp = 403$ m was used to remove the topographic phase. (d) The interferogram with $B_\perp = 690$ m and the DEM produced from the interferogram with $B_\wedge = 403$ m was used to remove the topographic phase. A full cycle of colors represents a phase change of 360 degrees.

was unwrapped (see Figure 5.4b), and a DEM was generated. We did not use the DEM from the tandem interferogram to simulate the topographic phase because the SRTM is far more accurate than the DEM based on the tandem pair. The DEM from the tandem pair was produced from an interferogram with a smaller baseline. Consequently, the interferometric phase is not very sensitive to topographic relief and any possible atmospheric delay anomalies in the data will significantly bias the DEM accuracy. If an existing DEM is not available, however, the DEM produced from the interferogram with smaller baseline can be used to simulate the topographic phase in the interferogram with larger baseline. Finally, the DEM produced using the interferogram with $B_\perp = 403$ m (see Figure 5.4b) was then used to assist unwrapping the October–November 1993 pair (with $B_\perp = 395$ m) (see Figure 5.4c) and the interferogram with $B_\perp = 690$ m (see Figure 5.4d). Two more DEMs were produced. A simple weighted approach was used to combine the four DEMs:

$$h = \frac{\sum_{i=1}^{4} h_i c_i B_{\perp i}^2}{\sum_{i=1}^{4} c_i B_{\perp i}^2}, \tag{5.5}$$

where h_i and c_i are height and coherence values from the four DEMs, and $B_{\perp i}$ is the perpendicular component of the baseline for each interferogram. The height value of each pixel in the final DEM results from the weighted average of the four DEMs, and height from the interferogram with larger baseline and higher coherence will be more heavily weighed. This procedure not only reduces the possible atmosphere-induced errors in each DEM but also improves accuracy of the final DEM. We used this procedure to generate a DEM depicting the topography of Okmok volcano before the 1997 eruption. A more sophisticated data-fusion technique, such as the wavelet method, can be used to combine DEMs from several interferograms with different spatial resolution, coherence, and vertical accuracy to generate the final DEM product (e.g., Ferretti et al., 1999; Lu et al., 2010b).

Figure 5.5a shows the thickness of the 1997 lava flows derived from the difference between the preeruption and the posteruption DEMs. We can see that the thickness of the lava is very heterogeneous. The thickest portion of the lava happens to be near the distal end (adjacent to Cone D) of the right arm of the Y-shaped flows and reaches almost 50 m. The flow is thickest here because there was a substantial preexisting depression, which caused the flow to pond. In fact, this depression hides the extreme thickness in this area because the closest measurements of the flow margin's height do not exceed 20 m (Moxey et al., 2001). If the preeruption surface is not flat, measurements at the edges are not representative of total thickness.

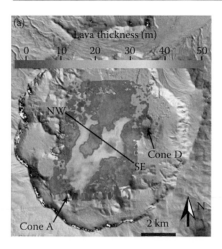

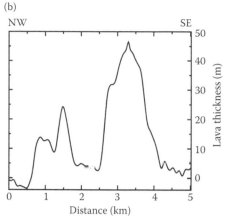

Figure 5.5 **(See color insert.)** (a) Thickness of lava flows emplaced during the 1997 eruption at Okmok volcano, Alaska. Flow thickness was derived from the height difference between the posteruption SRTM DEM and a preeruption DEM constructed from multi-temporal interferograms. (b) Lava thickness along profile northwest-southeast, reaching nearly 50 m in the thickest part of the flow. The red line represents the lava perimeter based on field data collected in August 2001.

Therefore, accurate DEMs are required to calculate more accurate values of lava thickness and eruption volume (e.g., Lu et al., 2003).

The standard deviation of our measurement can be estimated using the root mean-square values of DEM difference over the areas outside of the lava flows. We estimate the mean and 1 σ uncertainty of the DEM difference to be 1.8 m and 2.6 m, respectively, after spatial averaging with a 100 m moving window filter. This means the DEM produced using the four ERS interferograms has a relative vertical accuracy of about 5 m at the 95% confidence limit.

5.3 DEM FROM MULTITEMPORAL InSAR PROCESSING

Multitemporal InSAR (MTInSAR)—that is, persistent scatterer InSAR (PSInSAR) or small baseline subset (SBAS) InSAR (Ferretti et al., 2001; Berardino et al., 2002; Rocca, 2007; Hooper, 2008)—is one of the most significant recent advances in InSAR processing. "Multi" in this context refers to a series of InSAR observations in time, thus affording the opportunity to recognize spurious effects. The objective is to fuse multiple-interferogram measurements of the same area to characterize the spatial and temporal behaviors of the deformation signal, the topography signal, and various artifacts and noise sources (atmospheric delay anomalies, orbit errors), and then to remove the artifacts and anomalies to retrieve time-series deformation measurements as well as an accurate DEM height at the SAR pixel level.

Assuming we have more than about 20 repeat-pass SAR images, we can generate a stack of N coregistered multitemporal interferograms. Because our goal is to generate a new DEM or update an existing low-resolution or low-accuracy DEM, interferograms with short time separations and large baselines are preferred. This is critical to derive a high-accuracy DEM. If we generalize Equation 5.1 for a total of N interferograms, the phase value for the ith coherent pixel in the kth interferogram with a time separation of t^k, ϕ_i^k, can be expressed as the following (after the topographic contribution has been removed from an existing low-resolution DEM):

$$\phi_i^k = -\frac{4\pi}{\lambda}\left(v_i t^k + \mu_i^k\right) - \frac{4\pi}{\lambda}\frac{B_\perp^k\, h_i}{R^k \sin\theta^k} + \alpha_i^k + n_i^k. \tag{5.6}$$

The first term on the right side of Equation 5.6 represents the phase contribution related to ground surface deformation and consists of two components: one is due to the constant velocity (v_i), and the other is due to the nonlinear motion μ_i^k. The second term on the right side of Equation 5.6 represents the phase component due to DEM error (h_i) and is related to the perpendicular baseline ($B_\perp^k$), the distance from the master sensor to the scene center (R_k), and the SAR look angle (θ_k) at the scene center. α_i^k is the phase contribution due to atmospheric and baseline anomalies, and the last term (n_i^k) is the decorrelation phase. Because multitemporal InSAR mainly focuses on coherent points, the noise term is expected to be small and Gaussian in the selected interferograms. It is worth noting that the aforementioned phase terms have different spatial-temporal characteristics, allowing us to separate the interested signals (i.e., DEM error and deformation velocity). By examining the unknown parameters, it becomes apparent that the DEM error and constant velocity do not change with time for each coherent pixel, whereas the remaining terms usually vary from interferogram to interferogram. Therefore, Equation 5.6 can be grouped into a time-invariant part and a time-variant part:

$$\phi_i^k = \left[\left(-\frac{4\pi}{\lambda}t^k\right)v_i + \left(-\frac{4\pi}{\lambda}\frac{B^k}{R^k \sin\theta^k}\right)h_i\right] + \left[-\frac{4\pi}{\lambda}\mu_i^k + \alpha_i^k + n_i^k\right]. \tag{5.7}$$

Let us look into the spatial characteristic of the time-variant part of Equation 5.7 in more detail, which is helpful to construct an optimal observation model for parameter estimation. Generally speaking, the phase component due to baseline inaccuracy in α_i^k effects interferograms in the form of an almost-linear signal of long spatial wavelength. The other part of α_i^k, that is, the atmospheric phase, varies spatially with a typical scale of several kilometers. If we triangulate the coherent points in the study area and ensure that the length of the network edges (also called arcs) is less than a certain threshold (e.g., 1 km), the phase differences of α_i^k at the arcs will be limited to a low level (normally less than 0.1 rad^2). Because the motions of neighboring pixels are normally correlated, the variance of nonlinear motion (μ_i^k) is also very small. Consequently, for a given arc, the sum of the time-variant parts in Equation 5.7 (now represented by $w_{x,y}^k$) in Equation 5.8 should be very small. An observation model of multitemporal InSAR for a given arc constructed by two neighboring coherent pixels (x, y) is thus defined as follows:

$$\Delta\phi_{x,y}^k = \left[\left(-\frac{4\pi}{\lambda}t^k\right)\Delta v_{x,y} + \left(-\frac{4\pi}{\lambda}\frac{B^k}{R^k \sin\theta^k}\right)\Delta h_{x,y}\right] + w_{x,y}^k. \tag{5.8}$$

Equation 5.8 indicates the relationship between the differential phase ($\Delta\phi_{x,y}^k$) at the arc and the velocity difference ($\Delta v_{x,y}$) and height error difference ($\Delta h_{x,y}$), which is the basis for DEM refinement under the framework of MTInSAR. Because the observed differential phases are known only modulo 2π, the estimation of DEM error has to be performed either as a nonlinear inversion problem or as a linear problem, provided that the arcs with phase ambiguities can be reliably resolved or removed. Three approaches are introduced briefly here.

5.3.1 Two-dimensional solution search

As a nonlinear inversion problem, a search through the solution space must always be performed to estimate DEM errors as well as the deformation velocity from Equation 5.8. A temporal coherence index is commonly used for this inversion (Ferretti et al., 2001; Mora et al., 2003, Zhang et al., 2011b).

$$\gamma_{x,y} = \frac{1}{N}\left|\sum_{k=1}^{N}e^{-Jw_{x,y}^k}\right| = \frac{1}{N}\left|\sum_{k=1}^{N}e^{-J\left[\Delta\phi_{x,y}^k + \left(\frac{4\pi}{\lambda}t^k\right)\Delta v_{x,y} + \left(\frac{4\pi}{\lambda}\frac{B^k}{R^k \sin\theta^k}\right)\Delta h_{x,y}\right]}\right|, \tag{5.9}$$

where $J = \sqrt{-1}$. By setting proper variation ranges for velocity difference ($\Delta v_{x,y}$) and height error difference ($\Delta h_{x,y}$), one can search for the maximum coherence ($\gamma_{x,y}$) within the specified two-dimensional ranges using small sampling intervals. Then the optimum solutions of $\Delta v_{x,y}$ and $\Delta h_{x,y}$ can be found. After $\Delta v_{x,y}$ and $\Delta h_{x,y}$ between all neighboring pixels are determined, the absolute values of DEM error and linear deformation rate at each coherent pixel can be derived through spatial integration with respect to a reference point at which the DEM error and linear deformation rate are known or assumed to be zero. The solution search can be successfully performed only under the condition of $\left| \omega_{i,x,y}^{k} \right| < \pi$, which can be met in most cases.

5.3.2 Linear inversion with phase ambiguity detector

Although temporal coherence maximization (see Equation 5.9) has the ability to resolve DEM errors from wrapped phase data, the method might result in several local maxima during the search of parameters (i.e., $\Delta v_{x,y}$ and $\Delta h_{x,y}$), which means a unique solution cannot be guaranteed. Because of possible phase ambiguities at some arcs, parameters cannot be estimated using a linear inversion (Zhang et al., 2011a). If ambiguities at troublesome arcs can be reliably removed, estimating parameters ($\Delta v_{x,y}$ and $\Delta h_{x,y}$) can be simplified significantly. Following is Equation 5.8 rewritten in a simplified vector form:

$$\Delta \Phi = A \begin{bmatrix} \Delta h_{x,y} \\ \Delta v_{x,y} \end{bmatrix} + w, \tag{5.10}$$

where $\Delta \Phi$ is the differential phase vector at a given arc constructed by two pixels x and y, A is the design matrix containing the coefficients of unknowns (i.e., $\Delta v_{x,y}$ and $\Delta h_{x,y}$), and w is a stochastic vector with an expectation of zero. For any arc regardless of phase ambiguities, the least squares solution of unknowns is as follows:

$$\begin{bmatrix} \Delta \hat{h}_{x,y} \\ \Delta \hat{v} \end{bmatrix} = (A^{T} P_{x,y} A)^{-1} A^{T} P_{x,y} \Delta \Phi$$

$$r = \Delta \Phi - A (A^{T} P_{x,y} A)^{-1} A^{T} P_{x,y} \Delta \Phi, \tag{5.11}$$

where r is the least squares residual vector and $P_{x,y}$ is the weight matrix that can be obtained by taking the inverse of a prior variance matrix of the double-difference phases (Zhang et al., 2011a). It has been observed that the least squares residuals for an arc with and without phase ambiguities are quite different (see Figure 5.6), indicating that phase ambiguities can bias the parameter estimation significantly. An ambiguity detector therefore can be designed according to the least squares residuals. The details can be found in Zhang et al. (2011a). After removing the arcs with phase ambiguities, parameters at the remaining arcs are integrated to obtain the parameter estimates at all coherent points with respect to a reference point.

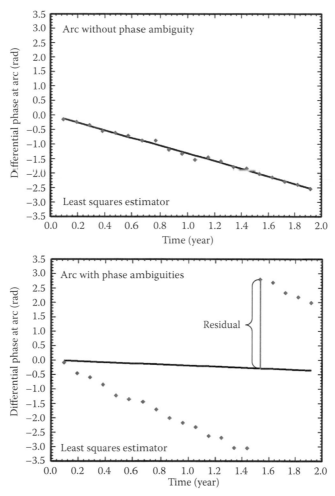

Figure 5.6 Least squares residuals at an arc with and without phase ambiguity. For illustrative purpose, only one parameter is estimated.

5.3.3 Linear inversion with phase unwrapping

Besides the removal of arcs with phase ambiguities, there is another way to linearly estimate DEM errors as well as deformation rates. Provided that two-dimensional phase unwrapping can be reliably performed on each interferogram, estimating parameters turns out to be a simple least squares problem where no ambiguity detector is needed:

$$
\begin{bmatrix} \Delta \hat{b}_{x,y} \\ \Delta \hat{v} \end{bmatrix} = (A^{\mathsf{T}} P_{x,y} A)^{-1} A^{\mathsf{T}} P_{x,y} \, \Delta \Phi_{\mathrm{unwrapped}}.
\tag{5.12}
$$

Parameters at pixels can be achieved by integration as done in the aforementioned approaches. Although the solution form of Equation 5.12 is similar to the one in small baseline subset SBAS InSAR method (Berardino et al., 2002), there is an essential difference: SBAS takes phases at pixels as observations whereas this model takes the differential phases at arcs as observations. The latter can better suppress the effects of atmospheric anomalies (Zhang et al., 2011a).

Figure 5.7 shows the DEM error map of Okmok volcano derived from MTInSAR processing. This DEM error map is based on a DEM constructed from a two-antenna airborne InSAR system (Lu et al., 2003). Because the airborne system typically images an area of about 10 km wide, multiple DEMs from several flight passes were used to generate the DEM mosaic. However, the original DEM patches have systematic errors of ±50m (Lu et al., 2003). These errors result in some artifacts in the DEM mosaic over the western flank of the volcano. The MTInSAR technique can correct the DEM error (see Figure 5.7) and produce a better DEM for Okmok volcano.

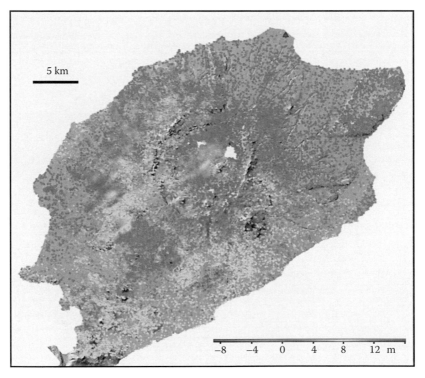

Figure 5.7 **(See color insert.)** DEM height update from MTInSAR processing technique for Okmok volcano, Alaska. The height difference is with respect to an airborne DEM mosaic.

5.4 DEM FROM ENVISAT/ERS-2 CROSS-PLATFORM INSAR

As discussed and illustrated in previous sections, a rule of thumb on generating DEMs from repeat-pass InSAR imagery is to select interferograms with the largest available baselines within the limit of correlation. That is because the DEM accuracy is inversely proportional to the baseline length (Equation 5.2). We have demonstrated that the vertical resolution of DEMs derived from repeat-pass InSAR imagery can reach to a few meters if multi-temporal interferograms are available. Therefore, it is generally impossible to generate a DEM with submeter accuracy, which requires the InSAR baseline to be larger than the critical baseline.

Generally speaking, a SAR system images the ground surface at a specific radar carrier frequency and incidence angle. A change in radar incidence angle (which translates to a change in baseline) or radar carrier frequency can result in a change in the radar reflectivity spectrum in range direction. The change in radar reflectivity spectrum (Δf) can be defined as follows (Gatelli et al., 1994):

$$\Delta f = -\frac{cB_{\perp}}{\lambda R \tan(\theta - \alpha)},$$

(5.13)

where c is speed of light and α is the terrain slope angle.

If two SAR images used to create an InSAR image are from SAR systems of the same carrier frequency, the change in radar reflectivity spectrum in range direction needs to be small so that the reflectivity spectra of the two signals overlap, which translates to the requirement of $B_{\perp}$ < the critical baseline. There is, however, a condition under which the effect of the baseline can be exactly compensated by a difference in radar frequency between two SAR acquisitions (Gatelli et al., 1994; Guarnieri and Prati, 2000; Colesanti et al., 2003). From Equation 5.13, for a given frequency difference, the perpendicular baseline component ($B_{\perp}$) required to compensate for the frequency difference can be defined as follows:

$$B_{\perp} \approx -\frac{(f_2 - f_1)\, R \tan(\theta - \alpha)}{f_1},$$

(5.14)

where f_1 and f_2 are radar carrier frequencies of two SAR acquisitions.

Acquisitions from C-band Envisat and ERS-2 SARs from ESA were on the same orbital plane with a 35-day repeat and a 28-minute time lag. The radar frequency of Envisat, however, is slightly different from that of ERS-2 by 31 megahertz (MHz). Because of this difference, Envisat SAR images generally cannot be combined with ERS-2 data for repeat-pass cross-platform InSAR processing. Fortunately, the 31 MHz frequency difference between ERS-2 and Envisat SAR images can be compensated by a perpendicular baseline of approximately 2 km over a flat surface (Equation 5.14). Consequently, Envisat and ERS-2 can be combined to preserve InSAR coherence in spite of a large baseline of about 2 km. For an interferogram of $B_\perp = 2$ km, one interferometric fringe corresponds to a topographic relief of ~4.8 m (Equation 5.3). Therefore, an Envisat/ERS-2 cross-platform InSAR image is capable of generating a submeter-accuracy DEM. Because the two SAR images are acquired about 28 minutes apart, temporal decorrelation as well as atmospheric artifacts are significantly reduced (Wegmuller et al., 2009).

We demonstrate the DEM generation using a pair of Envisat/ERS-2 images acquired on January 25, 2008, over northern Alaska (see Table 5.2). The exiting DEM from the USGS National Elevation Dataset (NED), with a posting of 2-arc-second and vertical accuracy of several meters, are also used in InSAR processing. We generated Envisat and ERS-2 SLC images using the same Doppler centroid and Doppler bandwidth calculated by the azimuth common band filtering to minimize the misregistration, and then we oversampled SLC images twice to reduce the phase unwrapping error. After DEM-assisted coregistration of Envisat and ERS-2 SLC images (Lee et al., 2010), the Envisat/ERS-2 cross-platform interferogram is created. The interferogram is unwrapped and converted into a topographic height map (see Figure 5.8a).

To assess the vertical accuracy of the Envisat/ERS-2 cross-platform InSAR-derived DEM, we compared it with an airborne DEM produced in July 2002, which has a spatial resolution of 5 m and vertical height specification of 10 cm. We selected an area in Figure 5.8a for accuracy assessment. The average height of the area is about 80 m. Figures 5.8b and 5.8c show the InSAR-derived DEM and the airborne DEM, and Figure 5.8d represents the height difference between the two DEMs along profile A-A'. The mean and standard deviation of the difference over the study area are 1 cm and 39 cm, respectively. Figure 5.8e shows the histogram of the height difference. The histogram generally follows a Gaussian distribution with no bias. From this comparison, we conclude the vertical accuracy of the DEM from Envisat/ERS-2 cross-platform InSAR is less than 40 cm.

This method of cross-platform InSAR utilizing the baseline difference to compensate the difference in SAR central frequency requires two similarly configured SAR systems on two separate platforms to generate interferograms with a baseline large enough to be sensitive to terrain height to

Table 5.2 Characteristics of ERS-2 and envisat pair for cross-platform InSAR

Parameters	ERS-2	Envisat
Central frequency (MHz) (GHz)	5.3	5.331
Acquisition date	January 25, 2008	
Time interval (min.)	28 minutes (Envisat – ERS2)	
$B_\perp$ (m)	2,400	

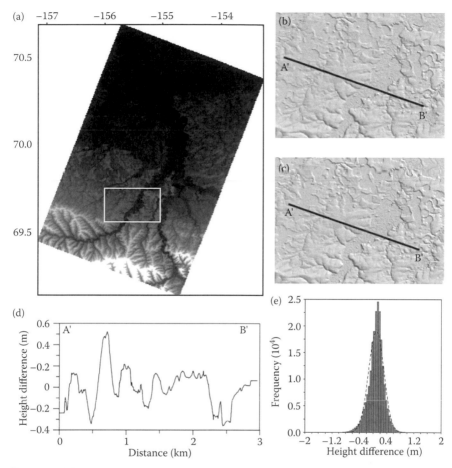

Figure 5.8 (a) Envisat/ERS-2 cross-platform InSAR-derived DEM over northern Alaska. Comparison of the InSAR-derived DEM (b) and a high-resolution airborne DEM (c) over an area outlined by the while rectangle in (a). (d) Height difference between the InSAR-derived DEM and the airborne DEM along profile of A'-B'. (e) Histogram of height differences between the Envisat/ERS-2 InSAR-derived DEM and the airborne DEM.

submeter accuracy. Such accurate DEM generation is the ultimate objective for the current DLR TanDEM-X mission.

5.5 DEM FROM TANDEM-X

The TerraSAR-X tandem mission for DEM measurements, TanDEM-X, was launched by DLR in 2010 (http://www.dlr.de/hr/en/desktopdefault.aspx/tabid-2317). TanDEM-X is a new high-resolution constellation InSAR mission that relies on an innovative flight formation of two tandem TerraSAR-X satellites to produce InSAR-derived DEMs on a global scale with accuracy better than SRTM (Krieger et al., 2007). In addition, TanDEM-X can enable precise mapping of ocean currents by fusing two SAR images steered in the along-track direction. The resulting product will be invaluable for monitoring extreme waves and ocean hazards. Furthermore, TanDEM-X will provide data to assess the utility of new methods, including bistatic multiangle SAR imaging, digital beam formation, and polarimetric InSAR, for monitoring landscape changes.

The two X-band SARs on two separate TerraSAR-X satellites record data synchronously and create a baseline ranging from ~200 m to ~500 m. The precise baseline determination and simultaneous data acquisitions can generate InSAR images that are immune to baseline errors, atmospheric contaminations, and temporal decorrelation that plague the accuracy of DEMs derived from repeat-pass InSAR. Therefore, the TanDEM-X satellite constellation allows the generation of global DEMs of an unprecedented accuracy, coverage, and quality. TanDEM-X DEMs have a specified relative vertical accuracy of 2 m and an absolute vertical accuracy of 10 m at a horizontal resolution of 12 m (Krieger et al., 2007).

5.6 CONCLUSION

The ideal SAR configuration for accurate DEM production is a single-pass (simultaneous) two-antenna system. Repeat-pass single-antenna satellite InSAR can be used to produce useful DEMs, particularly in areas where the photogrammetric approach to DEM generation is hindered by persistent clouds or other factors. There are many sources of errors in DEM construction from repeat-pass SAR images, including inaccurate determination of the InSAR baseline, atmospheric delay anomalies, and possible surface deformation resulting from tectonic, volcanic, or other sources during the time interval spanned by the images. To generate a high-quality DEM, these errors must be identified and corrected using a multi-interferogram approach. A data fusion technique such as PSInSAR or SBAS InSAR can be applied to a stack of repeat-pass InSAR images to generate a DEM accurate to meters. Ultimately, special InSAR image formation from constellation satellites, such as TanDEM-X, will continue advancing the spatial resolution and vertical accuracy of the global DEM.

ACKNOWLEDGMENTS

ERS-1, ERS-2, and Envisat SAR images are copyright © ESA and provided by ESA and the Alaska Satellite Facility. This work was supported by the USGS Volcano Hazards Program and the NASA's Earth Surface and

Interiors Program (2005–0021). We thank Dave Ramsey and Russ Rykhus for careful edits and constructive review.

REFERENCES

Berardino, P., G. Fornaro, R. Lanari, and E. Sansosti. 2002. A new algorithm for surface deformation monitoring based on small baseline differential SAR interferograms. *IEEE Transactions on Geoscience and Remote Sensing* 40: 2375–2383.

Colesanti, C., F. De Zan, A. Ferretti, C. Prati, and F. Rocca. 2003. Generation of DEM with sub-metric vertical accuracy from 30′ ERS–ENVISAT pairs. In *Proceedings from Fringe 2003 Workshop, ESA-ESRIN, Frascati, Italy.*

Costantini, M. 1998. A novel phase unwrapping method based on network programming. *IEEE Transactions on Geoscience and Remote Sensing* 36: 813–821.

Emardson, T. R., M. Simons, and F. H. Webb. 2003. Neutral atmospheric delay in interferometric synthetic aperture radar applications: statistical description and mitigation. *Journal of Geophysical Research* 108: 2231.

Farr, T. G., E. Caro, R. Crippen, R. Duren, S. Hensley, M. Kobrick, M. Paller, et al. 2007. The Shuttle Radar Topography Mission. *Reviews of Geophysics* 45: RG2004.

Ferretti, A., C. Prati, and F. Rocca. 1999. Multibaselime InSAR DEM Reconstruction: the wavelet approach. *IEEE Transactions on Geoscience and Remote Sensing* 37: 705–715.

Ferretti, A., C. Prati, and F. Rocca. 2001. Permanent scatterers in SAR interferometry. *IEEE Transactions on Geoscience and Remote Sensing* 39:8–20.

Foster, J., B. Brooks, T. Cherubini, C. Shacat, S. Businger, and C. L. Werner. 2006. Mitigating atmospheric noise for InSAR using a high-resolution weather model. *Geophysical Research Letters* 33: L16304.

Gabriel, A., R. Goldstein, and H. Zebker. 1989. Mapping small elevation changes over large areas: differential radar interferometry. *Journal of Geophysical Research* 94: 9183–9191.

Gatelli, F., A. Monti-Guarnieri, F. Parizzi, P. Pasquali, C. Prati, and F. Rocca. 1994. The wavenumber shift in SAR interferometry. *IEEE Transactions on Geoscience and Remote Sensing* 31: 855–865.

Goldstein, R. 1997. Atmospheric limitations to repeat-track radar interferometry. *Geophysical Research Letters* 22: 2517–2520.

Goldstein, R., H. Zebker, and C. Werner. 1988. Satellite radar interferometry: two-dimensional phase unwrapping. *Radio Science* 23: 713–720.

Gong, W., F. J. Meyer, P. Webley, and Z. Lu. 2011. Methods of InSAR atmosphere correction for volcano activity monitoring. *Geoscience and Remote Sensing Symposium (IGARSS), 2011 IEEE International, Vancouver, BC, Canada,* 1654–1657.

Gray, A. L., K. E. Mattar, and G. Sofko. 2000. Influence of ionospheric electron density fluctuations on satellite radar interferometry. *Geophysical Research Letters* 27: 1451–1454.

Guarnieri, A. M., and C. Prati. 2000. ERS-ENVISAT combination for interferometry and super-resolution. In *Proceedings of ERS-ENVISAT Symposium, European Space Agency, Frascati, Italy.*

Hensley, S., R. Munjy, and P. Rosen. 2001. Interferometric synthetic aperture radar (IFSAR). In *Digital Elevation Model Technologies and Applications: The DEM Users Manual*, edited by D. F. Maune, 143–206. Bethesda, MD: American Society for Photogrammetry and Remote Sensing.

Hooper, A. 2008. A multi-temporal InSAR method incorporating both persistent scatterer and small baseline approaches. *Geophysical Research Letters* 35: L16302.

Krieger, G., A. Moreira, H. Fiedler, I. Hajnsek, M. Werner, M. Younis, and M. Zink. 2007. TanDEM-X: a satellite formation for high-resolution SAR interferometry. *IEEE Transactions on Geoscience and Remote Sensing* 45: 3317–3341.

Lanari, R., G. Fornaro, D. Riccio, M. Migliaggio, K. P. Papathanassiou, J. R. Moreira, M. Schwäbisch, et al. 1996. Generation of digital elevation models by using SIR-C/X-SAR multifrequency two-pass interferometry: the Etna case study. *IEEE Transactions on Geoscience and Remote Sensing* 34: 1097–1114.

Lee, W. J., H. S. Jung, and Z. Lu. 2010. A study of high-precision DEM generation using ERS-Envisat SAR cross-interfeometry. *Journal of Korean Society of Surveying, Geodesy, Photogrammetry and Cartography* 28: 431–439.

Li, Z., J. P. Muller, and P. Cross. 2003. Comparison of precipitable water vapor derived from radiosonde, GPS, and Moderate-Resolution Imaging Spectroradiometer measurements. *Journal of Geophysical Research* 108: 4651.

Lu, Z. 2007. InSAR imaging of volcanic deformation over cloud-prone areas: Aleutian Islands. *Photogrammetric Engineering and Remote Sensing* 73: 245–257.

Lu, Z., and W. Danskin. 2001. InSAR analysis of natural recharge to define structure of a ground-water basin, San Bernardino, California. *Geophysical Research Letters* 28: 2661–2664.

Lu, Z., D. Dzurisin, J. Biggs, C. Wicks, and S. McNutt. 2010a. Ground surface deformation patterns, magma supply, and magma storage at Okmok volcano, Alaska, from InSAR analysis: 1. intereruption deformation, 1997–2008. *Journal of Geophysical Research-Solid Earth* 115(B00B02): doi:10.1029/2009JB006969.

Lu, Z., D. Dzurisin, H. S. Jung, J. X. Zhang, and Y. H. Zhang. 2010b. Radar image and data fusion for natural hazards characterization. *International Journal of Image and Data Fusion* 1: 217–242.

Lu, Z., E. Fielding, M. Patrick, and C. Trautwein. 2003. Estimating lava volume by precision combination of multiple baseline spaceborne and airborne interferometric synthetic aperture radar: the 1997 eruption of Okmok volcano, Alaska. *IEEE Transactions on Geoscience and Remote Sensing* 41: 1428–1436.

Lu, Z., and J. Freymueller. 1998. Synthetic aperture radar interferometry coherence analysis over Katmai volcano group, Alaska. *Journal of Geophysical Research* 103: 29887–29894.

Lu, Z., O. Kwoun, and R. Rykhus. 2007. Interferometric synthetic aperture radar (InSAR): its past, present and future. *Photogrammetric Engineering and Remote Sensing* 73: 217–221.

Lu, Z., D. Mann, and J. Freymueller. 1998. Satellite radar interferometry measures deformation at Okmok volcano. *EOS Transactions* 79: 461–468.

Lu, Z., D. Mann, J. Freymueller, and D. Meyer. 2000. Synthetic aperture radar interferometry of Okmok volcano, Alaska: radar observations. *Journal of Geophysical Research* 105: 10791–10806.

Lu, Z., T. Masterlark, and D. Dzurisin. 2005. Interferometric Synthetic Aperture Radar (InSAR) study of Okmok volcano, Alaska, 1992–2003: magma supply dynamics and post-emplacement lava flow deformation. *Journal of Geophysical Research* 110: B02403. doi:10.1029/2004JB003148.

Massmann, F. H. 1995. Information for ERS PRL/PRC Users. GeoForschungsZentrum Potsdam Technical Note.

Massonnet, D., and K. Feigl. 1998. Radar interferometry and its application to changes in the Earth's surface. *Reviews of Geophysics* 36: 441–500.

Maune, D. 2001. *Digital Elevation Model Technologies and Applications: The DEM Users Manual.* Bethesda, MD: American Society for Photogrammetry and Remote Sensing.

Miller, T. P., R. G. McGimsey, J. R. Richter, J. R. Riehle, C. J. Nye, M. E. Yount, and J. A. Dumoulin. 1998. Catalog of the historically active volcanoes of Alaska. USGS Open-File Report, 98–582.

Mogi, K. 1958. Relations between the eruptions of various volcanoes and the deformations of the ground surface around them. *Bulletin of the Earthquake Research Institute of the University of Tokyo* 36: 99–134.

Mora, O., J. J. Mallorqui, and A. Broquetas. 2003. Linear and nonlinear terrain deformation maps from a reduced set of interferometric SAR images. *IEEE Transactions on Geoscience and Remote Sensing* 41: 2243–2253.

Moxey, L., J. Dehn, K. R. Papp, M. R. Patrick, and R. Guritz. 2001. The 1997 eruption of Okmok volcano, Alaska, a synthesis of remotely sensed data. *EOS Transactions* 82: 47.

Rocca, F. 2007. Modeling interferogram stacks. *IEEE Transactions on Geoscience and Remote Sensing* 45: 3289–3299.

Rosen, P., S. Hensley, H. Zebker, F. H. Webb, and E. J. Fielding. 1996. Surface deformation and coherence measurements of Kilauea volcano, Hawaii, from SIR-C radar interferometry. *Journal of Geophysical Research* 101: 23109–23125.

Rosen, P., S. Hensley, I. R. Joughin, F. K. Li, S. N. Madsen, E. Rodriguez, and R. M. Goldstein. 2000. Synthetic aperture radar interferometry. *Proceedings of IEEE* 88: 333–380.

Ruffino, G., A. Moccia, and S. Esposito. 1998. DEM generation by means of ERS tandem data. *IEEE Transactions on Geoscience and Remote Sensing* 36: 1905–1912

Rykhus, R., and Z. Lu. 2008. InSAR detects possible thaw settlement in the Alaskan Arctic Coastal Plain. *Canadian Journal of Remote Sensing* 34: 100–112.

Sansosti, E., G. Fornaro, G. Franceschetti, M. Tesauro, G. Puglisi, and M. Coltelli. 1999. Digital elevation model generation using ascending and descending ERS-1/ERS-2 tandem data. *International Journal of Remote Sensing* 20: 1527–1547.

Small, D., C. Werner, and D. Muesch. 1995. Geocoding and validation of ERS-1 InSAR-derived digital elevation models. *ERSSeL Advanced Remote Sensing* 4: 26–39.

Wegmuller, U., S. Maurizio, C. Werner, and T. Strozzi. 2009. DEM generation using ERS-Envisat interferometry. *Journal of Applied Geophysics* 69: 51–58.

Zebker, H. A., S. N. Madsen, J. Martin, K. B. Wheeler, T. Miller, Y. Lou, G. Alberti, S. Vetrella, and A. Cucci. 1992. The TOPSAR interferometric radar topographic mapping instrument. *IEEE Transactions on Geoscience and Remote Sensing* 30: 933–940.

Zebker, H. A., C. L. Werner, P. A. Rosen, and S. Hensley. 1995. Accuracy of topographic maps derived from ERS-1 interferometric radar. *IEEE Transactions on Geoscience and Remote Sensing* 32: 823–836.

Zhang, L., X. L. Ding, and Z. Lu. 2011a. Modeling PSInSAR time series without phase unwrapping. *IEEE Transactions on Geoscience and Remote Sensing* 49: 547–556.

Zhang, Y. H., J. X. Zhang, H. A. Wu, Z. Lu, and G. T. Sun. 2011b. Monitoring of urban subsidence with SAR interferometric point target analysis: a case study in Suzhou, China. *International Journal of Applied Earth Observation and Geoinformation* 13: 812–818.

Chapter 6

Shoreline mapping

Christopher E. Parrish

CONTENTS

With coastal aerial imagery acquisition programs dating back to 1919 (Smith, 1981), shoreline mapping is one of the most established civilian applications of photogrammetry and remote sensing. In this chapter we first discuss the importance of shoreline mapping and some associated definitions. We then describe historical and current methods of mapping shorelines using field techniques, aerial imagery, satellite imagery, light detection and ranging (LiDAR), and other remotely sensed data. A pervasive concept is that the goal of mapping tidally referenced shorelines leads to unique challenges and considerations in planning, acquisition, and processing of remotely sensed data, including tide coordination, spectral band selection, and effects of beach slope. Uncertainty analysis is another important topic. Although it is typically required in any mapping application to assess and quantify the uncertainty (or "error") in output geospatial data products, it is especially critical in shoreline mapping because of the legal implications of the data and their use in informing complex policy decisions. We conclude with a look at the future of shoreline mapping using various remote-sensing technologies.

6.1 THE IMPORTANCE OF SHORELINE MAPPING

Discussions of the significance of mapping the world's shorelines often start with the fact that a large portion of the world's population lives along the coasts. Published figures vary, but frequently cited statistics are that approximately 40% the world's population lives within 100 kilometers (km) of the coast, and, in the United States (U.S.), approximately 53% of the population lives in coastal counties (Crossett et al., 2004; Agardy et al., 2005). When these figures are considered along with projections of global sea level rise in this century of 0.18–0.59 meters (m), or perhaps substantially greater with rapid melting of the Greenland and Antarctic ice sheets (Meehl et al., 2007; Grinsted et al., 2010), estimates of the number of people worldwide residing in areas vulnerable to flooding range in the tens of millions (Nicholls, 2004; National Research Council [NRC], 2010). This potential impact provides one strong motivating factor for mapping—and periodically remapping—the shoreline.

In the United States, the responsible organization for mapping the national shoreline depicted on the nation's nautical charts is the National Geodetic Survey (NGS), a program office of the National Oceanic and Atmospheric Administration (NOAA), National Ocean Service (NOS). This agency and its predecessors have been tasked with mapping the nation's shoreline since the original Survey of the Coast was authorized by Congress in 1807, following a recommendation from President Thomas Jefferson (Shalowitz, 1964). The shoreline produced by NGS is used to update NOAA nautical charts and, thus, to support safe marine navigation.

Beyond navigational safety, the national shoreline has legal significance. The government may be held liable for loss of life or property resulting from chart inaccuracies (Defense Mapping Agency [DMA], 1984), including inaccuracies in charted shoreline position. Additionally, as depicted in Figure 6.1, the position of the shoreline is used in defining the boundary between state-owned and federally owned submerged lands (Shalowitz, 1962). This demarcation can have great financial significance, for example, because of the ability of states to issue oil and gas leases on submerged lands for which they can claim ownership. Numerous boundary disputes in coastal areas have been resolved, in part, through reference to the shoreline depicted on the largest scale nautical charts of the area. Likewise, shoreline is used in defining the boundary between privately owned and state-owned waterfront property in coastal states. Setback requirements—which establish the boundary lines within which building is prohibited—for development in coastal areas, and even public rights to beach access in certain states, are related to shoreline position (Monmonier, 2008).

The legal significance of shoreline also extends to international maritime boundary determination. In particular, shorelines serve in defining the nation's territorial sea, contiguous zone, and exclusive economic zone (see Figure 6.1).

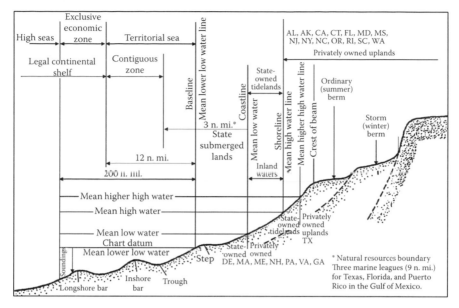

Figure 6.1 Legal significance of shoreline. Note that although this figure uses the terms "shoreline," "coastline," and "baseline" to indicate distinctions that arise in boundary determination, for purposes of this chapter, we consider each of these to be simply a different tidally referenced shoreline definition. (Reprinted from Gill, S. K., and J. R. Schultz, *Tidal Datums and Their Applications*, NOAA, 2001. Courtesy of NOAA.)

On the basis of international law, as reflected in the United Nations Convention on the Law of the Sea (UNCLOS), coastal nations have certain, defined rights in each of these sea zones, with respect to, for example, their ability to enforce regulations and exploit economic resources.

Other important uses of mapped shoreline abound in coastal science, engineering, and management. We will not attempt to provide a comprehensive list here, let alone a complete treatment of each. However, a few of the key uses include the following:

- Analysis and modeling coastal sedimentary processes
- Environmental protection
- Posthurricane damage assessment and improved coastal vulnerability models
- Storm surge and tsunami inundation modeling
- Analysis of wetlands and habitat loss
- Understanding and responding to threats of climate change, including coastal erosion

Imagery and other remotely sensed data used to map shoreline increasingly are being made publicly available to support a wide range of applications, as part of the U.S. interagency Integrated Ocean and Coastal Mapping (IOCM) initiative (Scott et al., 2009). This initiative aims to increase coordination among government, private sector, and academic partners engaged in ocean and coastal mapping to (a) reduce duplication of effort and (b) support the greatest possible range of ocean and coastal geospatial data users and applications. As just one example of the IOCM concept, light detection and ranging (LiDAR) data and imagery collected to support shoreline mapping for nautical charts currently are being used for mapping and monitoring environmentally sensitive areas, such as mangroves, coral reefs, and wetlands.

6.2 SHORELINE DEFINITIONS

Although the discussion thus far has addressed the question of why shoreline mapping is important, an even more fundamental question is, "What is shoreline?" The answer may seem obvious, as most of us feel we have a good, intuitive sense of what shoreline is. Many texts give succinct definitions of shoreline, such as the line of intersection between a water body and land.

To understand the limitations of this seemingly straightforward definition, it may be helpful to view the brief video clip showing a short stretch of shoreline on Assateague Island, Maryland (see http://ccom.unh.edu/sites/default/files/videos/misc/waves_at_assateague.avi). Is it possible to identify precisely "the line of intersection between the water and land" in this video? The obvious point is that this line is continuously changing. Wave swash, as shown in the video clip, is just one cause of the change in the position of the land-water interface over extremely short time periods: tides, currents, storms, seasonal beach changes, sediment transport, and sea level change are some of the other causes of shoreline change over time scales ranging from minutes to centuries. In addition, there are human-induced changes, such as coastal development and engineering, including such activities as beach nourishment and construction of jetties, groins, seawalls, and breakwaters.

The challenges in defining the shoreline are further discussed in Boak and Turner (2005). As these authors noted, it is, of course, possible to define an "instantaneous shoreline": the position of the land-water interface at one instant in time. The land-water interface delineated in aerial or satellite imagery acquired at an arbitrary—and quite possibly unknown—stage of tide is an example of an instantaneous shoreline. Unfortunately, instantaneous shoreline is of relatively little value in charting applications or in coastal change analysis, as it lacks repeatability of measurement and is difficult to compare with other mapped shorelines in a meaningful way. This difficulty has led to the use of a number of shoreline "indicators" or "proxies" (Boak and Turner, 2005) that can be readily and consistently identified in the field or in aerial imagery and that can serve as a surrogate for the "true" shoreline position. Examples include the following:

- Bluff line
- Vegetation line
- High water line (HWL), often defined as the wet-dry line from the previous high tide
- Wrack line, defined as the line of seaweed and debris washed ashore

An alternative to these types of visually identifiable shoreline indicators is datum-based shoreline, defined as the intersection of a vertical datum—usually a tidal datum—with the coastal topographic surface. Although detailed discussion of tidal datums is beyond the scope of this chapter, briefly, a tidal datum is a vertical reference defined in terms of a certain phase of tide. Tidal datums are referenced to a specific National Tidal Datum Epoch (NTDE), a 19-year period over which tide observations are obtained and averaged. (The length of the NTDE is based on 18.6-year cycle of the moon's nodes, rounded to the nearest integer.) The latest NTDE was 1983–2001. The two tidal datums most commonly used in defining tide-datum-based shoreline are mean high water (MHW), the arithmetic mean of all the high water heights observed over the NTDE, and mean lower low water (MLLW), the arithmetic mean of the lower low water height of each tidal day observed over the NTDE (Gill and Schultz, 2001). Shoreline mapped by NOAA NGS is tide datum based (specifically, MHW and MLLW), and the remainder of this chapter will focus on tide-datum-based shoreline, except where otherwise noted.

6.3 BRIEF HISTORY OF SHORELINE MAPPING IN THE UNITED STATES

As shoreline mapping campaigns started in the United States in the early 19th century—well before even the earliest documented acquisition of aerial photography from a hot air balloon in 1858—the initial techniques were all field based. The field survey procedures entailed using a plane table (level mapping surface mounted on a tripod), alidade (instrument composed of a straightedge and telescopic sight), and stadia rod (graduated rod used in measuring distances). An observer set up and leveled the plane table with a map sheet at a location from which a stretch of shoreline could be observed (see Figure 6.2). An assistant, known as a rodman, then walked the stretch of shoreline with the stadia rod, pausing at each prominent location or salient change in shoreline direction so that distances and directions could be observed and recorded on the sheet (Graham et al., 2003; Monmonier, 2008). Not including uncertainty introduced in sketching between surveyed

Figure 6.2 Field-based shoreline mapping procedures, using an alidade and plane table. (Courtesy of NOAA.)

points, Shalowitz (1964) estimated the maximum positional uncertainty (error) of the mapped shoreline to be 10 m when using these methods. The techniques, however, were clearly laborious and time consuming.

As in many mapping applications, the motivating factors in the transition from field surveys to photogrammetry included (a) the desire to map large areas quickly; (b) the difficulty—or, in some cases, outright impossibility—of accessing certain areas from the ground; and (c) the wealth of detail available in aerial photographs. Regarding the latter point, U.S. Coast and Geodetic Survey (USC&GS) cartographers quickly realized they could, in some cases, produce much more detailed shoreline maps using aerial photography. One of the initial test projects was conducted in coastal New Jersey in 1919 (see Figure 6.3). Although the accuracy of the photogrammetric methods initially was questioned, over a period of several years, techniques were developed and refined that enabled high accuracy as well as efficiency. By the early 1940s, a joint program between the U.S. Coast Guard and USC&GS allowed for the use of Coast Guard aircraft (see Figure 6.4), with USC&GS supplying cameras and personnel (Smith, 1981).

6.4 PHOTOGRAMMETRIC METHODS

Photogrammetric data continue to serve as the primary source for mapping shoreline depicted on U.S. nautical charts (Graham et al., 2003). Advantages of the photogrammetric methods include high spatial resolution, ability to

Figure 6.3 Photo mosaic of shoreline between Corson Inlet, New Jersey (just left of left edge of image) and Ocean City, New Jersey (right side of image) from 1919 U.S. Coast & Geodetic Survey (USC&GS) aerial photography. (Courtesy of NOAA.)

plan acquisition for a certain stage of tide, and the relative maturity of photogrammetric tools and techniques, as compared with other airborne and spaceborne remote-sensing technologies.

Aerial imagery acquisition for shoreline mapping is, in general, subject to many more constraints than is acquisition for other applications. The biggest constraint is typically tide coordination. To map a tide-datum-based shoreline on the basis of the position of the land-water interface in aerial imagery,

Figure 6.4 Aerial survey aircraft for coastal mapping, 1941. (Courtesy of NOAA.)

the imagery is acquired when the instantaneous water level in the project site coincides with the tide-datum surface. The times at which this occurs can be estimated from water levels that are either observed (via concurrent data from tide gauges) or predicted. When based on observed water levels, the imagery is referred to as "tide coordinated"; when based on predicted waver levels, it is referred to as "tide predicted" (Graham et al., 2003).

Even with extensive water-level data from gauges in or near the project site, acquisition of the imagery at the exact instant that the water level reaches the tide-datum surface is impossible in practice because of the following: (a) uncertainty in the water-level observations and (b) inability to acquire imagery for the entire project site instantaneously, no matter how fast the survey aircraft or how large the image footprint on the ground. A practical means of overcoming the latter challenge is to establish a "tide window," which is a small range of water levels, specified in terms of an allowable tolerance about the desired level. The tolerance is established as a compromise between the competing interests of maximizing the shoreline positional accuracy and providing sufficient time to enable the aircraft to acquire the imagery. Since 1982, this tolerance has been defined as a (discontinuous) function of mean tide range, r, for the project site:

$$T(r) = \begin{cases} 0.3 \text{ ft } (0.09 \text{ m}), \ r < 5 \text{ ft } (1.5 \text{ m}) \\ 0.1 \ r, \ r \geq 5 \text{ ft } (1.5 \text{ m}) \end{cases}. \tag{6.1}$$

A potential alternative to this one-size-fits-all threshold is a project-specific threshold based on beach slopes in the project site and a priori estimates of measurement uncertainties. This proposed approach would enable International Hydrographic Organization (IHO) shoreline positional uncertainty standards to be met as efficiently as possible, by customizing the tide window size to the project site characteristics, predicted measurement uncertainties, and shoreline uncertainty specifications. A drawback, however, is that this method requires that reliable beach slope data be available during the project planning stage.

Sun angle poses another constraint in image acquisition. Project instructions typically require a sun angle of 30° or greater. Furthermore, to avoid sun glint over water, the maximum sun angle is typically set to 45°, so that the allowable range of sun angles for acquisition can be expressed as follows: $30° \leq \emptyset \leq 45°$. Although this requirement can be fairly restrictive, the technological transition from aerial film to digital cameras may relax sun-angle restrictions because of digital cameras' greater dynamic range.

The spectral band(s) and spectral sensitivity range of the imagery are further considerations when acquiring aerial imagery for shoreline mapping. Near infrared (NIR) imagery provides strong contrast between land and water surfaces because NIR radiation is strongly absorbed by water, which results in submerged areas being dark in the imagery. Meanwhile, the Earth materials that make up beaches are typically bright NIR reflectors. This contrast greatly facilitates delineating a land-water interface, regardless of whether this delineation is done by a human compiler or an auto-feature extraction algorithm.

The ability to more easily demarcate the land-water interface led the USC&GS to begin using black-and-white infrared (B&WIR) film for aerial photography starting sometime in the 1950s or early 1960s, depending on which historical account one follows (Swanson, 1960; Smedley, 1986). The B&WIR emulsion also enabled better haze penetration. It quickly became apparent, however, that use of this emulsion alone was not always sufficient to guarantee sharp contrast between land and water. In areas of clear, shallow water and bright, sandy bottom, NIR water penetration may make it difficult to distinguish between subaerial (land) and subaqueous (underwater) regions. Empirical analysis in the late 1960s and 1970s led to the development and purchase of optical filters: specifically, long-wave pass filters, aimed to block radiation below a specified cut-on in the NIR (Parrish et al., 2005). Because of more recent research in this area and the larger spectral sensitivity ranges achievable with modern digital aerial cameras, the digital aerial camera currently operated by NGS uses an optical filter with an 850 nm cut-on wavelength, enabling acquisition in an NIR band of ~850–1,000 nm.

The importance of spectral band selection for shoreline mapping can be understood with reference to Figure 6.5. The left image (Figure 6.5a) shows a small bay in Fort De Soto Park, Florida, imaged in the visible portion of the electromagnetic (EM) spectrum (~400–700 nm). The center

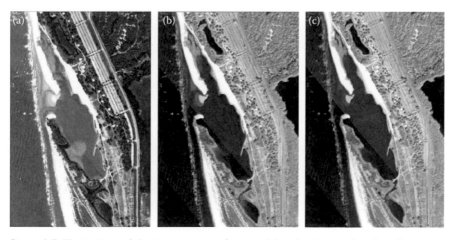

Figure 6.5 Illustration of the importance of spectral band selection for shoreline mapping. A small bay in Fort Desoto County Park, Florida, is shown imaged in (a) the visible portion of the EM spectrum; (b) a NIR band of ~700–1,000 nm; and (c) an NIR band of ~850–1,000 nm.

image (Figure 6.5b) shows the same area imaged in the NIR (~700–1,000 nm) and the right image (Figure 6.5c) shows the area again imaged in the NIR, but this time with a cut-on wavelength of ~850 nm (i.e., wavelengths shorter than 850 nm are blocked). Focusing on the small, roughly triangular feature near the north end of the inlet and the narrow, northern portion of the bay, it is difficult to distinguish between submerged and subaerial (i.e., above-water) areas in imagery acquired in the visible portion of the EM spectrum (Figure 6.5a). It is substantially easier in the NIR (Figure 6.5b). The water body on the right-hand side of the image provides another clear example of the superiority of NIR over visible bands for distinguishing between land and water. It is difficult to pick out this water body in Figure 6.5a, whereas in image Figure 6.5b, there is sharp contrast between the water body and surrounding vegetation. The differences between Figure 6.5b and Figure 6.5c are subtler, but careful examination shows that the land-water interface is clearest in the latter, illustrating the importance not only of imaging in the NIR, but also of using a long-wave pass optical filter to block all wavelengths below a specified cut-on. (The slight, but perceptible, changes in tone across diagonal lines running northwest-southeast are seamlines between adjacent aerial images.)

Additional requirements for image acquisition include the following (NOAA, 2011a):

- There shall not be any clouds or cloud shadows appearing on the photographs.
- Visibility at time of acquisition should be ≥ 8 statute miles (13 km).
- Camera tilt shall not exceed 3° for any image.
- Endlap (overlap between successive exposures) shall normally be 60%, except over areas of primarily water, in which case it should be 80%.
- Sidelap (strip overlap between adjacent flight lines) is normally 30%.

The combination of these requirements can make it challenging and time consuming to acquire imagery for mapping tide-datum-based shoreline. All of the requirements were developed over a period of decades, however, to ensure that the final shoreline data product meets specifications.

Although aerial imagery has been used in shoreline mapping in NOS and its predecessor agencies since the 1920s, technological advancements continue to be made. From the early 1980s through early 2000s, advancements have included the transition from analog to analytical and finally softcopy photogrammetry. The first of these phases, analog, involved using hardcopy photographs and large, optical-mechanical instruments. In analytical photogrammetry, hardcopy images were still used, but the three-dimensional (3D) modeling was done mathematically using computers. Finally, in softcopy photogrammetry, images are digital (not hardcopy) and all of the display and mathematical modeling is handled on a computer. Another advancement that began during the latter part of this period was the transition from film-based aerial cameras to digital cameras. One of the more recent advancements involves the use of automatic feature extraction (AFE) tools for delineating the shoreline. A common approach entails running an object-oriented image classification algorithm to segment the image—at appropriate, user-specified scale(s)—followed by classifying and aggregating the image objects to output a binary classification (land-water) image. The boundaries between the two classes are then vectorized, followed by editing. AFE can lead to great time savings by dramatically reducing the amount of manual compilation required in a project.

6.5 LIDAR-BASED METHODS

Although tide-coordinated aerial imagery remains the mainstay for mapping tide-datum-based shoreline in NOAA, LiDAR-based methods, where applicable, can be advantageous. LiDAR technology is covered elsewhere in this book, but, briefly, an airborne LiDAR system uses a pulsed laser in an aircraft to measure ranges to the Earth's surface (ground or elevated objects, such as canopy). Most mapping systems employ a scanning mechanism, such as an oscillating or rotating mirror, to deflect outgoing pulses through a range of scan angles (e.g., $\pm 15 - 35°$), thereby creating a swath on the ground. To obtain georeferenced point clouds, the measured ranges and scan angles are combined with sensor position and orientation information obtained from postprocessed airborne kinematic GPS, inertial measurement unit (IMU) data, and sensor calibration data. A point cloud is a collection of points in 3D space, with each point containing 3D spatial coordinates (e.g., latitude, longitude, height)—and possibly intensity and other attributes—for each point from which a laser pulse was reflected as the survey aircraft passed over. If desired for the end user's application, the point cloud data can be filtered to "bare Earth" and gridded to create digital elevation models (DEMs) and other derived terrain data products, such as contours. Topographic LiDAR systems typically use NIR lasers for measuring subaerial topography, whereas topographic-bathymetric systems use green (532 nm) laser beams that can penetrate the water column to also obtain estimates of seafloor elevation. Some topographic-bathymetric systems use both NIR and green laser beams, whereas others, such as the U.S. Geological Survey (USGS) Experimental Advanced Airborne Research LiDAR (EAARL) system, are green-only.

One benefit of the use of LiDAR for shoreline mapping is that it allows some of the data acquisition constraints to be relaxed. LiDAR is an active remote-sensing technology, meaning that it uses its own source of EM radiation (the laser), as opposed to relying on reflected sunlight. Thus, LiDAR data can be collected without regard to sun angle (i.e., even at night). In theory, elimination of the sun angle requirement may increase the amount of time available per year for acquisition by a factor of ~3.5. In practice, however, project instructions often call for imagery to be collected concurrently with the LiDAR data, effectively negating this benefit. Yet, even with requirements for simultaneous imagery acquisition, LiDAR acquisition windows are still much larger than those required for tide-coordinated aerial imagery. The reason is that LiDAR data for shoreline mapping typically need to be acquired only when the instantaneous water level in the project site, z, is *below* some specified level (i.e., $z < T$). This is a less stringent requirement than for tide-coordinated imagery, which must be acquired within a narrow time window (i.e., $T_1 < z < T_2$), centered on the time at which the instantaneous water level in the project area is predicted or observed to coincide with the MHW or MLLW tide-datum surface. Other requirements for coastal imagery acquisition are also not applicable in LiDAR acquisition, including the following: (a) the restriction on cloud shadows (NIR LiDAR radiation cannot penetrate clouds, but clouds above the survey aircraft typically do not pose a problem); (b) the 8 mile (13 km) visibility requirement; and (c) the requirement that the sun be over the water.

Another advantage is that LiDAR-based shoreline extraction methods are usually highly automated, as compared with photogrammetric methods, as the latter often entail manual delineation of the shoreline on the tide-coordinated aerial imagery. This increase in automation not only decreases the required human time, but also makes the process more repeatable. Applying the same algorithms to the same LiDAR data set should yield the same output (plus or minus very small round-off errors that may occur on different computers or operating systems), regardless of the day, time, or person who launches the computer program. This does not necessarily mean that the absolute accuracy of LiDAR-derived shoreline is better than that of photogrammetrically mapped shoreline. In fact, it is not possible to make blanket statements about which method yields more accurate shoreline, as the component uncertainties are different for each process, and their magnitudes can vary from project to project. Nevertheless, although increased repeatability (or precision) does not necessarily imply increased accuracy, it is a desirable characteristic.

Coastal LiDAR data also have the capability to support a wide range of applications, from regional sediment management, to storm vulnerability analysis, to habitat mapping. Of course, this is not a comprehensive list; any application requiring high-accuracy 3D spatial data in the coastal zone stands to benefit from coastal LiDAR. These combined uses have led to LiDAR data becoming available for increasingly larger portions of the coastal United States. The U.S. Army Corps of Engineers (USACE), through the Joint Airborne LiDAR Bathymetry Technical Center of Expertise (JALBTCX), administers a National Coastal Mapping Program that includes coastal data acquisition with the Compact Hydrographic Airborne Rapid Total Survey (CHARTS) sensor suite (Wozencraft and Millar, 2005). CHARTS includes a bathymetric-topographic LiDAR, multispectral aerial camera, and hyperspectral imager. The survey specifications for the JALBTCX program are to acquire bathymetric data from the shoreline to 1 km offshore and topographic data from the shoreline to 0.5 km onshore. The USACE and JALBTCX are currently funding development of a new system, known as the Coastal Zone Mapping and Imaging LiDAR (CZMIL) (Wozencraft, 2010).

Other federally funded or administered LiDAR coastal mapping efforts include the USGS National Geospatial Program's (NGP) LiDAR acquisition along portions of the U.S. coastline in support of the National Map. NGP coastal areas of focus include the coasts of the northern Gulf of Mexico, the Chesapeake Bay, and portions of the coast of California, including San Francisco and San Pablo Bays. Coastal LiDAR acquisition for other science projects is carried out with the EAARL sensor, owned and operated by the USGS Coastal and Marine Geology Program. Originally developed at the NASA Wallops Flight Facility, the EAARL is a short-pulse, green-wavelength topographic-bathymetric LiDAR system that has been used in mapping coral reefs, rivers, barrier islands, wetlands, and coastal vegetation, as well as shoreline (Brock et al., 2004; Nayegandhi et al., 2006; Kinzel et al., 2007; White, 2007). The EAARL is currently being upgraded to improve measurement of bathymetry in shallow and turbid water environments. LiDAR data from these federally funded programs is available through NOAA's Digital Coast web portal at http://www.csc.noaa.gov/digitalcoast (NOAA, 2011b), the USGS Center for LIDAR Information Coordination and Knowledge at http://LiDAR.cr.usgs.gov (USGS, 2011a), and the USGS Date Series and Open Access Reports (USGS, 2011b).

The method used by NOAA NGS to map the national shoreline from LiDAR is based on NOAA's vertical-datum transformation utility, VDatum (Myers et al., 2007), which is described in detail in White (2007) and White et al. (2011). The input to the process consists of coastal LiDAR point clouds in LAS file format. LAS is a publicly available point cloud format maintained by the American Society for Photogrammetry and Remote Sensing (ASPRS) and is described in Graham (2005). The input LiDAR point clouds are referenced to the North American Datum of 1983 (NAD 83) and contain continuous data across the shoreline of interest. Topographic-bathymetric LiDAR data with accurate, continuous coverage across the entire intertidal zone and adjacent land and water areas are ideal. Topographic-only data are also viable, however, if collected when the instantaneous water level is below the tide-datum corresponding to the shoreline to be mapped (e.g., MHW) by a defined threshold.

Briefly, the first step is to clean the LiDAR point cloud in the immediate vicinity of the shoreline to remove returns from vegetation, birds, and so on,

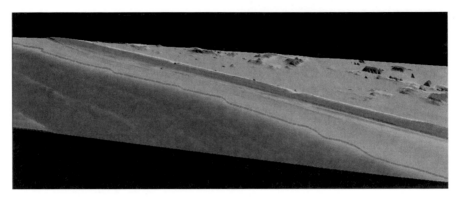

Figure 6.6 **(See color insert.)** LiDAR-derived shoreline (magenta line).

or to classify these returns in the LAS file, such that they can be excluded in subsequent processing steps. Next, the entire point cloud is processed through VDatum to transform from NAD 83 ellipsoid heights to the tide datum corresponding to the shoreline to be mapped (we will assume MHW for this discussion). The MHW-referenced point cloud is then interpolated to a regular grid, commonly referred to as a DEM. Typical grid spacing is on the order of 1 m. An autocontouring routine is then run on the DEM and a zero-elevation contour is extracted (see Figure 6.6). At this point in the process, the zero-elevation contour represents the modeled intersection of the MHW tidal datum and topographic surface, which is how the tide-datum-based shoreline is defined. To complete the process, this shoreline must be edited and attributed (i.e., a shoreline classification code must be assigned). Finally, quality assurance and quality control (QA/QC) are performed. The latter steps (i.e., editing, attribution, and QA/QC) are ideally performed with the aid of concurrently collected, stereo-aerial imagery, although nonconcurrent, monoscopic imagery can be used.

Another approach to mapping shoreline from LiDAR is based on extracting a set of cross-shore profiles (Stockdon et al., 2002). A linear or higher degree polynomial fit to each elevation profile is then performed, after which the shoreline position—defined by a specified elevation—along the profile can be computed. This results in N nodes (where N is the number of profiles) that can be connected to generate a shoreline vector. When applying this method, the spacing of the profiles should be considered in relation to the fractal dimension (or other measure of jaggedness) of the shoreline, as well as the desired mapping resolution.

Although these LiDAR-based approaches to shoreline mapping are proving to be effective and efficient, there are some challenges. In particular, LiDAR shoreline mapping requires obtaining good, continuous coverage in the intertidal zone, which is arguably *the* most difficult area in which to acquire data. By definition, the intertidal zone is intermittently inundated and exposed. Even when exposed, it is typically wet from the previous high tide, which can reduce signal-to-noise ratio (SNR) because of the absorption of the laser energy by the wet surface. Wave swash poses another problem. Additionally, the substrate in the intertidal zone can be very dark at NIR wavelengths. With topographic LiDAR systems, these factors can lead to sparse, noisy data. Additionally, even if the coverage in the intertidal zone is adequate, the shoreline extraction algorithm is often forced to operate on the very edge of the data, where many algorithms have difficulty.

Topographic-bathymetric LiDAR systems, which can acquire both submerged and subaerial topography, would seem to inherently overcome the challenge of acquiring seamless data across the land-water interface. Although topographic-bathymetric systems are, in fact, extremely well-suited for shoreline mapping, the use of such a system does not guarantee good, continuous coverage across the land-water interface. One reason is that it can be very difficult to resolve the bottom return in LiDAR waveforms acquired over shallow, turbid areas, which sometimes can lead to nearshore data gaps. Furthermore, some topographic-bathymetric LiDAR-processing software requires an estimate of the position of the land-water interface, such that different waveform-processing algorithms can be applied on either side of this interface. If the estimate of the land-water interface is poor, this can compromise the data in the intertidal zone. In general, the "ideal" LiDAR data set for shoreline mapping includes topographic LiDAR acquired at low tide and bathymetric LiDAR acquired at high tide. Such a merged data set, especially when carefully edited, tends to work very well for shoreline extraction.

Another challenge with topographic LiDAR is that when used alone, it does not enable identification of slightly submerged, alongshore rocks, as is possible with natural-color aerial imagery. This problem, however, is easily overcome with concurrently collected imagery. Notwithstanding the other noted challenges, LiDAR is already proving extremely beneficial in coastal mapping, and its use in the coastal zone is likely to continue to grow exponentially in the coming years.

6.6 SATELLITE IMAGERY

In the 21st century, commercial high-resolution satellite imagery that can be used for mapping applications is readily available. Table 6.1 contains a list of commercial satellites providing better than 1 m spatial resolution.

One attractive feature of satellite imagery is cost: In many cases, satellite imagery covering a project site can be obtained for a fraction of the cost of dedicated aerial imagery acquisition. (An important clarification here is that, in this cost comparison, we are referring only to the cost to end users of obtaining existing satellite imagery from distributers. If we were

Table 6.1 High-resolution commercial satellite imagery

Satellite	Company	Launch year	Spatial resolution
IKONOS	GeoEye (formerly Space Imaging)	1999 (first submeter commercial satellite)	0.80-m panchromatic and 3.2-m multispectral
QuickBird	DigitalGlobe	2001	0.60-m panchromatic and 2.4-m multispectral
WorldView-1	DigitalGlobe	2007	0.50-m panchromatic
GeoEye-1	GeoEye	2008	0.41-m panchromatic and 1.65-m multispectral
WorldView-2	DigitalGlobe	2009	0.46-m panchromatic and 1.84-m, 8-band multispectral

to account for the life-cycle costs of the satellites themselves, this would alter the comparison.) Additionally, satellites offer the advantage of repeat passes, providing multitemporal imagery, which is highly advantageous in analyzing coastal processes. Some challenges, however, are associated with using satellite imagery for mapping tide-datum-based shoreline, including the following:

- It is extremely difficult to obtain tide-coordinated commercial satellite imagery (i.e., imagery acquired at a specified stage of tide).
- Although high-resolution commercial satellite imagery can be obtained as stereo pairs, it is much more commonly available in the form of monoscopic images. Hence, photogrammetric procedures for shoreline extraction that rely on stereoscopic measurement are often not amenable to satellite imagery.

Because of the lack of tide coordination, use of satellite imagery for mapping shoreline is often limited to (a) nontidal areas, or areas of diminished tidal influence, defined by NOS as areas with a mean tide range < 0.2 ft. (6 cm) (Graham et al., 2003); (b) ports and other areas of manmade shoreline; or (c) areas for which contemporary shoreline does not exist and are so difficult to map that instantaneous shoreline is the most practical option. Satellite imagery also can be used for analyzing changes to shoreline features (e.g., piers, bulkheads, jetties, and groins), as is done in NGS's Coast and Shoreline Change Analysis Program (CSCAP). Methods of mapping tidally referenced shoreline from a time-series of instantaneous shorelines extracted from satellite imagery (Li et al., 2002) also may lead to increased use of satellite imagery for mapping tide-datum-based shoreline. The general concept is that from a time-series of instantaneous shorelines bracketing MHW or MLLW, an appropriate interpolation technique can be applied to obtain the MHW or MLLW shoreline.

6.7 OTHER REMOTE-SENSING TECHNOLOGIES FOR SHORELINE MAPPING

Operating in the microwave portion of the EM spectrum, synthetic aperture radar (SAR) systems provide the advantage of being able to penetrate a variety of types of weather, from clouds to rain, snow, ice, and fog. This makes SAR particularly well suited for coastal areas of persistent cloud cover, such as portions of Alaska. Like LiDAR, SAR is an active sensor technology, meaning acquisition is also independent of time of day. Tests conducted by NGS in the late 1990s and early 2000s revealed that commercially available SAR imagery available at that time did not provide sufficient image resolution for mapping the national shoreline (Graham et al., 2003). Since then, however, high-resolution (up to 1 m in "spotlight" modes) SAR imagery has become available from TerraSAR-X, Cosmo-SkyMed, and Radarsat-2, leading to renewed interest.

Other possible deterrents to wider use of radar for shoreline mapping are that (a) radar images are generally more difficult to (manually) interpret than optical images, and (b) fewer people are trained in SAR analysis, as compared with optical imagery. The all-weather capabilities and commercially available, high-resolution SAR data make the technology increasingly attractive for shoreline mapping. It is likely that the use of SAR for shoreline mapping will become increasingly prevalent.

Hyperspectral imagery (HSI) is another promising data source for shoreline mapping. As noted, the spectral band(s) in which one images are a strong factor in the ability to distinguish land from water. By imaging in dozens to hundreds of narrow, contiguous spectral bands across a wide spectral sensitivity range, HSI sensors, combined with suitable signal processing algorithms, are readily able to distinguish between land and water. Furthermore, HSI enables analysts to go beyond this simple, binary (land-water) classification and automatically assign shoreline attributes (e.g., mangrove or cypress, marsh, manmade, etc.) through image classification routines. There are, however, some drawbacks to the use of HSI for shoreline mapping. The first is that there is a general trade-off between spectral resolution and spatial resolution and accuracy that occurs with multispectral images and HSI: Greater spectral sensitivity generally comes at the expense of decreased spatial resolution and accuracy. Because positional accuracy is such a strong consideration in shoreline mapping (see Section 6.8, Uncertainty Analysis), spatial resolution and accuracy are typically prioritized over spectral resolution. Furthermore, software and tools for efficient, productive use of HSI data have, arguably, lagged a bit behind those for working with multispectral (NIR, R, G, B) digital camera imagery. Notwithstanding these issues, hyperspectral imagery nicely complements other sensor data when working in the coastal zone. For example, the fusion of HSI and LiDAR data used in the CHARTS and CZMIL systems enables autogeneration of a variety of products (Park et al., 2010; Park and Tuell, 2010). In particular, this combination of sensor technologies can provide enhanced information on both *where* things are (high-accuracy spatial coordinates) and *what* they are (image-feature classification).

Thermal imagery is yet another potential data source for shoreline mapping. The use of thermal imagery for this application relies on being able to distinguish land from water based on radiant temperature differences (see Figure 6.7). Because water surface temperature is typically lower than that of the adjacent land during the day and higher than that of the land at night, the land-water interface can be extracted, based on this temperature contrast. Beyond the usual tide coordination requirements, the time of image acquisition is a critical factor, as there are generally two times per day when contrast between land and water is low (the so-called thermal crossover points). Other factors that can affect the contrast include currents (ebb vs. flow), season, moisture condition of the terrain surface (wet vs. dry), and atmospheric conditions. Generally, however, availability of suitable thermal imagery is a greater concern than the technical challenges related to being able to extract the land-water interface. Many Earth-observing satellites

Figure 6.7 **(See color insert.)** (a) Thermal image of a stretch of shoreline in Monterey, California, acquired with an Itres TABI. (b) RGB imagery of same area acquired with an Applanix DSS digital aerial camera. These data were acquired in the late afternoon. (A seamline from the image mosaicing process is evident in the bottom image.)

have thermal infrared bands (e.g., Landsat TM and ETM+ band 6; ASTER bands 10–14; and MODIS bands 20–36), but their spatial resolution is far too coarse for shoreline mapping for nautical charting purposes. Mapping-grade airborne thermal imagers do exist and are used for applications ranging from wildfire mapping to urban energy efficiency studies. They are much less common, however, than aerial mapping cameras that sense reflected radiation in the visible and near infrared, and the achievable spatial resolution and accuracy often are not as good. As with HSI, thermal imagery is of interest in fusion-based approaches to coastal mapping.

6.8 UNCERTAINTY ANALYSIS

It is nearly always required to assess and report the spatial accuracy—or "uncertainty," to use the preferred term—of geospatial data products. Uncertainties are provided to end users to assist in evaluating the suitability of the data for a particular purpose. For shorelines, reliable estimates of uncertainty are vital because of the legal significance of the data. Specifically, with private, state, federal property rights and international maritime boundaries at stake, estimates of uncertainty must accompany the data products. Furthermore, the use of the shoreline in supporting safety of navigation necessitates that applicable standards be satisfied and that uncertainties be published in the mandatory metadata.

The IHO S-44: *Standards for Hydrographic Surveys* (IHO, 2008) contains minimum standards for uncertainty of "coastline" (shoreline) positioned as part of a hydrographic survey. This document states that "a statistical method, combining all uncertainty sources, for determining positioning uncertainty should be adopted." The positional uncertainty standard, at the 95% confidence level, for coastlines is listed as 10 m for Special Order surveys and 20 m for all other survey orders. In practice, the total propagated uncertainty (TPU) can be assessed using analytical uncertainty propagation, or when the analytical approach is unfeasible, Monte Carlo simulations (White et al., 2011). In the LiDAR shoreline mapping workflow, component uncertainties in the process include uncertainties in sensor position and orientation, range, and scan angle, as well as uncertainties introduced in the vertical datum transformation, gridding, and contouring steps. In the photogrammetric shoreline mapping workflow, component uncertainties include those in the camera exterior orientation parameters, image measurement, and water-level data, as well as uncertainty introduced in the manual (or auto) shoreline delineation. In both cases, these component uncertainties must be propagated through the end-to-end shoreline mapping process to obtain estimates of uncertainty in the final, mapped shoreline position.

Beach slope is an important factor in the positional uncertainty of mapped shoreline, regardless of the mapping technologies or methods.

Shallow-sloped beaches—mudflats being an extreme case—lead to higher uncertainty in shoreline position. To understand this, consider our earlier definition of tide-datum-based shoreline: the intersection of the topographic surface and the tide-datum surface. When beach slopes are shallow (i.e., the topographic surface is nearly flat), even small errors in the topographic surface or tidal-datum surface propagate to large horizontal errors. This can be quantified as follows: an elevation uncertainty of σ_z propagates to an uncertainty in the position of the shoreline (in the aspect direction), σ_s, as follows:

$$\sigma_s = \sigma_z \cot \theta. \tag{6.2}$$

In Equation 6.2 we assume the beach slope, σ_s, to be constant over short distances in the intertidal zone. For example, assuming an elevation uncertainty of 30 cm leads to the shoreline positional uncertainties shown in Table 6.2 for various beach slopes. The relationships between shoreline positional uncertainty, beach slope, and elevation uncertainty are further illustrated in Figure 6.8.

6.9 SUMMARY AND FUTURE VISION

Given the importance of shoreline mapping in marine navigation, coastal science, climate change studies, and legal boundary determination—and the extent of coastal areas to be mapped worldwide—it is little wonder that the use of remote-sensing data and tools for this application remains a topic of great interest. The evolution of shoreline mapping technology has been remarkable, from the plane tables and alidades used in the field throughout the 19th century, to early aerial photography techniques developed in

Table 6.2 Propagated horizontal position uncertainty of shoreline for various beach slopes, assuming a constant 30 cm vertical uncertainty

Beach slope (°)	Reference beach type	Uncertainty in mapped shoreline position (m)
0.25	Mudflat	68.8
1.00	Sandy beach	17.2
3.00	Sandy beach	5.7
4.00	Sandy beach	4.3
6.00	Steep, sandy beach	2.9
8.00	Steep, rocky beach	2.1
15.00	Very steep, rocky	1.1
∞	Vertical seawall or vertical rock face	0

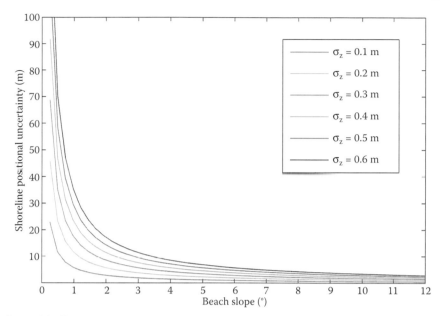

Figure 6.8 (**See color insert.**) Plots of shoreline positional uncertainty as a function of beach slope and elevation uncertainty, σ_z.

first half of the 20th century, to the advanced photogrammetry, LiDAR, and other remote-sensing tools and techniques in use in the 21st century. Current trends suggest that this rapid development will continue into the foreseeable future.

One ongoing area of research is enhanced efficiency in shoreline mapping. This is especially critical because of the number of considerations and constraints in acquiring and processing remotely sensed data for shoreline mapping, several of which were described in this chapter. Extraction of tidally referenced shoreline from multitemporal, but non-tide-coordinated, satellite imagery is one promising approach. Although photogrammetric and LiDAR-based shoreline mapping methods are operational, further investigation and use of radar, HSI, and broadband thermal imagery may lead to gains in efficiency and enhanced information about the coastal zone. Multisensor fusion-based operational approaches are especially attractive, not only for improving efficiency but also for supporting the goals of the IOCM initiative. This effort aims to provide multiuse data for a range of coastal mapping, science, and management applications. Use of remotely sensed data collected primarily for nautical charting to simultaneously support coastal ecosystems studies, change analysis, wetlands conservation efforts, and coastal vulnerability modeling are just a few examples.

The final topic discussed in this chapter, uncertainty analysis, is another area in which continuing research is needed. All of the applications of shoreline mapping data discussed in this chapter require reliable estimates of the uncertainty in the mapped shoreline position. The legal implications of the charted shoreline and navigational safety aspects are among the biggest factors motivating the focus on TPU. The challenge, of course, lies in the fact that with so many different sensors, algorithms, and workflows being used or investigated for shoreline mapping, it is difficult to generate reliable TPU estimates for each technique. Yet, there is little doubt that ongoing work will lead to progress on all of these fronts. We look forward to a future in which shorelines around the world can be mapped efficiently, accurately, and frequently enough to support detailed change analysis using the latest developments in remote sensing.

REFERENCES

Agardy, T., J. Alder, P. Dayton, S. Curran, A. Kitchingman, M. Wilson, A. Catenazzi, et al. 2005. Coastal systems in current state & trends assessment. *Millennium Assessment Report Series: Global Assessment Reports 1*. Washington, DC: Island Press.

Boak, E. H., and I. L. Turner. 2005. Shoreline definition and detection: a review. *Journal of Coastal Research* 21(4): 688–703.

Brock, J. C., C. W. Wright, T. D. Clayton, and A. Nayegandhi. 2004. LIDAR optical rugosity of coral reefs in Biscayne National Park, Florida. *Coral Reefs* 23: 48–59.

Crossett, K. M., T. J. Culliton, P. C. Wiley, and T. R. Goodspeed. 2004. *Population Trends along the Coastal United States: 1980–2008*. National Oceanic and Atmospheric Administration (NOAA), National Ocean Service. http://oceanservice.noaa.gov/programs/mb/pdfs/coastal_pop_trends_complete.pdf (accessed December 21, 2011).

Defense Mapping Agency (DMA). 1984. *Nautical Charting with Remotely Sensed Imagery 1*. Washington, DC: Defense Mapping Agency.

Gill, S. K., and J. R. Schultz. 2001. *Tidal Datums and Their Applications*. National Oceanic and Atmospheric Administration (NOAA) Special Publication NOS CO-OPS 1. http://tidesandcurrents.noaa.gov/publications/tidal_datums_and_their_applications.pdf (accessed December 21, 2011).

Graham, D., M. Sault, and J. Bailey. 2003. National Ocean Service shoreline: past, present, and future. *Shoreline Mapping and Change Analysis: Technical Considerations and Management Implications: A Special Issue of Journal of Coastal Research* 38(Special Issue): 14–32.

Graham, L. 2005. The LAS 1.1 standard. *Photogrammetric Engineering and Remote Sensing* 71(7): 777–780.

Grinsted, A., J. C. Moore, and S. Jevrejeva. 2010. Reconstructing sea level from paleo and projected temperatures 200 to 2100 AD. *Climate Dynamics* 34(4): 461–472.

International Hydrographic Organization (IHO). 2008. *IHO Standards for Hydrographic Surveys*, 5th ed., 36. Monaco: International Hydrographic Bureau.

Kinzel, P. J., C. W. Wright, J. M. Nelson, and A. R. Burman. 2007. Evaluation of an experimental LiDAR for surveying a shallow, braided, sand-bedded river. *Journal of Hydraulic Engineering* 133(7): 838–842.

Li, R., R. Ma, and K. Di. 2002. Digital tide-coordinated shoreline. *Marine Geodesy* 25: 27–36.

Meehl, G. A., T. F. Stocker, W. D. Collins, P. Friedlingstein, A. T. Gaye, J. M. Gregory, A. Kitoh, R. Knutti, et al. 2007. Global climate projections. In *Climate Change 2007: The Physical Science Basis. Contribution of Working Group I to the Fourth Assessment Report of the Intergovernmental Panel on Climate Change*, edited by S. Solomon, D. Qin, M. Manning, Z. Chen, M. Marquis, K. B. Averyt, M. Tignor, et al. Cambridge and New York: Cambridge University Press. http://www.ipcc .ch/publications_and_data/publications_ipcc_fourth_assessment_report_wg1_ report_the_physical_science_basis.htm

Monmonier, M., 2008. *Coastlines: How Mapmakers Frame the World and Chart Environmental Change*, 228. Chicago: The University of Chicago Press.

Myers, E., K. Hess, Z. Yang, J. Xu, A. Wong, D. Doyle, J. Woolard, et al. 2007. VDatum and strategies for national coverage. In *Proceedings of the Marine Technology Society/IEEE OCEANS Conference, Vancouver, BC, Canada*. http://ieeexplore.ieee.org/stamp/stamp.jsp?arnumber=04449348.

National Oceanic and Atmospheric Administration (NOAA). 2011a. *Scope of Work: Shoreline Mapping, Version 14A*. NOAA Coastal Mapping Program, Remote Sensing Division, National Geodetic Survey, U.S. Department Of Commerce. http://www.ngs.noaa.gov/ContractingOpportunities/CMPSOWV14.pdf (accessed December 21, 2011).

National Oceanic and Atmospheric Administration (NOAA). 2011b. *The Digital Coast*. http://www.csc.noaa.gov/digitalcoast (accessed December 21, 2011).

National Research Council (NRC). 2010. *Advancing the Science of Climate Change*, 528. Washington, DC: National Academies Press.

Nayegandhi, A., J. C. Brock, C. W. Wright, and M. J. O'Connell. 2006. Evaluating a small footprint, waveform-resolving LiDAR over coastal vegetation communities. *Photogrammetric Engineering and Remote Sensing* 72(12): 1407–1417.

Nicholls, R. J. 2004. Coastal flooding and wetland loss in the 21st century: changes under the SRES climate and socio-economic scenarios. *Global Environmental Change-Human and Policy Dimensions* 14(1): 69–86.

Park, J. Y., V. Ramnath, V. Feygels, M. Kim, A. Mathur, J. Aitken, and G. Tuell. 2010. Active-passive data fusion algorithms for seafloor imaging and classification from CZMIL data. In *Algorithms and Technologies for Multispectral, Hyperspectral, and Ultraspectral Imagery XVI, Proceedings of SPIE*, edited by S. S. Shen and P. E. Lewis, 7695, 769515.

Park, J. Y., and G. Tuell. 2010. Conceptual design of the CZMIL Data Processing System (DPS): algorithms and software for fusing LiDAR, hyperspectral data, and digital images. In *Algorithms and Technologies for Multispectral, Hyperspectral, and Ultraspectral Imagery XVI, Proceedings of SPIE*, edited by S. S. Shen and P. E. Lewis, 7695, 769510.

Parrish, C. E., M. Sault, S. A. White, and J. Sellars. 2005. Empirical analysis of aerial camera filters for shoreline mapping. In *Proceedings of American Society for Photogrammetry and Remote Sensing Annual Conference, March 7–11,*

Baltimore, Maryland. Bethesda, MD: American Society of Photogrammetry and Remote Sensing.

Scott, G., N. Wijekoon, and S. White. 2009. Multi-sensor mapping: integrating data streams for coastal science and management. *Hydro International* 13(2): 20–23.

Shalowitz, A. L. 1962. *Shore and Sea Boundaries: Volume 1, Boundary Problems Associated with the Submerged Lands Cases and the Submerged Lands Acts.* Publication 10-1. Washington, DC: U.S. Department of Commerce.

Shalowitz, A. L. 1964. *Shore and Sea Boundaries: Volume 2, Interpretation and Use of Coast and Geodetic Survey Data.* Publication 10-1. Washington, DC: U.S. Department of Commerce.

Smedley, K. G. 1986. Imaging land/water demarcation lines for coastal mapping. Bachelor's thesis, 47, Rochester Institute of Technology. Rochester, NY: Center for Imaging Science.

Smith, J. T., Jr. 1981. *A History of Flying and Photography in the Photogrammetry Division of the National Ocean Survey, 1919–79.* Chicago: University of Chicago Press.

Stockdon, H. F., A. H. Sallenger, Jr., J. H. List, and R. A. Holman. 2002. Estimation of shoreline position and change using airborne topographic LiDAR data. *Journal of Coastal Research* 18(3): 502–513.

Swanson, L. W. 1960. Photogrammetric surveys for nautical charting, use of color and infrared photography. *Photogrammetric Engineering* 26(1): 137–141.

U.S. Geological Survey (USGS). 2011a. Center for LIDAR Information Coordination and Knowledge (CLICK). http://LiDAR.cr.usgs.gov (accessed December 21, 2011).

USGS. 2011b. Data Series and Open File Reports. http://ngom.usgs.gov/dsp/data/products_region.php (accessed December 21, 2011).

White, S. 2007. Utilization of LIDAR and NOAA's vertical datum transformation tool (VDatum) for shoreline delineation. In *Proceedings of the Marine Technology Society/IEEE OCEANS Conference, Vancouver, BC, Canada.*

White, S. A., C. E. Parrish, B. R. Calder, S. Pe'eri, and Y. Rzhanov. 2011. LiDAR-derived national shoreline: empirical and stochastic uncertainty analyses. *Journal of Coastal Research* 62(Special Issue): 62–74.

Wozencraft, J. M. 2010. Requirements for the Coastal Zone Mapping and Imaging LiDAR (CZMIL). In *Algorithms and Technologies for Multispectral, Hyperspectral, and Ultraspectral Imagery XVI, Proceedings of SPIE*, edited by S. S. Shen and P. E. Lewis, 7695.

Wozencraft, J. M., and D. Millar. 2005. Airborne LiDAR and integrated technologies for coastal mapping and charting. *Marine Technology Society Journal* 39(3): 27–35.

Chapter 7

Seeing residential buildings from remotely sensed imagery

An object-oriented approach

Xuelian Meng, Nate Currit, Le Wang, and Xiaojun Yang

CONTENTS

Automatically extracting land use information from remotely sensed imagery is an active yet challenging topic. As the urban unit that is related most closely to the spatial distribution of population, the buildings for residential land uses are of special interests for broad applications. This paper presents a three-step approach to identify residential buildings using light detection and ranging (LiDAR) data, aerial photographs, and road maps. A multidirectional ground-filtering algorithm first separates ground from LiDAR data to produce a digital surface model, a digital terrain model, and the height of objects above ground. Then, a morphology-based building-detection method extracts buildings by gradually removing other objects (especially trees) based on the difference in the first and last returns of LiDAR data, building height, vegetation indexes from aerial photograph, and the morphological characteristics of building footprints. Finally, residential buildings are separated through the classification based on seven land use indicators: area, height and compactness of buildings, the distance to major roads, the percentage of green space and parking space surrounding a building, and building density within a block. The method was tested in an area in Austin, Texas. The results showed that the method successfully extracted buildings from LiDAR and aerial photographs and identified 81.1% of the residential buildings.

7.1 INTRODUCTION

The accelerated development of human activities over the last century has altered approximately one-third of the Earth's surface (Vitousek et al., 1997). In addition to a larger global population, more than 50% of the Earth's population now lives in urban areas (Lutz et al., 2004; United Nations, 2007). Characterizing the spatial distribution of residential land uses is a critical topic in remote sensing because it contributes to broader research areas such as population estimation, industrial or commercial site selection, urban sprawl modeling, and intraurban physical structure analysis (Mesev, 2003; Wu, 2006; Meng, 2010; Meng et al., 2012). Accurately identifying residential land uses directly from remotely sensed images, however, has proven to be a challenging issue because land use functions reflect social applications of buildings or lands instead of the physical materials (Gong and Howarth, 1992; Herold et al., 2003; Lu and Weng, 2006). For example, some researchers have classified residential pixels based on the assumption that the tilted roofs of residential buildings usually reflect differently from nonresidential buildings with flat and concrete roofs (Barnsley and Barr, 1996; Tang et al., 2007). This assumption, however, may lead to the over- or underestimation of residential land uses because many buildings do not meet this assumption. For example, high-rise buildings may actually be multiresidential buildings and single-story buildings actually may house small businesses. Therefore, relying on physical reflectance alone cannot accurately be used to identify residential buildings.

Recent advancements in image resolution, the frequency of image acquisition, and the areal coverage of acquired imagery allow for improved quantitative analysis of the spatial distribution of urban land use patterns and their relationship with environmental change and other issues occurring at various scales (Jensen, 2007). Within the scope of urban remote sensing, three main topics are of special significance: (a) the use of multiscale images to characterize the spatial distribution of human activities and their changes over time (e.g., land cover and land use change); (b) feature detection from high-resolution data to depict specific elements of the urban landscape (e.g., building arrangements); and (c) determination of the drivers of a changing urban landscape (e.g., urban sprawl, environmental change, land use planning, and resource consumption).

Characterizing residential land use distribution, which is important for a wide range of applications, remains an unsolved challenge in urban remote sensing. Here, we introduce methods that provide an object-oriented approach to characterizing urban residential buildings using light detection and ranging (LiDAR) data and aerial photographs. Land use classification at the building level is a challenging task in urban studies, in part, because of the difficulty of identifying individual buildings, but also because of the need to identify the land use category of buildings. Several studies have explored methods to observe how buildings with different land uses differ from each other on the basis of study sites with buildings of one land use type by comparing such characteristics as area and compactness of building footprints, node degree (number of edges to a node represented by a building), and distance between adjacent buildings (Barnsley et al., 2003; Barnsley and Barr, 1997). At the time of this writing, however, no other groups have reported usage of those characteristics to classify land use categories of buildings. In this chapter, we present a three-step approach to detect all buildings and to specifically identify residential buildings based on seven spatial statistical indicators. Section 7.2 presents the theoretical logic of the three methods: (a) ground surface detection, (b) building extraction, and (c) residential building identification using LiDAR data, aerial photographs, and a road vector file; Section 7.3 presents validation results for a study site in Austin, Texas; and Section 7.4 assesses the overall performance of the proposed methods and discusses their potential applications in urban areas.

7.2 AN OBJECT-ORIENTED APPROACH TO IDENTIFY RESIDENTIAL BUILDINGS FROM LiDAR AND AERIAL PHOTOGRAPHS

Current methods for urban land use classification can be grouped into three main categories: pixel-based, parcel-based, and object-oriented classification (Wu, 2006; Meng et al., 2012). Pixel-based classification methods usually apply supervised or unsupervised algorithms to classify each individual pixel as a particular land cover or land use type. Parcel-based methods first group adjacent pixels based on shared common characteristics—such as homogeneity, smoothness, and texture—and then classify each group as a particular land cover or land use type. In contrast, object-oriented methods first detect features of interest (such as buildings) as objects and then classify each object into a land use category based on object-associated land use indicators. A land use parcel may contain different types of urban objects, but an urban object usually corresponds to a feature of interest (e.g., building). For example, a commercial land use parcel may contain commercial buildings, grass, trees, parking lots, and roads. Object-oriented classification methods aim to identify each of those objects individually. The advantages of the object-oriented land use classification approach include its effectiveness in defining object properties, such as area, height, and shape compactness; its ability to analyze the spatial relationships of objects to their surrounding objects; and its similarity to human visual interpretation of land uses.

In this research, our goal is to detect residential buildings directly from remotely sensed imagery. Three data sources are required: LiDAR data to determine building footprints and heights, aerial photographs to determine the amount of green space surrounding buildings, and a road map to determine the distance to major roads. Raw LiDAR data include a collection of points with three-dimensional (3D) coordinates from multiple returns of LiDAR signals. The first return represents points reflected from the top layer of the imaged surface, including bare ground, building roofs, or tree crowns (Vosselman, 2000; Zhang et al., 2003; Sithole and Vosselman, 2004; Meng et al., 2010). The last return records the lowest place the laser beams can reach. Therefore, the largest differences between the first and last returns normally occur in vegetated areas, which is useful for separating trees from other features, such as buildings and ground area. The following content introduces a three-step object-oriented method to utilize the three data sources and data characteristics to separate residential buildings. We first use a multidirectional ground filtering algorithm to separate ground areas to generate a digital elevation model (DEM) and to identify object heights above the ground. Then a morphology-based building-detection method further removes other nonground areas to derive the outline of building footprints. Finally, the building-based land use classification method calculates seven statistics as land use indicators through spatial analysis tools and separate residential buildings through a supervised decision tree classification method.

7.2.1 Multidirectional ground filtering from LiDAR data

The purpose of LiDAR ground filtering is to identify and separate LiDAR returns originating from the ground surface from those originating from a nonground surface. The algorithm presented here has been developed based on the observation that elevation differences between nonground points (e.g., vegetation or buildings) and neighboring ground points are generally more abrupt than elevation differences between neighboring ground points. Specifically, we use the first return for ground filtering, DEM generation, and building detection (Meng et al., 2009a, 2009b). The last return is used to identify and remove vegetation and thus assists in building detection.

Compared with the abrupt elevation changes on the boundaries of objects above ground, ground slopes are usually relatively flat and change in a gradual fashion with a smooth texture (Shan and Sampath, 2005; Meng et al., 2009a). Therefore, many ground-filtering algorithms utilize these characteristics to separate ground areas based on slope and the elevation difference in reference to its local environment (Vosselman, 2000; Zhang and Whitman, 2005). The introduced multidirectional ground-filtering algorithm is an improved version of the labeling algorithm developed by Shan and Sampath (2005). The algorithm labels each pixel as uncertain, ground, or nonground based not only on the slopes along the labeling direction and elevations to its local minimum values but also on the label of the previous pixel along the labeling direction and the elevation differences to its local minimum elevation and nearby ground elevation (see Figure 7.1). This labeling process is repeated from different combination of the four possible directions: left to right, right to left, top to bottom, and bottom to top. As a result, more identified ground pixels are gradually added to the ground collection, and chances to find a closer ground point are significantly improved in the next labeling process, thus leading to more accurate labeling.

Current LiDAR ground-filtering algorithms commonly work on two types of input: the raw LiDAR point clouds and the interpolated LiDAR data in raster format. Advantages and disadvantages of both input types have been discussed in Meng et al. (2009a). The ground-filtering method in this research uses interpolated raster data derived by nearest neighbor point sampling (Zhang et al., 2003; Meng et al., 2009a). A pixel of the ground sample is automatically selected using the lowest location in a specified

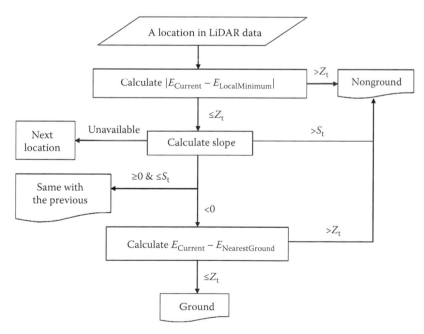

Figure 7.1 Flowchart of the multidirectional ground filtering algorithm. (Modified from Meng, X. L., L. Wang, J. L. Silván-Cárdenas, and N. Currit, *ISPRS Journal of Photogrammetry and Remote Sensing* 64: 117–124, 2009. Courtesy ISPRS.)

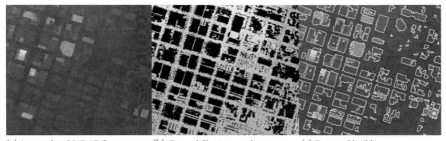

(a) Interpolated LiDAR first return   (b) Ground filtering result   (c) Detected buildings

Figure 7.2 An example of (a) interpolated LiDAR first return, (b) ground-filtering result, and (c) building-detection result. The black color in (b) represents nonground and the detected building footprint in white is overlaid on the interpolated LiDAR data in (c).

window to initiate the labeling process. The flowchart in Figure 7.1 illustrates the logic to determine the label of each pixel along a selected direction. For each pixel, the algorithm first compares its elevation with its minimum elevation in a predefined neighborhood and labels it as nonground if the elevation difference is larger than the user-specified elevation threshold. If the elevation difference is smaller than the elevation threshold, the algorithm will calculate the slope along the labeling direction. If the slope is not available (e.g., the first point of a line along the labeling direction), the algorithm skips to the next pixel. If the slope is larger than the threshold, it indicates that the slope is heading upward and the pixel is labeled as nonground. If the slope is positive but not larger than the threshold, the pixel is labeled the same as the previous pixel. If the slope is negative (i.e., descending) and its elevation to the nearest ground is smaller than the elevation threshold, the pixel is labeled as ground; otherwise, it is labeled a nonground pixel. Figure 7.2b shows an example of the results of ground filtering applied to the area shown in Figure 7.2a.

7.2.2 Morphology-based building detection

After ground filtering, the remaining task for building detection is to remove other above-ground objects, such as trees, shrubs, and outdoor facilities (Meng et al., 2009b). Buildings differ from other objects in height, texture, shape, and color (Cho et al., 2004; Zhang et al., 2006). The morphology-based building-detection method uses these morphological differences as well as other characteristics to gradually remove other objects from building candidates. The early development and evaluation of this method is published in Meng et al. (2009b). In this section, we introduce the general theory of building detection for a complete understanding of the three-step land use classification.

The flowchart in Figure 7.3 illustrates the analytical strategies to gradually remove other objects after ground filtering. The main challenging issue

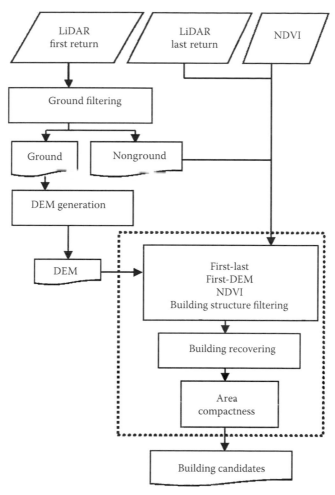

Figure 7.3 Flowchart of the morphology-based building-detection method. (Meng, X. L, N. Currit, and L. Wang, and X. Yang, *Photogrammetric Engineering & Remote Sensing* 78: 35–44, 2010. Courtesy PE&RS.)

in this process is to separate buildings from trees of similar height and size to the building. Because laser beams can penetrate vegetation and reach deeper areas in a tree, the elevation differences between the first and last returns are generally larger for vegetation than for buildings. As a result, some returns within the tree crown can be removed using a difference threshold between the first and last returns. We observe small elevation differences, however, between the first and last returns in building areas and used 0.2 meters (m) as the threshold to remove vegetation in our previous research (Meng et al., 2012). Objects lower than the threshold of minimum building height are removed, eliminating shrubs, short walls along walkways, and equipment, for example. The calculated elevation differences between the first return and the DEM interpolated from the ground-filtering results indicate the height of aboveground objects. When aerial photographs with red and near-infrared bands are available, the normalized difference vegetation index (NDVI; Jensen, 2007) can be calculated to further remove vegetation as illustrated in Figure 7.3. We used a 0.2 m threshold in Austin, Texas (Meng et al., 2009b; Meng, 2010). After removing all of the possible vegetated areas and short objects, the remaining objects are mainly composed of vegetated areas in fragmented shapes and building footprints.

Because many remaining vegetation areas are linear and thin fragments, a building structure filtering process is proposed to further remove pixels in vegetated areas. The process examines the four-neighborhood window and removes the pixel if both the pixels above and below or the pixels on the left and right do not belong to buildings. Because a building is rarely one-pixel wide in a meter- or submeter resolution data set, this process breaks vegetation areas into smaller pieces. After removing the above processes at a pixel level, a few random pixels may be removed erroneously from actual buildings and need to be replaced. The hole-shaped pixels within a building might result from vegetation or antenna-like linear objects on top of buildings (e.g., open chimney or flag poles) or to anomalies from the interpolation process of the LiDAR point cloud. Therefore, the small holes are filled by the average value of the surrounding building pixels. Finally, the blocks of building candidates are converted into polygons and further removed based on area and the compactness calculated using area and perimeter of a building polygon as illustrated in Equation 7.1. Figure 7.2c provides an example of building-detection results using morphology-based building detection.

$$\text{Compactness} = \text{area} \times \text{perimeter}^2/(4\pi) \qquad (7.1)$$

7.2.3 Object-oriented building land use classification

Land use classification is a challenging issue in urban remote sensing. Current classification methods are mainly based on pixel-based and parcel-based methods and relatively few are based on objected-oriented methods that identify individual buildings (Wu et al., 2006; Meng et al., 2009b). Furthermore, few studies examine building characteristics for typical building land use types (e.g., residential, industrial, commercial, etc.) (Barnsley and Barr, 1997; Barnsley et al., 2003; Barr et al., 2004; Wu et al., 2006). In addition, automatically identifying the relationship between buildings and their surrounding environments is technically difficult. Alternatively, some researchers have developed a land use classification system based on manual inputs of a set of statistics from visual interpretation.

However, individually and manually estimating statistics for classification is time consuming and costly.

This section introduces an approach to derive land use indicators based on three data resources and identifies residential buildings using a classification method. Table 7.1 lists typical characteristics of single-family and multifamily residential buildings and nonresidential buildings in industrial countries. Some features (e.g., fences) or characteristics (e.g., garage space

Table 7.1 Characteristics of different building land use types in developed countries

Single-family residential	Single-family house	• A single driveway, distance to major roads • One front sidewalk leading to the front door • A front yard, a backyard (often fenced) • A garage or carport, roof types, size • Usually ≤ 3 stories in height • Space between houses, building density
	Small mobile home	• Much smaller than other single-family homes • Rarely have a garage (but may have a carport) • May have a paved driveway or sidewalk • Roofs can be flat or pitched
Multifamily residential	Multifamily residential	• Large collective above- or below-ground parking garages, parking lots, or concrete ground covers • May be ≥ 2 stories in height • Share front or back yards, some have fences • Green spaces, outside swimming pools • Recreation courts • Similar building shape and size • More than one sidewalk • Some are in residential areas and are usually surrounded by residential buildings • Distance to major roads
Nonresidential	Large	• Large building sizes • Next to large or small nonresidential buildings • Along major roads • Huge open parking spaces in rural areas • High-rise buildings in downtown • May have multiple air-conditioning units
	Smaller	• Size may equal or larger than single-family house • Some storage or equipment room maybe smaller • Along major roads or at intersections • Next to nonresidential buildings • Shared parking lots or no parking lots in downtown

Source: Meng, X. L, N. Currit, and L. Wang, and X. Yang, *Photogrammetric Engineering & Remote Sensing* 78: 35–44, 2010. Courtesy PE&RS.

Table 7.2 Elements and land use indicative characteristics that can be derived from the three LiDAR data sets, aerial photographs, and road maps

Target elements	Source	Land use indicative characteristics
Buildings	Detected*	Height, area, roof type, distance to the closest building, census block building density, and building size diversity within a block
Green spaces	Detected	Size: small, middle, and large grass or trees
Water bodies	Detected	Size, shape
Parking lots or concrete ground covers	Detected	Size, area
Roads	GIS data	Classified roads from GIS data, two or three classes

Source: Meng, X. L, N. Currit, and L. Wang, and X. Yang, *Photogrammetric Engineering & Remote Sensing* 78: 35–44, 2010. Courtesy PE&RS.

*"Detected" means that the elements need to be generated or extracted from the three data resources.

in a building), however, may be difficult or impossible to automatically detect using machine-learning algorithms. Hence, we simplify the features and characteristics that are useful and detectable for building land use identification as shown in Table 7.2. Finally, seven statistics are designed for building land use classification: the area, height, and compactness of buildings; the distance to major roads; percentage of green space within a buffer zone of a building; percentage of parking space within the buffer; and building density in a block.

The seven building-based statistics can be analyzed using spatial analysis and zonal statistics tools. Figure 7.4 shows how to derive these statistics in three tiers: the input data, spatial analysis and zonal statistics, and the seven statistics. Within the seven statistics, three can be calculated directly from the building candidates after building detection: area, height, and compactness. The other four statistics reflect the spatial relationship between buildings and adjacent objects in the local environment. To obtain the distance to major roads, a spatial distance analysis function first generates a raster file of the distance to the major roads. Then a zonal statistic tool calculates the smallest value within the zone of a building as the distance to major roads for each building. Calculating the percentage of green space requires two steps. First, the NDVI is calculated from the aerial photograph and values larger than 0.2 are identified as green space. Then the zonal statistics of the green space within a building's buffer zone divided by the area of the buffer zone represent the percentage of green space around a building. The calculation of the percentage of parking distance is similar to the calculation of green space. Identifying parking garages from images

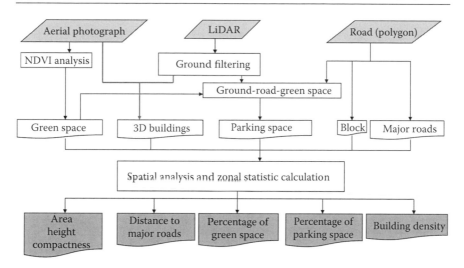

Figure 7.4 A flowchart showing the process to calculate the seven statistics from the three data sources using spatial analysis and zonal statistic tools. (From Meng, X. L, N. Currit, and L. Wang, and X. Yang, *Photogrammetric Engineering & Remote Sensing* 78: 35–44, 2010. Courtesy PE&RS.)

is difficult and hence not considered in the this calculation of parking statistics. The parking area is derived by removing road and green space from the ground calculated during the ground-filtering process. In our previous research, we used 10 m as the buffer distance for buildings. For building density in a block, the method defines a block as the area enclosed by roads and calculates the building density using the number of buildings within a block divided by the area of the block.

The spatial analysis and zonal statistics described in the previous paragraph produce a building data set with seven variables for each building polygon. To complete this step, many statistical classification methods are available for building land use classification. Among these methods, we select C4.5 decision tree classification, which has been widely used in land cover and land use classification and has produced promising results (Friedl et al., 1999; Homer et al., 2004; Rogan et al., 2008). This decision tree classification method learns from training samples to produce a series of rules that define particular classes (see Quinlan, 1993). Because defining expert-knowledge-based systems requires comprehensive knowledge of and familiarity with the local environment, a supervised classification method is preferred to allow for flexibility to adapt to various environments by automatically learning from selected training samples. As the purpose of the research is to separate residential buildings, separate sets of training and assessment building samples will be randomly selected, and each sample set will contain two types of buildings: residential and other buildings. The

C4.5 decision tree classification method then uses these training samples to formulate the classification rules and the assessment samples for accuracy assessment in quality control.

7.3 METHOD VALIDATIONS AND PERFORMANCE

The object-oriented land use classification method has been validated and tested in an 8.24 square kilometer (km²) study site located in Austin, Texas, as illustrated in Figure 7.5 (Meng, 2010, Meng et al., 2012). The features in the site include part of the downtown area with complicated roofs and shapes of commercial buildings, elevated highways, and a section of residential areas with dense trees that are challenging for building detection. For the ground-filtering process, we first validated the ground-filtering method in a 0.016 km² local site using ground reference data. Similarly, we validated the building-detection method in a local site of 0.67 km² using building vector data downloaded from the City of Austin. The parameters used for validation purposes were applied to the entire 8.24 km² study site in Austin, Texas. Finally, the seven land use indicators were calculated using spatial analysis tools. For supervised building classification, we randomly selected approximately one-third of the buildings for the land use survey, which included 169 residential buildings (43.1%), 213 nonresidential buildings (54.35%), and 10 trees (2.55%) that failed to be removed. The training data set contained 94 residential buildings and 109 nonresidential buildings, and the testing data set contained 75 residential buildings and 104 nonresidential buildings. The results from the C4.5 decision tree classification method yielded classification accuracies of 81.1% for residential buildings and 85.7% for nonresidential buildings.

7.4 CONCLUSION

Detecting urban features from high-resolution remotely sensed imagery has significant scientific and engineering benefits. Especially important findings include how to automatically extract social indicators, such as land uses, that are not directly related to the physical reflectance measured by remote sensors (Gong and Howarth, 1992; Herold et al., 2003). In this chapter, we presented an object-oriented land use classification approach to characterize the spatial distribution of residential buildings based on LiDAR, aerial photographs, and road maps. In urban studies, residential buildings are the smallest human dwelling units that are accessible from remotely sensed images, and they represent the spatial distribution of the human population (Meng, 2010; Meng et al., 2012). To make building

Legend

■ Training ■ Testing □ Other

Figure 7.5 (a) Interpolated gray-scale image of the LiDAR first return, (b) ground-filtering results, (c) building-detection result overlaid on top of interpolated image, and (d) training and testing sets for land use classification. (From Meng, X. L, N. Currit, and L. Wang, and X. Yang, *Photogrammetric Engineering & Remote Sensing* 78: 35–44, 2010. Courtesy PE&RS.)

classification possible, three consecutive methods have been developed to extract buildings and associated statistics from remotely sensed images: a multidirectional ground-filtering algorithm, a morphology-based building-detection method, and a building land use classification method based on seven statistics derived from spatial analysis and the C4.5 decision tree classification algorithm. The method has been tested and evaluated in the 8.24 km² study site in Austin, Texas. The results showed that the proposed approach successfully detected buildings and identified 81.1% of the residential buildings.

Our previous research indicates that the multidirectional ground-filtering algorithm has the merits of being fast, easy to understand, robust to parameter selection, and sensitive in detecting short and small aboveground objects, such as shrubs and short walls (Meng et al., 2009a, 2010). The morphology-based building-detection method effectively detected buildings in both developed downtown commercial zones and residential zones, even with dense trees surrounding the buildings. The building classification method could likely be modified to detect buildings using only LiDAR data, but it has been demonstrated that high-resolution optical images (e.g., aerial photograph) can be used effectively to reduce the confusion between buildings and trees (Meng et al., 2009b). Finally, previous building-based land use studies mainly focus on broad land use zones, such as commercial and residential zones. The developed residential building-detection method is the first method that directly detects buildings and their associated land use indicators from a few remotely sensed images and identifies residential buildings using classification methods (Meng et al., 2012). The method is designed such that it does not require many preexisting data sets to operate, which makes it available to a wide range of countries lacking data for existing urban structures. Potential applications and future research for the residential building-detection method include sub-block-level population estimation based on residential buildings, census population data redistribution, and analysis of changes to the inner structure of urban areas caused by economic development.

ACKNOWLEDGMENTS

This study was supported by grants to Le Wang from the National Science Foundation (BCS-0822489).

REFERENCES

Barnsley, M. J., and S. L. Barr. 1996. Inferring urban land use from satellite sensor image using Kernel-based spatial reclassification. *Photogrammetric Engineering & Remote Sensing* 62(8): 949–958.

Barnsley, M. J., and S. L. Barr. 1997. Distinguishing urban land-use categories in fine spatial resolution land-cover data using a graph-based, structural pattern recognition system. *Computers, Environment and Urban Systems* 21(3/4): 209–225.

Barnsley, M. J., A. M. Steel, and S. L. Barr. 2003. Determining urban land use through an analysis of the spatial composition of buildings identified in LiDAR and multispectral image data. In *Remotely Sensed Cities*, edited by V. Mesev, 83–108. New York: Taylor & Frances.

Barr, S., M. J. Barnsley, and A. Steel. 2004. On the separability of urban land-use categories in fine spatial scale land-cover data using structural pattern recognition. *Environment and Planning B: Planning and Design* 31: 397–418.

Cho, W., Y. Jwa, H. Chang, and S. Lee. 2004. Pseudo-grid based building extraction using airborne LiDAR data. *International Archives of Photogrammetry and Remote Sensing* 35(B3): 378–381.

Friedl, M. A., C. E. Brodley, and A. H. Strahler. 1999. Maximizing land-cover classification accuracies produced by decision trees at continental to global scales. *IEEE Transactions on Geoscience and Remote Sensing* 37(2): 969–977.

Gong, P., and P. J. Howarth. 1992. Frequency-based contextual classification and gray-level vector reduction for land-use identification. *Photogrammetric Engineering & Remote Sensing* 58(4): 423–437.

Herold, M., X. Liu, and K. Clarke. 2003. Spatial metrics and image texture for mapping urban land use. *Photogrammetric Engineering & Remote Sensing* 69(6): 991–1001.

Homer, C., C. Huang, L. Yang, B. Wylie, and M. Coan. 2004. Development of a 2001 national land-cover database for the United States. *Photogrammetric Engineering & Remote Sensing* 70(3): 829–840.

Jensen, J. 2007. *Remote Sensing of the Environment: An Earth Resource Perspective.* Saddle River, NJ: Prentice-Hall.

Lu, D., and Q. Weng. 2006. Use of impervious surface in urban land-use classification. *Remote Sensing of Environment* 102: 146–160.

Lutz, W., W. C. Sanderson, and S. Scherbov. 2004. *The End of World Population Growth in the 21st Century: New Challenges for Human Capital Formation and Sustainable Development.* London: Earthscan.

Meng, X. 2010. Determining urban land uses through building-associated element attributes derived from LiDAR and aerial photographs. PhD diss., Texas State University-San Marcos.

Meng, X., N. Currit, L. Wang, and X. Yang. 2012. Detect residential buildings from LiDAR and aerial photographs through object-oriented land-use classification. *Photogrammetric Engineering & Remote Sensing* 78(1): 35–44.

Meng, X., N. Currit, and K. Zhao. 2010. Ground filtering algorithms for airborne LiDAR data: a review of critical issues. *Remote Sensing* 2(3): 833–860.

Meng, X., L. Wang, and N. Currit. 2009b. Morphology-based building detection from airborne LiDAR data. *Photogrammetric Engineering & Remote Sensing* 75(4): 427–442.

Meng, X. L., L. Wang, J. L. Silván-Cárdenas, and N. Currit. 2009a. A multi-directional ground filtering algorithm for airborne LiDAR. *ISPRS Journal of Photogrammetry and Remote Sensing* 64(1): 117–124.

Mesev, V. 2003. *Remotely Sensed Cities*. London and New York: Taylor & Francis.

Quinlan, J. R. 1993. *C4.5: Programs for Machine Learning*. San Mateo, CA: Morgan Kaufmann.

Rogan, J., J. Franklin, D. Stow, J. Miller, C. Woodcock, and D. Roberts. 2008. Mapping land-cover modifications over large areas: a comparison of machine learning algorithms. *Remote Sensing of Environment* 112(5): 2272–2283.

Shan, J., and A. Sampath. 2005. Urban DEM generation from raw LiDAR data: a labeling algorithm and its performance. *Photogrammetric Engineering & Remote Sensing* 71(2): 217–226.

Sithole, G., and G. Vosselman. 2004. Experimental comparison of filter algorithms for bare Earth extraction from airborne laser scanning point clouds. *ISPRS Journal of Photogrammetry and Remote Sensin* 59(1-2): 85–101.

Tang, J., L. Wang, and S. W. Myint. 2007. Improving urban classification through fuzzy supervised classification and spectral mixture analysis. *International Journal of Remote Sensing* 28(18): 4047–4063.

United Nations. 2007. State of world population 2007: unleashing the potential of urban growth. http://www.unfpa.org/swp/2007/english/ introduction.html (accessed March 3, 2008).

Vitousek, P. M., H. A. Mooney, J. Lubchenco, and J. M. Melillo. 1997. Human domination of Earth's ecosystems. *Science* 277(5325): 494–499.

Vosselman, G. 2000. Slope based filtering of Laser altimetry data. *International Archives of Photogrammetry, Remote Sensing and Spatial Information Sciences* 33(Part B3-2): 935–942.

Wu, S., B. Xu, and L. Wang. 2006. Urban land-use classification using variogram-based analysis with an aerial photograph. *Photogrammetric Engineering & Remote Sensing* 72(7): 813–822.

Zhang, K., S. Chen, D. Whitman, M. Shyu, J. Yan, and C. Zhang. 2003. A progressive morphological filter for removing nonground measurements from airborne LiDAR data. *IEEE Transactions on Geoscience and Remote Sensing* 41(4): 872–882.

Zhang, K., and D. Whitman. 2005. Comparison of three algorithms for filtering airborne LiDAR data. *Photogrammetric Engineering & Remote Sensing* 71(3): 313–324.

Zhang, K., J. Yan, and S. Chen. 2006. Automatic construction of building footprints from airborne LiDAR data. *IEEE Transactions on Geoscience and Remote Sensing* 44(9): 2523–2533.

Chapter 8

Assessment of urbanization patterns and trends in the Gulf of Mexico region of the southeast United States with Landsat and nighttime lights imagery

George Xian and Collin Homer

CONTENTS

Urban land cover change between 2001 and 2006 was quantified by comparing the National Land Cover Database (NLCD) 2001 and NLCD 2006 percent impervious surface (PIS) products in the Gulf of Mexico coastal and inland areas of the United States. The 2006 PIS was estimated using an improved method with both coarse resolution nighttime lights imagery and medium resolution Landsat thematic mapper imagery. The nighttime lights imagery was used to mask nonurban areas, and Landsat imagery was used to estimate impervious surface growth in 30-meter pixels in urban areas. Thresholds were applied to the 2006 imperviousness products to identify classes of urban land cover. The regional impervious surface increases approximately 839.2 square kilometers (km^2), or 5.3%, over the study region. A total of 2081 km^2 of nonurban land was converted to urban land cover and the regional urban land cover increased about 2249.6 km^2 with an annual growth of 449.9 km^2. The improved method has been shown to produce percent impervious product in an efficient and cost-effective way.

8.1 INTRODUCTION

Urban land cover is an important component of regional and global environmental change and has significant implications for a range of ecological, biophysical, social, and climate consequences (Seto and Shepherd, 2009; DeFries et al., 2010; McCarthy et al., 2010). Urban growth has become one of the most important trends on Earth because the growth usually lasts decades and often results in the deterioration of natural vegetation and environmental conditions associated with complex development forces. As a consequence of urban growth in the past several decades, a significant amount of natural landscape has been converted into anthropogenic impervious surface around the world. Impervious surface has been recognized as the most prominent stressor for the ecological conditions in watersheds (Grimm et al., 2000; Alberti et al., 2007). The runoff from urban built-up areas, which can be accelerated by increasing the level of imperviousness, usually increases pollutant loadings and further deteriorates aquatic ecosystems (Xian et al., 2007; Kaushal et al., 2008; Thompson et al., 2008). In addition, the warming trends associated with urban areas have long been recognized (Quattrochi and Ridd, 1994; Owen et al., 1998; Voogt and Oke, 2003). Trends in urban heat islands in some locations have similar magnitudes or are greater than those from greenhouse gas–forced climate change (Stone, 2007). The magnitude of the urban heat island is a function of urban morphology and physical characteristics, urban extent, waste heat release, and regional climate factors. Previous studies have suggested that impervious surface with other urban landscape information can be used as an indicator for urban environmental conditions, including urban heat island effect (Xian and Crane, 2006; Xian et al., 2007).

The speed and growth patterns of urban development are determined mainly by population growth and economic conditions. Previous studies have found that the population of the United States has migrated toward the coasts, concentrating along the earthquake-prone Pacific Coast and the hurricane-prone Atlantic and Gulf Coasts, with associated urban land cover and population increasing substantially (Iwan et al., 1999; Crowell et al., 2007). Approximately 3% of the U.S. population lives in areas subject to the 1% annual chance coastal flood hazard (Crowell et al., 2010). In Florida, for instance, the population has increased fivefold since 1950, with 80% of the population living within 35 kilometers (km) of the coast (Iwan et al., 1999). The growths in both population and built-up land across the coastal region may further stress ecosystems by converting natural or cultivated crop land into urban land (Duh et al., 2008). More than 84,000 acres of wetland have been lost to development since 1990 in Florida alone (Pittman and Waite, 2009). Accurate spatial extent of impervious surface and its temporal change information is critical for assessing variations of urban land cover and associated ecological and climatic effects.

Satellite remote-sensing data acquired from medium-resolution satellites (e.g., Landsat) have provided repetitive and consistent observations of the terrestrial surface across large areas over time. These data sets have been widely used to monitor urban land cover change at local and regional scales (Small, 2003; Maktav et al., 2005; Lu and Weng, 2009; Potere et al., 2009). Urban extents and structures cannot be clearly determined by using discrete classification methods along with medium-resolution remote-sensing data in part because of highly heterogeneous features of urban land cover. Most urban areas, especially in single-family residential development areas, exhibit subpixel characteristics that mix impervious surface with other land covers (e.g., grass) in medium-resolution satellite imagery. The urban landscape can be treated as a continuum, however, while using modeling techniques to extract urban characteristics. Percent impervious surface (PIS) estimated from satellite data can serve as a surrogate to determine urban extent and infrastructure and to assess changes in the urban environment (Civco et al., 2002; Powell et al., 2007; Xian et al., 2008; Prasada and Wu, 2010). The U.S. Geological Survey (USGS) National Land Cover Database (NLCD) 2001 impervious surface product has been used to assess the extent of urban development and associated ecological effects in the conterminous United States (Imhoff et al., 2010). Additionally, remote sensing of nocturnal light provides a straightforward view of the global distribution and density of developed areas (Elvidge et al., 2007). The nighttime lights imagery collected from the U.S. Air Force Defense Meteorological Satellite Program (DMSP) has been used to estimate urban development and population distribution at regional and global scales (Imhoff et al., 1997; Small et al., 2005; Elvidge et al., 2007; Nghiem et al., 2009; Townsend and Bruce, 2010). The annual cloud-free nighttime lights composite from the DMSP provides a consistent review of urban development and can discriminate nonurban biases that often are produced from models that map urban land cover. Furthermore, the Landsat satellite record now spans more than three decades and can be used to generate high–temporal frequency information about urban landscape condition. The two data sets can be combined to monitor rapid growth of urban land through frequent updates for a large geographic area.

The goals of this chapter are to (a) describe an improved method evolved from an original NLCD 2006 impervious surface updating prototype (Xian and Homer, 2010) by using both Landsat and nighttime lights imagery to estimate impervious surface; (b) implement the method to quantify impervious surface variation and associated land cover change between 2001 and 2006 in the coastal and inland areas in the Gulf of Mexico region of the United States; and (c) analyze land cover transition associated with urban developments during the period in the region.

8.2 STUDY AREA AND DATA

We implemented the approach in a study area of 866,740 square kilometers (km²) that encompasses both inland and coastal areas of Texas, Louisiana, Mississippi, Alabama, Georgia, and Florida in the United States. The study area also included an area of 311,682 km² within 100 km off the coast (see Figure 8.1). This region had experienced signification degradation of natural habitats and the associated loss of ecological services because of population growth, changes in land use, and other natural disturbances, including hurricanes, sea level rising, and warming trends, which have exacerbated degradation trends in the region (Karl et al., 2009). We divided the study area into three regions—eastern, central, and western—to compare urbanization patterns and trends, as well as the overall and regional landscape effects of urban land cover change, especially within 100 km off the coast zone.

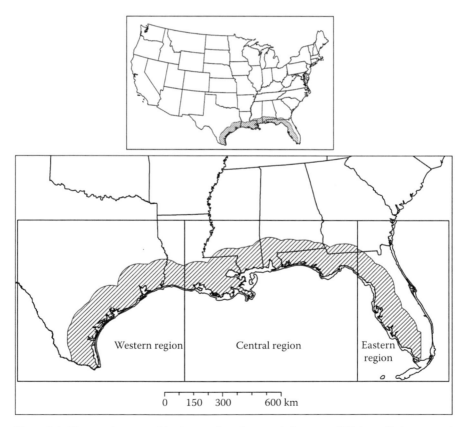

Figure 8.1 The study area with three subregions and the zone 100 km off the coastal line.

According to the NLCD 2001 land cover classification, the largest and second-most-dominant cover types in the region were shrub and evergreen forestlands, which covered approximately 19.7% and 13.6% of the land area, respectively. The extent of urban land cover was about 8.5% of the land area. Impervious surface area in the region was 15,213 km², or 1.8% of the regional land extent. Within the 100 km off the coast zone, the two most dominant types of land cover were woody wetland and evergreen forest, which count about 19.3% and 13.8% of the total land cover. Total urban land cover was about 11.1% of the land area in the 100 km zone. Impervious surface was about 7,146 km², or 2.4%, of the land area.

Landsat imagery between 2001 and 2006 was used as the primary data source to estimate impervious surface change. The DMSP nighttime lights imagery in 2001 and 2006 was used to help create training data and modify the final product. Table 8.1 lists path and rows and acquisition dates of all

Table 8.1 Path and row numbers and acquisition dates of landsat images used to estimate

Path	Row	2001 Date	2006 Date	Path	Row	2001 Date	2006 Date
15	41	05/14/01	05/04/06	23	38	05/14/01	05/12/06
15	42	02/21/00	05/04/06	23	39	06/23/01	06/13/06
15	43	04/09/00	05/04/06	23	40	04/26/03	06/13/06
16	38	10/23/99	10/15/05	24	38	04/08/00	05/19/06
16	39	04/27/01	03/08/06	24	39	11/16/99	10/23/05
16	40	04/03/01	03/05/05	25	38	10/06/99	11/18/06
16	41	04/03/01	03/05/05	25	39	10/06/99	06/27/06
16	42	05/18/00	03/05/05	25	40	05/07/99	04/08/06
16	43	04/19/01	03/08/06	26	38	02/23/02	03/11/05
17	38	05/09/00	05/02/06	26	39	02/23/02	03/11/05
17	39	06/16/02	05/02/06	26	40	07/22/01	05/17/06
17	40	03/28/02	05/02/06	26	41	02/07/02	12/27/06
17	41	04/10/01	05/02/06	26	42	05/09/03	04/18/07
18	38	04/30/00	06/10/06	27	38	10/25/01	10/28/05
18	39	02/10/00	03/25/07	27	39	10/25/01	09/26/05
19	38	04/29/00	04/27/05	27	40	09/04/00	09/26/05
19	39	04/05/00	04/27/05	27	41	09/04/00	09/26/05
20	38	09/17/99	09/28/06	27	42	07/21/01	07/08/05
20	39	06/07/00	04/05/06	28	38	10/16/01	10/22/06
21	38	06/17/01	05/27/05	28	39	10/16/01	10/19/05
21	39	10/15/01	10/18/05	28	40	07/12/01	07/18/06
21	40	10/15/01	10/18/05	28	41	07/12/01	07/18/06
22	38	07/15/00	06/06/06	29	38	07/12/01	07/18/06
22	39	01/21/00	02/11/05	29	39	09/24/02	09/24/05
22	40	11/07/01	10/09/05	29	40	05/03/02	04/01/05

Landsat images used for the study area. In each Landsat image, all bands except band 6 were used. The nighttime lights images used for the area are taken from average visible, stable lights, and cloud-free coverages of Operational Linescan System instruments smooth data from 2001 and 2006.

8.3 METHODS

We developed an algorithm that evolved from an early prototype proposed for updating the USGS NLCD 2006 impervious surface product (Xian and Homer, 2010). The original approach relied on image change detection to locate changes between two times and to label these changes with specific land cover types. To mask nonurban land, the land cover data were then applied to the impervious surface product estimated from the regression tree models. The accuracy of the PIS product depended on the land cover classification and the impervious surface estimation. Moreover, the pixel-based thematic land cover classification did not capture urban pixels very well, especially on fringes of existing urban areas where most new urban developments emerge. To improve the accuracy of estimated PIS and further use it to determine urban land cover characteristics, it is preferable to have a method that depends on only one mapping step. Therefore, a new approach was developed with the requirement that the percent imperviousness estimated by regression tree models be accurate and consistent with the previous NLCD product. The method should be feasible for the nationwide impervious surface mapping effort. Figure 8.2 illustrates three major steps involved in the method. In step 1, the DMSP nighttime lights imagery was first resampled to 30 meters (m) and was subset to each Landsat path and row. We then intersected 2001 DMSP imagery with the NLCD 2001 impervious surface product to exclude low-density imperviousness outside urban and suburban centers. These impervious zones in the urban core areas were considered stable and were used as reliable training data sets. Two training data sets, one having a relatively larger urban extent and one having a relatively smaller extent, were produced using two different thresholds of city light imagery with the 2001 impervious product. In step 2, the two training data sets combined with 2001 Landsat imagery were applied separately with regression tree algorithms to build up regression tree models (Xian and Homer, 2010). Two sets of regression tree models were produced and used to estimate PIS for the time period 2001–2006. After that, two 2001 synthetic impervious surface products were produced. Similarly, the same two training data sets were used with 2006 Landsat and DMSP images to create two sets of regression tree models to produce two 2006 synthetic impervious surfaces. The regression tree models estimated PIS according to spectral signatures of input satellite imagery as well as the information from ancillary data sets. Some nonurban areas that have

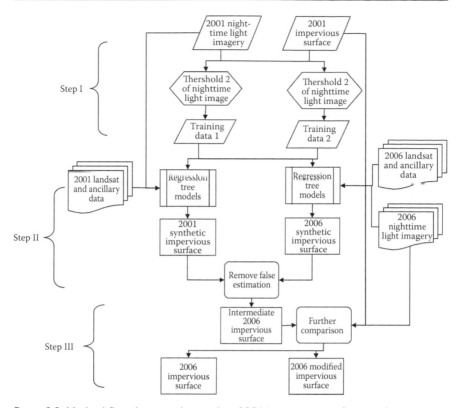

Figure 8.2 Method flowchart used to update 2006 impervious surface product.

strong reflectance around the existing urban area could be misrepresented as urban land, resulting in biases from estimated impervious surface. The two synthetic products were used to remove these false estimates with the following constraints:

$$PIS_{i,n}(2006) = \begin{cases} 0, \ S_n(2001) > S_n(2006) \ \ and \ \ S_n(2001) \not\subset PIS(2001) \\ S_n(2006), \ S_n(2001) = 0 \ \ and \ \ S_n(2006) > 0 \end{cases}, \quad (8.1)$$

where $PIS(2001)$ is the 2001 impervious surface from the NLCD 2001 product, $S_n(2001)$ and $S_n(2006)$ are synthetic imperviousness estimates in 2001 and 2006, respectively, n represents the synthetic products 1 and 2, and $PIS_{i,n}(2006)$ is the 2006 intermediate estimate. Two 2006 synthetic products were produced to ensure that only the stable prediction was chosen in the two intermediate products. In step 3, the 2001 PIS product was used to retain the 2001 impervious surface in the unchanged areas. The 2006 DMSP image was employed to ensure that nonimperviousness areas were not included and that new impervious surfaces emerged in the nightlight lights imagery extent. The final $PIS(2006)$ was determined by comparing the two 2006 intermediate pairs with constraints of $PIS(2006) = 0$ if $PIS_{i,n}(2006) = 0$, or $PIS(2006) = PIS_{i,1}(2006)$ if $PIS_{i,1}(2006) > PIS_{i,2}(2006)$, or $PIS(2006) = PIS_{i,2}(2006)$ if $PIS_{i,2}(2006) > PIS_{i,1}(2006)$. After that, estimations were produced of the 2006 new impervious surface and the 2006 modified impervious surface that had a higher percent of imperviousness in 2006 than the 2001 existing impervious surface area. The latter product was created with the same 40% threshold in the original approach (Xian and Homer, 2010).

We quantified the total impervious surface area and its change between 2001 and 2006 by directly comparing the PIS products from the two time periods. Urban land cover categorized by PIS as 1%–20% for open space development, 21%–50% for low-intensity development, 51%–80% for medium-intensity development, and 81%–100% for high-intensity development (Homer et al., 2007) were measured, along with associated land cover transition. Two additional variables were also estimated. The first one was the annual growth rate (GR) defined as the annual transition from a nonurban land cover type to an urban land cover type between 2001 and 2006:

$$GR = \frac{\Delta A}{\Delta t} = \frac{U(2006) - U(2001)}{\Delta t} = \frac{U(2006)_{new}}{5}. \tag{8.2}$$

The second variable that was estimated was the growth ratio or transition rate (r) defined as the rate of transition from a nonurban land cover type to any type of urban land cover between 2001 and 2006:

$$r_k = \frac{L(2006)_k}{L(2001)_k}, \tag{8.3}$$

where L is the extent of a land cover type k.

8.4 RESULTS

8.4.1 Impervious surface

Increases of impervious surface area associated with urban land cover expansion in the region were directly estimated from the updated PIS product after training data sets obtained by applying two thresholds of 10 and 20 to 2001 nighttime lights imagery and using 2001 impervious surface product were implemented to create regression tree models. Table 8.2 summarizes areal increases and growth rates of impervious surface from 2001 to 2006 in different regions. The increments of impervious surface were 265.4 km², 125.1 km², and 438.6 km² in the eastern, central, and western

Table 8.2 Impervious surface growth in km² between 2001 and 2006, growth ratios (%), and annual growth rate (km²/yr)

Region	Eastern (km²) (%)	Central (km²) (%)	Western (km²) (%)	Total area (km²) and growth rate (km²/yr)
100 km	107.5 (5.6%)	88.1 (4.3%)	220.4 (7.0%)	416, 83.2
Total area	265.4 (5.7%)	135.1 (4.2%)	438.6 (6.0%)	839.2, 167.8

regions, respectively, resulting in growths of 5.7%, 4.2%, and 6.0% from 2001 to 2006 in the three regions. The overall growth rate for the entire study area was 167.8 km²/yr, and the growth ratio between 2001 and 2006 reached 839.2 km², or 167.8 km² annually. The 100 km zone in the western region remained the largest impervious surface area and had the highest growth rate. Overall, impervious surface grew about 416 km², or 83.2 km² annually, in the 100 km zone.

The increase of impervious surface varies with different imperviousness categories in different regions. Figure 8.3a displays areas of 2001 and 2006 new impervious surfaces in different percent categories within the 100 km zones over the three regions. In the eastern region, growth peaks appear around the 61%–70% cover range, and the 2001 PIS peak is around the 41%–50% category. The distribution pattern of 2006 new impervious surface is smoother than that in 2001 in the central region. Additionally, the 2006 new PIS in the western region has an apparent peak around 70%–80%

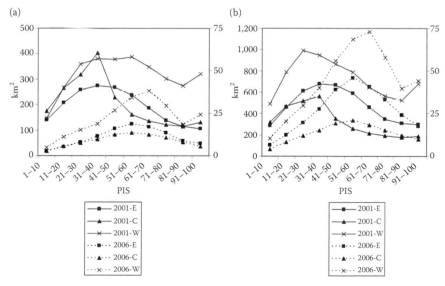

Figure 8.3 PIS in the (a) 100 km zone and (b) entire area of each three study regions. Labels of E, C, and W are for eastern, central, and western regions, respectively.

and 2001 PIS has a flat peak from 20% to 60%. The distributions of PIS in the three regions (see Figure 8.3b) exhibit similar patterns as in the 100 km zone, except in the western region where PIS has an apparent peak around 21%–30%.

8.4.2 Urban land cover change

Associated with the expansion of 2006 impervious surface is an overall increase in urban land cover by 2,249.6 km², or an annual increase of 449.9 km², from the region's 2001 baseline. Figure 8.4 shows spatial distributions of 2001 and 2006 new urban land in the study region. Most new urban land emerged around the existing urban centers. The extents of urban land cover in Houston, Texas, and Tampa Bay, Florida, are enlarged to exhibit development patterns in the two metropolitan areas. Most new urban growth emerged on city developments demonstrating apparent "ring" growth patterns in the two areas. Our analysis indicates that among the three regions, the western region had the largest urban land cover increase (1,135.3 km²), and the eastern region had the highest growth ratio (3.7%) (see Table 8.3). Similar growth patterns were observed in the 100 km zone: The western region had the largest spatial increment (559.6 km²), and the eastern region had the highest growth ratio (4.6%). Overall, more than 1,110 km² of new urban land was added in the 100 km zone.

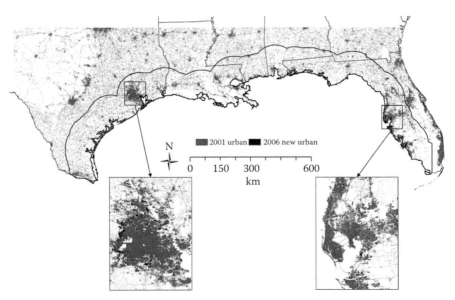

Figure 8.4 **(See color insert.)** Distributions of 2001 and 2006 new urban land cover in the region. The lower left and right panels are for Houston, Texas, and Tampa Bay, Florida.

Table 8.3 Urban land cover change in km² between 2001 and 2006, growth ratios (%), and annual growth rate (km²/yr)

Region	Eastern (km²) (%)	Central (km²) (%)	Western (km²) (%)	Total area (km²) and growth rate (km²/yr)
100 km	293.5 (4.6%)	257 (2.7%)	559.6 (3.3%)	1,110.1, 222
Total Area	713.4 (3.7%)	400.8 (2.4%)	1,135.3 (3.6%)	2,249.6, 449.9

Total urban land reached 8.5% of the total land area, which represented an increase of 0.3% from 2001 to 2006 in the region. The 2006 urban land was 11.5% of the total land area in the 100 km zone, which is about a 0.4% increase from 2001.

8.4.3 Urban land cover transition

Spatial distributions of nonurban to urban land cover conversion between 2001 and 2006 were analyzed using transition matrices to quantify land cover transition trajectories in the region. Figure 8.5 reveals extents of land cover transition in the region and the 100 km zone. Overall, the leading urban land cover transitions are from woody wetland (219.8 km²) in the eastern region, hay and pasture (80.7 km²) in the central region, and herbaceous (221.2 km²) in the western region (see Figure 8.5a). The land cover transitions in the 100 km zone are different in different regions (see Figure 8.5b). In the eastern region, the most prevalent land cover transitions are from woody wetlands (82.1 km²). In the central and western regions, hay and pasture (50.5 km² and 143.0 km²) are the dominant transition type. The urban land cover transition rates (see Table 8.4) are different from land cover areal changes. Barren land has the largest transition rates in the 100 km zone (2.19%) and the overall region (1.58%). Herbaceous, hay and pasture, and mixed forest have the second-largest transition rates in eastern, central, and western regions, respectively.

Table 8.4 Two land cover types having the highest urban land cover transitions and their transition rates (%) between 2001 and 2006

Region	Eastern	Central	Western	Total area
100 km	Barren (4.08), herbaceous (1.21)	Barren (1.49), hay (0.59)	Barren (1.89), mixed forest (1.51)	Barren (2.19), herbaceous (0.72)
Total area	Barren (4.24), herbaceous (1.04)	Barren (0.97), hay (0.44)	Barren (1.12), mixed forest (0.47)	Barren (1.58), herbaceous (0.47)

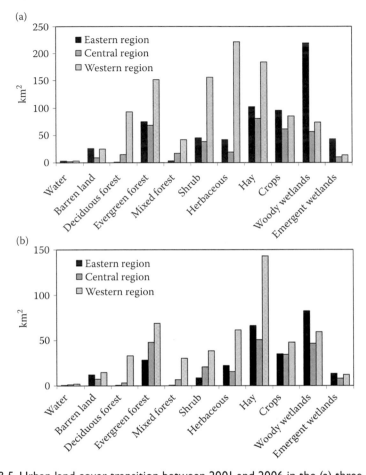

Figure 8.5 Urban land cover transition between 2001 and 2006 in the (a) three regions and (b) 100 km zone in the region.

8.5 DISCUSSION

Understanding the spatial and temporal dynamics of land cover change is critical to sustainable societal development. Because urban land usually lasts a long time after being built up, most urban developments are relatively stable over decadal time periods. It is advantageous to map urban land cover change using remote-sensing information by focusing on stable areas. The implemented method employed in this study combines the advantages of medium-resolution Landsat imagery and coarse resolution nighttime lights imagery to estimate impervious surface changes in a fast and reliable way. The DMSP images provide a relatively stable extent of potential urban land cover for a large spatial extent. The imagery helps reduce biases of urban land cover from training data sets and defines the potential urban boundary by setting thresholds for the intensities of city lights. The use of Landsat imagery with regression algorithms ensured that subpixel characteristics of urban land cover can be captured through the estimation of PIS. The latter product is then used to quantify the spatial extent and development intensity of urban land cover.

By conducting land use and land cover change analysis and comparing the results with the baseline information, transitions of how many nonurban lands are converted to urban land and even land cover change occurring within urban areas, as well as how fast and in what areas the changes have occurred, can be determined in a large area. Such analysis can be implemented in urban-water-balance and energy-balance models for optimal urban design options considering the mitigation of climate change impacts (Mitchell et al., 2008). To this end, the most up-to-date land cover status and frequent updates of its change are necessary.

The NLCD 2006 impervious surface updating was designed to provide the most current urban land cover distributions for the conterminous United States. To illustrate the algorithm and demonstrate the usefulness of this change product, we used the updated impervious surface products to quantify the type and spatial extent of urban land cover and to estimate land cover transition in both inland and coastal areas of the Gulf of Mexico region where many coastal wetlands have been deteriorated and damaged by intensive human activities. For example, in Tampa Bay, Florida, because the population was estimated to have increased nearly 10% between 2000 and 2006 (U.S. Census Bureau, 2007), many original woody wetlands in northeastern Tampa Bay transitioned into residential areas. Our analysis

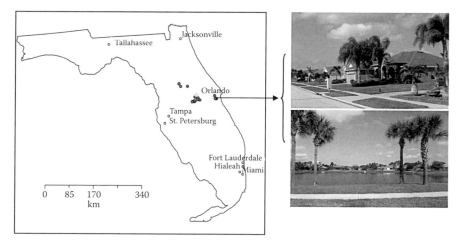

Figure 8.6 Field sites visited in 2010 in the Orlando area. The dots are visited sites. The photos on the right show residential housing built up after 2001.

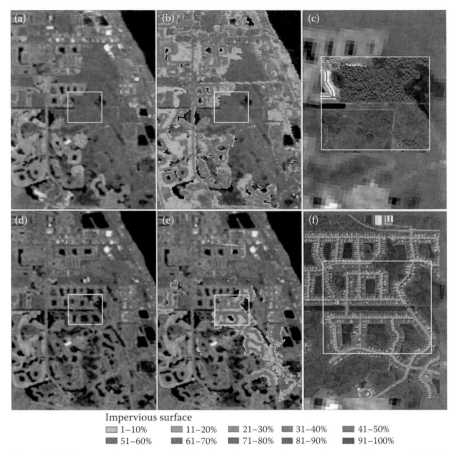

Impervious surface

▨ 1–10%	▨ 11–20%	▨ 21–30%	▨ 31–40%	▨ 41–50%
▨ 51–60%	▨ 61–70%	▨ 71–80%	▨ 81–90%	▨ 91–100%

Figure 8.7 **(See color insert.)** Landsat imagery, impervious surface, and high-resolution orthoimagery for the site where the photos shown in Figure 8.6 were taken. (a) Landsat image in 2001, (b) 2001 impervious surface, and (c) 1 m orthoimagery in 2002 show the land cover in nonurban areas within the white rectangle. (d) Landsat image in 2006 shows the land cover has been transformed into urban, and (e) 2006 impervious surface captures these changes. (f) The orthoimagery in 2006 displays details of urban land cover growth in the area.

confirmed that the largest land transition occurred in the woody wetland in the eastern region where Tampa Bay is located.

To validate the mapped urban land cover change, a visual comparison was performed using photos and field observations conducted in 2010 by visiting 114 randomly selected sites in Florida. Landsat images in 2001 and 2006, land cover change maps between 2001 and 2006 from NLCD products, and 1 m high-resolution orthoimages acquired in 2002 and 2006 were collected to compare with field verifications. The field observations verified that 107 sites (93% of the total) could confirm that the mapped urban land cover change had occurred. In seven other sites (7% of the total), changes were misclassified because these developed areas existed before 2001 but were missed in the 2001 map. Figure 8.6 demonstrates the field sites visited in 2010 near the Orlando area. The photos show residential housings built after 2001. Figure 8.7 displays Landsat images in 2001 and 2006, PIS in 2001 and new PIS in 2006, and orthoimages in 2002 and 2006 for the areas where onsite photos were taken (shown in Figure 8.6). The 2001 Landsat image shows that the area outlined by the white rectangle was not developed. The orthoimage indicates the area was covered by trees and grass in 2002. The 2001 impervious surface map also shows pervious land cover in the area. The 2006 Landsat image demonstrates that the area has been converted to developed land. The 2006 orthoimage shows that residential housings have been built and impervious surface has replaced most of the pervious land. The 2006 impervious surface map captures such land cover transformation well. The field validation, however, was not used to conduct the systematic accuracy assessment that quantifies differences between mapped and real PIS products. Such a task requires more resources and is beyond the scope of this research.

8.6 CONCLUSION

Our analysis for the Gulf of Mexico regions revealed that urban land cover transition between 2001 and 2006 involved almost all land cover types in both coastal and inland areas. We found that total impervious surface increased approximately 416 km², or 5.6%, from 2001 to 2006 in the 100 km zone over the three regions. Most new impervious surfaces were added to categories between 40% and 80% levels, indicating that most of new growth was medium- to high-intensity residential housing developments. The regional annual growth levels reached 3.7%, 2.4%, and 3.6% in the eastern, central, and western regions, respectively. Growth rates differ by region, and the overall annual growth rate was 449.9 km².

The area of urban land cover is larger than impervious surface area because the former was calculated for any land having PIS > 0, and the latter was totaled by the proportion of PIS. We found that approximately 2,081 km² of nonurban land was converted to urban land cover from 2001 to 2006, resulting in a 0.4% transition rate over the three regions. The two most dominant land cover types that transformed to urban land were hay and pasture and woody wetlands, with areas of 367.7 km² and 349.6 km², respectively. About 168 km² of new urban land was transferred from previous urban land by increasing imperviousness within previously low-intensity impervious surface areas. In the 100 km zone, the land cover transition patterns were similar to those in the entire region and the transition rate was 0.3%. About 86 km² of new urban land was added from previous low-intensity urban land. The most significant transition from nonurban to urban land cover was barren land, which reached about 1.58% of its 2001 coverage in the region. The second-largest land transition was herbaceous, which reached about 0.5% of the 2001 baseline cover. Similarly, nearly 2.2% of barren land and 0.7% of herbaceous in the 100 km zone were transferred to urban land. The two land cover types have high probabilities to be converted to urban land cover in the region. The probability of land cover transition also depends on geographic locations, however. In the eastern region, the highest transition rates appeared in barren and herbaceous. In the central region, the highest rates are in barren land and hay and pasture. In the western region, barren and mixed forest had the highest transition rates. Bare land has a high conversion rate because the land was likely in early development condition in 2001.

Our study reveals that urban development has converted a variety of land covers into a relatively unique urban land cover type and has substantially changed the landscape features in both inland and coastal areas. Hay and pasture and woody wetland were the two most dominant types of and cover being converted to urban land over the entire region. Furthermore, land cover transition has likely added substantial pressure on the regional ecosystems, especially within the areas closes to the coast. Beyond increasing surface runoff associated with increases in impervious surface, other physical properties of the regional landscape, such as surface albedo, emissivity, and surface temperature, also have been altered. The newly developed urban land cover change information will enable monitoring for regional environmental health and sustainability assessments and will be useful for evaluating potential ecological and hydroclimatic impacts on coastal landscape and ecosystems.

REFERENCES

Alberti, M., D. Booth, K. Hill, B. Coburn, C. Avolio, S. Coe., and D. Spirandelli. 2007. The impact of urban patterns on aquatic ecosystems: an empirical analysis in Puget lowland sub-basins. *Landscape and Urban Planning* 80(4): 345–361.

Civco, D., J. D. Hurd, E. H. Wilson, C. L. Arnold, M. P. Prisloe, Jr. 2002. Quantifying and describing urbanizing landscape in the northeast United States. *Photogrammetric Engineering and Remote Sensing* 68(10): 1083–1091.

Crowell, M., K. Coulton, C. Johnson, J. Westcott, D. Bellomo, S. Edelman, and E. Hirsch. 2010. An estimate of the U.S. population living in 100-year coastal flood hazard areas. *Journal of Coastal Research* 26(2): 201–211.

Crowell, M., S. Edelman, K. Coulton, and S. McAfee. 2007. How many people live in coastal areas? *Journal of Coastal Research* 23(5): iii–vi.

DeFries, R. S., T. Rudel, M. Uriarte, and M. Hansen. 2010. Deforestation driven by urban population growth and agricultural trade in the twenty-first century. *Nature Geoscience* 3: 178–181.

Duh, J., V. Shandas, H. Chang, and L. George. 2008. Rates of urbanisation and the resiliency of air and water quality. *Science of the Total Environment* 400(1–3): 238–256.

Elvidge, C. D., P. Cinzano, D. R. Pettit, J. Arvesen, P. Sutton, C. Small, R. Nemani, T. Longcore, C. Rich, J. Safran, J. Weeks, and S. Ebeber. 2007. The nightsat mission concept. *International Journal of Remote Sensing* 28(12): 2645–2670.

Grimm, N. B., J. M. Grove, S. T. A. Pickett, and C. L. Redman. 2000. Integrated approaches to long-term studies of urban ecological systems. *BioScience* 50(7): 571–584.

Homer, C., J. Dewitz, J. Fry, M. Coan, H. Hossain, C. Larson, H. Herold, et al. 2007. Completion of the 2001 National Land Cover Database for the conterminous United States. *Photogrammetric Engineering & Remote Sensing* 73: 337–341.

Imhoff, M. L., P. Zhang, R. E. Wolfe, and L. Bounoua. 2010. Remote sensing of the urban heat island effect across biomes in the continental USA. *Remote Sensing of Environment* 114(3): 504–513.

Iwan, W. D., L. S. Cluff, J. F. Kimpel, H. Kunreuther, S. H. Masakischatz, J. M. Nigg, R. S. Roth, Sr., E. Stanley, and F. H. Thomas. 1999. Mitigation emerges as major strategy for reducing losses caused by natural disasters. *Science* 284: 1943–1947.

Karl, T. R., J. M. Melillo, T. C. Oeterson, S. J. Hassol, eds. 2009. *Global Climate Change Impacts in the United States.* Cambridge and New York: Cambridge University Press.

Kaushal, S. S., P. M. Groffman, L. E. Band, C. A. Shields, R. A. Morgan, M. A. Palmer, K. T. Belt, C. M. Swan, S. E. G. Findlay, and G. T. Fisher. 2008. Interaction between urbanization and climate variability amplifies watershed nitrate export in Maryland. *Environmental Science & Technology* 42: 5872–5878.

Lu, D., and Q. Weng. 2009. Extraction of urban impervious surfaces from an IKONOS image. *International Journal of Remote Sensing* 30: 1297–1311.

Maktav, D., F. S. Erbek, and C. Jürgens. 2005. Remote sensing of urban areas. *International Journal of Remote Sensing* 26: 655–659.

McCarthy, M. P., M. J. Best, and R. A. Betts. 2010. Climate change in cities due to global warming and urban effects. *Geophysical Research Letters* 37: L09705. doi:10.1029/2010GL042845.

Mitchell, V. G., H. A. Cleugh, C. S. B. Grimmond, and J. Xu. 2008. Linking urban water balance and energy balance models to analyse urban design options. *Hydrological Processes* 22(16): 2891–2900.

Nghiem, S. V., D. Balk, G. Neumann, A. Sorichetta, C. Small, and C. D. Elvidge. 2009. Observations of urban and suburban environments with global satellite scatterometer data. *ISPRS Journal of Photogrammetry and Remote Sensing* 64(4): 367–380.

Owen, T. W., T. N. Carlson, and R. R. Gillies. 1998. An assessment of satellite remotely sensed land cover parameters in quantitatively describing the climatic effect of urbanization. *International Journal of Remote Sensing* 19(9): 1663–1681.

Pittman, C., and M. Waite. 2009. *Paving Paradise: Florida's Vanishing Wetlands and the Failure of No Net Loss.* Gainesville: University Press of Florida.

Potere, D., A. Schneider, S. Angel, and D. L. Civco. 2009. Mapping urban areas on a global scale: which of the eight maps now available is more accurate? *International Journal of Remote Sensing* 30: 6531–6558.

Powell, R., D. Roberts, P. Dennison, and L. Hess. 2007. Sub-pixel mapping of urban land cover using multiple endmember spectral mixture analysis: Manaus, Brazil. *Remote Sensing of Environment* 106(2): 253–267.

Prasada, M. R., and C. Wu. 2010. High resolution impervious surface estimation: An integration of Ikonos and Landsat-7 ETM+ imagery. *Photogrammetric Engineering and Remote Sensing* 76(12): 1329–1341.

Quattrochi, D., and M. K. Ridd. 1994. Measurement of thermal energy properties of common urban surfaces using the thermal infrared multispectral scanner. *International Journal of Remote Sensing* 15(10): 1991–2022.

Seto, K. C., and J. M. Shepherd. 2009. Global urban land-use trends and climate impacts. *Current Opinion in Environmental Sustainability* 1: 89–95.

Small, C. 2003. High spatial resolution spectral mixture analysis of urban reflectance. *Remote Sensing of Environment* 88: 170–186.

Small, C., F. Pozzi, and C. D. Elvidge. 2005. Spatial analysis of global urban extent from DMSP-OLS night lights. *Remote Sensing of Environment* 96: 277–291.

Stone, B., Jr., 2007. Urban and rural temperature trends in proximity to large US cities: 1951–2000. *International Journal Climatology* 27: 1801–1807.

Thompson, A. M., K. Kim, and A. J. Vandermuss. 2008. Thermal characteristics of stormwater runoff from asphalt and sod surfaces. *Journal of the American Water Resources Association* 44: 1325–1336.

Townsend, A., and D. Bruce. 2010. The use of night-time lights satellite imagery as a measure of Australia's regional electricity consumption and population distribution. *International Journal of Remote Sensing* 31(16): 4459–4480.

U.S. Census Bureau. 2007. 2006 American community survey fact sheet. U.S. Census Bureau. http://factfinder.census.gov (accessed December 6, 2010).

Voogt, J. A., and T. R. Oke. 2003. Thermal remote sensing of urban climates. *Remote Sensing Environment* 86: 370–384.

Xian, G., and M. Crane. 2006. An analysis of urban thermal characteristics and associated land cover in Tampa Bay and Las Vegas using Landsat satellite data. *Remote Sensing of Environment* 104: 147–156.

Xian, G., M. Crane, and C. McMahon. 2008. Quantifying multi-temporal urban development characteristics in Las Vegas from Landsat and ASTER data. *Photogrammetric Engineering & Remote Sensing* 74: 473–481.

Xian, G., M. Crane, and J. Su. 2007. An analysis of urban development and its environmental impact on the tampa bay watershed. *Journal of Environmental Management* 85(4): 965–976.

Xian, G., and C. Homer. 2010. Updating the 2001 National Land Cover Database impervious surface products to 2006 using Landsat imagery change detection methods. *Remote Sensing of Environment* 114(2): 1676–1686.

Chapter 9

Fractional vegetation cover mapping from the HJ-I small satellite hyperspectral data

Xianfeng Zhang and Chunhua Liao

CONTENTS

This chapter evaluates the usefulness of the hyperspectral imagery (HSI) in vegetation mapping onboard a Chinese HJ-1A small satellite. Fractional vegetation cover (FVC) is an important surface microclimate parameter for characterizing land surface vegetation cover as well as the most effective indicator for assessing desertification and crop growth condition. The HJ-1/HSI data were used to calculate the narrow-band vegetation index by using the Moderate Resolution Imaging Spectroradiometer leaf area index product as supporting data, which were then applied to a subpixel

203

decomposition model for the FVC estimation, namely the dimidiate pixel model (DPM). This model was used to retrieve the FVC information in the Shihezi area, Xinjiang, China. Cross-checked with the in situ measured FVC data, a correlation coefficient square of 0.86 and a root mean square error of 10.9% were statistically achieved. The verification indicates that the FVC result retrieved from the HJ-1/HSI data is well correlated with the in situ measurements, demonstrating that the HJ-1/HSI data are promising for studying the potential impacts of global climate change on the arid and semiarid landscapes.

9.1 INTRODUCTION

A hyperspectral sensor can detect detailed objects on the Earth's surface for continuous spectral information and hundreds of wavelength bands (Landgrebe et al., 2001). In recent years, hyperspectral remote sensing has been widely used in mineral mapping, vegetation classification, and environmental analysis. The use of spaceborne hyperspectral data in fractional vegetation cover (FVC) mapping will be discussed in this chapter.

9.1.1 Use of hyperspectral data in vegetation mapping

The applications of hyperspectral remote sensing in vegetation mapping are divided into two main areas. The first area is the estimation of various biophysical and biochemical vegetation variables, such as leaf area index (LAI) (Haboudane et al., 2004; Li et al., 2007; Darvishzadeh et al., 2008a; Liu et al., 2008; Wu et al., 2008), chlorophyll content (Blackburn, 1998; Zarco-Tejada et al., 2001; Darvishzadeh et al., 2008b; Haboudane et al., 2008), biomass (Liu et al., 2004; Cho et al., 2007), and FVC (McGwire et al., 2000), by building correlation relationship between these variables and hyperspectral vegetation indices (Haboudane et al., 2004; Darvishzadeh et al., 2008a,b; Wu et al., 2008). These parameters reflect the growth conditions of vegetation, and the vegetation information can be extracted based on these parameters. The second area is the classification of vegetation types (Goel et al., 2003; Tømmervik et al., 2003; Yuan and Niu, 2007; Luo and Chanussot, 2009; Oldeland et al., 2010; Tarabalka et al., 2010). In this field, especially for precision agriculture, some progress has been already achieved.

9.1.2 Preprocessing of hyperspectral data

Unlike the multispectral remote-sensing data, hyperspectral data have abundant and continuous spectral information, which means there is redundancy of data. It is not always practical to calculate all of the possible vegetation indices from the narrow bands. Similarly, the optimum narrow-band Normalized Difference Vegetation Index (NDVI) identified for one data set might not be the best for another. Therefore, the processing of hyperspectral data and selection of the proper combination of bands to calculate the vegetation index are important in vegetation mapping.

9.1.2.1 Atmospheric correction

Atmospheric correction is used to remove the atmospheric effects and transform the at-sensor radiance to surface reflectance. This correction is a prerequisite to most hyperspectral imagery (HSI) data analysis, especially when curve-shape matching is made with laboratory or field spectra (Lewis et al., 2001; Kruse et al., 2003). Three relevant atmospheric correction software packages are used to correct hyperspectral radiance data at sensor to surface reflectance: the Atmospheric REMoval program (ATREM), Atmospheric CORrection Now (ACORN), and the Fast Line-of-sight Atmospheric Analysis of Spectral Hypercubes (FLAASH). Atmospheric correction methods basically follow the radiative transfer model (Gao and Goetz, 1990), and each of the applications uses a slightly different version of the algorithm. ACORN utilizes MODTRAN4 to estimate atmospheric parameters and calculate water vapor on a per-pixel basis (Kruse, 2004). ACORN currently is used for correction of both airborne and satellite hyperspectral data and can produce high-quality surface reflectance without ground measurements (Kruse et al., 2003).

9.1.2.2 Dimension reduction

Principal component analysis (PCA) is a method routinely used in data extraction and compression from hyperspectral imagery, and each principal component is a linear combination of the original variables (Wang and Chang, 2006). Using PCA, the uncorrelated or the principal components are orthogonal to each other so that redundant data can thus be reduced by assigning scores (weighted sums of the original variables) without much loss of useful information, and the representation of the original data be maximized. The number of components is normally far less than that of original variables (Vogt and Tacke, 2001). PCA can be used in hyperspectral data processing mainly for the study of homogeneity (Ariana et al., 2006) and texture (Qiao et al., 2007) as well as for the assessment of quality parameters in leaves (Lelong et al., 1998) and fruits (Zou et al., 2010).

The minimum noise fraction (MNF) transformation proposed by Green et al. (1988) is an effective method to determine the inherent dimensionality of image data, to separate signal from noise, and to improve the efficiency of data processing by compressing spectral information to a few bands (Boardman and Kruse, 1994). The method includes two separate principal component transformation (PCT) procedures, assuming that each pixel contains both signal and noise. The first step separates white noise by creating a noise covariance matrix, which is used to decrease correlation among the spectral bands and to rescale the noise in the data. The second step recombines these bands into new composite bands that account for most of the variance in the original data. The examination of the results from the PCT determines the dimensionality of the image data into two parts, in which the first part is associated with those eigenvectors having large eigenvalues and their coherent eigenimages, and the second part is associated with near-unity eigenvalue and noise-dominated images. By using only the part with the coherent eigenimages, the noise can be separated from the signal in the image data (Underwood et al., 2003; Zhang et al., 2003).

9.1.2.3 Band selection methods

These two PCT methods are used for dimension reduction. On the basis of dimension reduction transformation, the high-dimensional data can be transformed directly down to a few dimensions or even to one dimension with a high speed, which is the advantage; the disadvantage is that it changes the original features of the image.

Another way to select optimal bands is based on a nontransformation method, such as spectral selection. This method overcomes the noted disadvantage and tends to maintain the image's original features. Therefore, studies using a spectral selection method to reduce the dimension of hyperspectral data have important applications.

Currently, the spectral selection criteria, from three different perspectives, can be summarized as follows: (a) from information theory point of view, the most informative band or band combination should be selected; (b) from the statistical point of view, the correlation between the selected bands should be minimal and thus ensure the independence and effectiveness of each band; and (c) from the spectral point of view, differences in the spectral characteristics of the selected bands should be the largest to improve the surface features of the separability.

9.2 ADVANCED HYPERSPECTRAL IMAGING SYSTEMS AND APPLICATIONS

9.2.1 Hyperspectral imaging systems

Since the first imaging spectrometer, Aero Image Spectrometer-1 (AIS-1), was successfully developed, more and more advanced HSI systems have been developed and tested. U.S. airborne HSI AVIRIS, spaceborne hyperspectral imager Moderate Resolution Imaging Spectroradiometer (MODIS), and Hyperion are

Table 9.1 Technical characteristics of the hyperspectral sensors

Name	Platforms	Number of bands	Spectral range	Spatial resolution	Average spectral resolution
Hyperion	EO-I (NASA)	242	400–2,500 nm	30 m	<10 nm
CHRIS	CHRIS/PROBA (ESA)	62	410–1,050 nm	18 m	10 nm
AVIRIS	Aircraft platform (NASA)	224	400–2,500 nm	20 m	10 nm
HSI	HJ-IA (China)	115	459–956 nm	100 m	5 nm

among the most popular hyperspectral imaging systems. The Compact High-Resolution Imaging Spectrometer (CHRIS)/Project for On-Board Autonomy (PROBA), launched by European Space Agency (ESA) in 2001, is a multiangled spectrometer, and provides hyperspectral observations from five angles. The Chinese Pushbroom Hyperspectral Imager (PHI) and Operative Modular Imaging Spectrometer (OMIS) are two representative hyperspectral imagers developed by the Chinese Academy of Sciences. In September 2008, China launched HJ-1A and HJ-1B small satellites, and a new hyperspectrometer HSI is onboard the HJ-1A satellite. In addition, some other typical hyperspectral imagers, including the Australian Hymap, Australian Resource Information and Environmental Satellite, thermal infrared line profiling spectrometer, and the Canadian Compact Airborne Spectrographic Imager (CASI), have been used in hyperspectral data acquisition and regional vegetation mapping.

The HJ-1/HSI hyperspectral sensor is carried by the HJ-1-A small satellite, conducting a repeated global monitoring at a ±30° side-viewing angle, with a 96 h revisiting cycle, a 100 m spatial resolution, 50 km swath, 115 working bands covering a spectral range of 0.45–0.95 µm, and an average spectral resolution of 4.32 nm. Although narrower in the spectral range as compared with NASA's widely used Earth Observing System MODIS (EOS-MODIS) and EO-1 Hyperion, the HJ-1/HSI spectrometer has improved spectral resolution for better ground feature identification and information extraction. It provides another valuable tool for developing quantitative research and application, such as atmospheric composition detection, water environment monitoring, and vegetation growth monitoring. Table 9.2 summarizes the specifications of the HJ-1/HSI sensor.

9.2.2 Hyperspectral data in vegetation mapping

In the field of vegetation study, especially for precise agriculture, some progress has been achieved (Tong et al., 2004). In the early 1980s and 1990s, derivative spectral analysis models were used for background noise removal, red-edge determination, and biochemical parameter estimation

Table 9.2 Specification of HJ-1/HSI data

Attribute	Value
Orbital altitude	649.093 km
Orbital inclination	97.9486°
Launch time	2008-9-18
Number of bands	115
Spectral range	459–956 nm
Spatial resolution	100 m
Average spectral resolution	4.32 nm
Swath width	50 km
Off-nadir angle	±30°

(Horler et al., 1983; Tanvir and Michael, 1990; Penuelas, 1994). Underwood et al. (2003) investigated the use of AVIRIS imagery to detect the invasive species iceplant (*Carpobrotus edulis*) and jubata grass (*Cortaderia jubata*). The green LAI of crop canopies was predicted by hyperspectral vegetation indices and novel algorithms using hyperspectral images were acquired by CASI (Calgary, Canada) (Haboudane et al., 2004). Wu et al. (2008) estimated chlorophyll content from hyperspectral vegetation indices using Hyperion data. McGwire et al. (2000) quantified sparse vegetation cover in arid environments with hyperspectral mixture modeling (Du and Chang, 2004). Zare et al. (2008) studied vegetation mapping for landmine detection using long-wave HSI. Yuan and Niu (2007) had a classification of tropical vegetation in Menghai County, Yunnan province, China using EO-1 Hyperion hyperspectral and Enhanced Thematic Mapper Plus (ETM+) data. Filippi and Jensen (2006) studied the fuzzy learning vector quantization for hyperspectral coastal vegetation classification. In recent years, the applications of spectrodirectional remote sensing have begun to take shape in the field of vegetation study, as multiangle, high-resolution imaging spectrometer CHRIS data can be used.

Not many studies on FVC estimation using hyperspectral image data have been reported. The next section presents a case study in which an improved mapping approach for FVC in Xinjiang, China, is proposed and tested using the HJ-1/HSI hyperspectral data and in situ measurements.

9.3 ESTIMATION OF FVC USING HJ-1/HSI HYPERSPECTRAL DATA

9.3.1 Dimidiate pixel model for FVC estimation

DPM assumes that the remote-sensing response in a pixel consists of soil and vegetation (Leprieur, 1994; Zribi et al., 2003), thus, the information S as captured by the remote sensor can then be expressed as consisting of

S_v as contributed by the green vegetation and S_s as contributed by the soil. Linearly decomposing S into S_v and S_s, the proportion of vegetation area in the pixel is the FVC, f_c, of that pixel, and, accordingly, the proportion of soil area will be $1 - f_c$. Assuming the spectral response information received by the all-vegetation pure pixel is S_{veg}, the information contributed by vegetation in the mixed pixel is S_v, and the spectral response contributed by vegetation in the mixed pixels can then be expressed as a product of S_{veg} and f_c:

$$S_v = f_c \times S_{veg}. \tag{9.1}$$

Similarly, assuming the remotely sensed information received by the all-pure soil pixel is S_{soil}, the information S_s as contributed by soil in the mixed pixel can then be expressed as a product of S_{soil} and $1 - f_c$:

$$S_s = (1 - f_c) \times S_{soil}. \tag{9.2}$$

Based on Equations 9.1 and 9.2, the spectral response of a mixed pixel can be derived as follows:

$$S = f_c \times S_{veg} + (1 - f_c) \times S_{soil}. \tag{9.3}$$

Equation 9.3 can be understood as linearly decomposing S into S_{veg} and S_{soil}, whose weights are the proportion of area taken by them respectively in the pixel, that is, f_c and $1 - f_c$.

For pixels containing more than two components, Equation 9.3 will have to be modified. In the case that elements other than vegetation and soil are included, such as water body, Equation 9.3 should be modified by the multicomponent mixed model. In the case of a mixture of only vegetation and soil, the FVC can be derived by modifying Equation 9.3 as follows:

$$f_c = \frac{(S - S_{soil})}{(S_{veg} - S_{soil})}, \tag{9.4}$$

where S_{soil} and S_{veg} are the spectral responses from pure soil and pure vegetation pixels, respectively. The model has a fairly sound theoretical basis and is widely applicable regardless of the geographic constraints. In addition, a major advantage of the DPM is that the impacts from atmosphere, soil background, and vegetation type are reduced. S_{soil} contains the soil information, including the contribution to remotely sensed data by elements such as the type, color, brightness, and moisture of soil, whereas S_{veg} contains the vegetation information, including the contribution to the remotely sensed data by elements such as the type and structure of vegetation. In fact, the DPM is a linear stretch based on the two regulatory factors of S_{soil}

and S_{veg}, whereby the impacts on remotely sensed data by atmosphere, soil background, and vegetation type are reduced to the minimum, keeping only the FVC information. Therefore, FVC can be estimated by Equation 9.4.

In the DPM, remotely sensed spectral response is well related linearly to the FVC. For instance, the most widely applied remote-sensing response is NDVI, which, being the best indicator for plant growth and density of vegetation spatial distribution, is linearly correlated to vegetation distribution density. The value of NDVI, which comprehensively reflects the vegetation type, canopy pattern, and growth status in per-unit pixel, is determined by elements like the FVC (horizontal density) and LAI (vertical density) and is notably correlated to the FVC. Inserting the NDVI into Equation 9.3, we can have the following approximation:

$$NDVI = f_c \times NDVI_{veg} + (1 - f_c) \times NDVI_{soil}. \tag{9.5}$$

Rearranging Equation 9.5, the FVC can be derived using the values of $NDVI_{soil}$ and $NDVI_{veg}$ as follows:

$$f_c = \frac{NDVI - NDVI_{soil}}{NDVI_{veg} - NDVI_{soil}} \times 100. \tag{9.6}$$

In Equation 9.6, the NDVI data, the NDVI for "pure" vegetation pixel, and the NDVI for "pure" bare soil need to be determined to retrieve the FVC information from the remotely sensed data.

9.3.2 Study area and data collection

In the middle section of the northern slope of the Tianshan Mountains in China's Xinjiang Uygur Autonomous Region, the Shihezi area is situated on the southern border of the Junggar Basin, bordered by the Gurbantunggut Desert in the north, with the geographic coordinates being 84°40′ by 86°43′E, 43°15′ by 45°23′N (see Figure 9.1). With an altitude of 500 to 800 m above sea level, and varying greatly, declining from southeast to northwest, the area consists of mountains, plains, and deserts. Its temperate continental climate features a long and cold winter and a short and hot summer, dry and scare rainfall, high evaporation, rapid rise and drop of temperature in spring and fall, respectively, and sharp contrast temperatures between daytime and night. The annual average temperature is 7.5°C to 8.2°C, with 2,318 to 2,732 hours of sunshine, 147 to 191 days of frostless period, annual precipitation of 180 to 270 mm, and an annual evaporation of 1,000 to 1,500 mm. The area is in a typical arid climate environment with a full range of ecological types, including snow cover, meadows and forests of the alpine ecosystem, hilly piedmont grassland, oasis agricultural system, and the desert system in the center of the basin. The natural vegetation

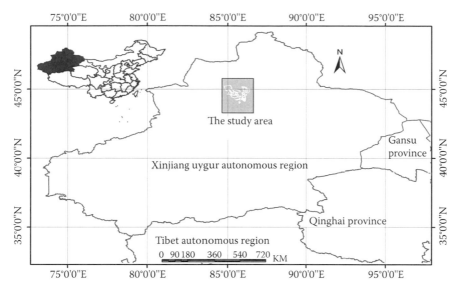

Figure 9.1 Location of the study area.

is highly dependent on rainfall, resulting in certain interannual fluctuations. The sparsely populated large area with a dry climate that seldom rains is perfect for application of the remote-sensing technique and is favorable for research on the interaction between human activity and the natural environment.

The HJ-1/HSI image data covering the entire Shihezi area in Xinjiang were acquired on July 24, 2009. Three scenes of the radiometrically calibrated HSI images were used in this study. For the purpose of comparative study, the MODIS LAI data products and Landsat Thematic Mapper (TM) imagery of the same period in the area were also obtained. In addition, the FVC data of 155 in situ plant quadrate data in the study area were also collected. The quadrate values falling in the same pixel were averaged to get the in situ measured value of that pixel. In the ground measuring, photographs taken by a digital camera Canon EOS500D were classified to extract the FVC (White et al., 2000; Zhou and Robson, 2001). The ground-measured quadrate FVC data were used mainly to assess the accuracy of the FVC data retrieved from remotely sensed data and to calibrate the parameters for the DPM.

9.3.3 Calibration and atmospheric radiation correction

In this study, the FLAASH module in the Environment for Visualizing Images (ENVI) package was used in the atmospheric radiometric correction of the HJ-1/HSI imagery. FLAASH uses the MODTRAN4 radiometric transmission model code and is one of the most accurate atmospheric radiometric correction models. Because no parameter is embedded in the FLAASH module for direct correction of the HJ-1/HSI data, the atmospheric radiometric correction is conducted on the basis of the HJ-1/HSI sensor's own parameters and characteristics. In the FLAASH atmospheric correction, the main parameters to be input include the central position of image; type of sensor; altitude; imaging time; mean ground elevation; image spatial resolution; atmospheric-, aerosol-, and water-vapor models; method for extraction of aerosol parameters; atmospheric visibility; and whether spectral smoothing and wavelength remodification are required to run the module.

In preprocessing, the HJ-1/HSI data is first radiometrically calibrated according to the radiometric calibration formula, and then the data is converted to the BIL (band interleaved by line) or BIP (band interleaved by pixel) format. At the same time, the central wavelength of each band and its full width at half maximum (FWHM) are specified. The image central position, sensor altitude, imaging time, and resolution are available in the header file. The average elevation of the land surface can be calculated from the digital elevation model (DEM) of the study area. In water-vapor retrieval, the water-vapor removal model was used to retrieve the water-vapor content in each pixel of the images. Because the HJ-1/HSI data have a spectral range from 450 nm to 950 nm, the water-vapor characteristic band should be 820 nm. When the model's parameters are correctly set, the correction of the HJ-1/HSI data in the study area can be conducted by running the FLAASH module. Figure 9.2 shows a comparison of the typical surface feature spectral curve before and after the correction. On finishing the atmospheric radiometric correction, the surface reflectance data was obtained and ready for extraction of the FVC information.

9.3.4 Selection of the narrow-band vegetation index

The vegetation index calculated from remotely sensed data is required when retrieving the vegetation coverage using the DPM. The HSI has a number of red bands and near-infrared bands, resulting in many combinations for calculating narrow-band NDVI. Many recent studies that used hyperspectral data have focused on improving the commonly and widely used NDVI and on developing new indices aiming to compensate for soil background influences and atmospheric effects (Verstraete et al., 1993; Broge and Leblanc, 2001; Stagakis et al., 2010). Thus, there is a need to determine a more effective combination of bands. A number of methods are available for selecting the bands, including stepwise regression, simple band autocorrelation, and weighted principal component analysis (Quamby et al., 1992).

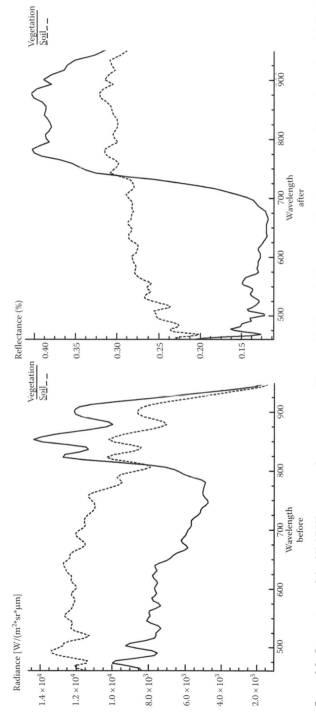

Figure 9.2 Comparison of the HJ-1/HSI spectra of green grass and bare soil before and after atmospheric correction (the radiance is multiplied a scale of 100).

Table 9.3 Comparison of the five best fitting pairs between LAI and NDVI

NIR/Red band (nm) combination	r^2 for linear regression	r^2 for exponential regression
877/682	0.7654	0.8836
900/682	0.7219	0.7999
782/660	0.6627	0.7989
782/692	0.6174	0.7449
759/631	0.5872	0.7244

Comparing the NDVI and LAI obtained by all possible band pairs, the best band combination for the FVC estimation can be determined. In the HJ-1/HSI data, there are a total of 22 bands in the near infrared region (760–900 nm) and 15 bands in the red band wavelength region (630–690 nm), and the possible pairs can give a total of 330 NDVI values.

The study shows that a considerable positive correlation exists between FVC and LAI. Given the LAI data of the same period of time, the correlation between NDVI and LAI is compared, and the more correlated NDVI and LAI are, the more suitable the NDVI retrieved from that band combination for the retrieval of fractional vegetation cover. In this chapter, 330 NDVI image data are evaluated one by one with the MODIS LAI data from the same time. Table 9.3 gives the fitting results between NDVI from several typical band combinations and the MODIS/LAI data. As shown in Table 9.3, when the 877 nm near-infrared and 682 nm red bands are used for calculation of NDVI, the correlation of simple linear regression and exponential regression between NDVI and LAI is the highest, with the correlation coefficient of exponential regression being above 0.88. Therefore the 877/682 band combination is the best for calculating the hyperspectral narrow-band NDVI for FVC retrieval. This band-selection method can be easily and quickly implemented in the ENVI/Interactive Data Language (IDL) environment.

9.3.5 Determining the $NDVI_{veg}$ and $NDVI_{soil}$

In the study area, there are "pure" pixels for forests and agricultural areas because of the dense vegetation. In most areas of this region, however, mixed pixels are widespread because of the sparse vegetation and the 100 m resolution of HJ-1/HSI data, and a mixed pixel often contains multiple spectra of elements like soil, vegetation, and shadows. Because this study used the DPM for the FVC extraction, the challenge is to identify the correct NDVI values for "pure" vegetation and soil pixels, respectively.

Ideally, $NDVI_{soil}$ is not to change with time and should be around zero for the bare surface. Because of the atmospheric impact and the changes in surface soil moisture, however, $NDVI_{soil}$ may change with time. In addition, $NDVI_{soil}$ also may change with space depending on conditions like soil moisture, roughness, soil type, and color. Therefore, it is not practical to think of a fixed and ideal $NDVI_{soil}$ value, as the value will have to change even for the same scene of imagery. For easy adjustment, it is not necessary to know the actual $NDVI_{soil}$ value; instead, it can be extracted based on the image data.

$NDVI_{veg}$ represents the maximum value of all the vegetation pixels. Depending on the vegetation type, the seasonal change of the canopy, the interference of the foliage background, as well as wet ground, snow, and fallen leaves, the $NDVI_{veg}$ value is determined similarly to that of the $NDVI_{soil}$, as the $NDVI_{veg}$ value will also change with time and space. Therefore, it also is not advisable to think of an ideal $NDVI_{veg}$ value.

To further improve the prediction of the model, the parameters $NDVI_{veg}$ and $NDVI_{soil}$ were determined for the four main land cover types in the study area: farming land, grassland, forest, and water body–snow packs. With the aid of ground measurements of FVC in the study area, an approach was proposed and used to resolve the suitable $NDVI_{veg}$ and $NDVI_{soil}$ values for the three main vegetation cover types. For each pixels in which the in situ measurements are located, the relationships between the FVC and NDVI can be depicted using the DPM:

$$\begin{bmatrix} fc_1 & 1-fc_1 \\ \vdots & \vdots \\ fc_n & 1-fc_n \end{bmatrix} \begin{bmatrix} NDVI_{veg} \\ NDVI_{soil} \end{bmatrix} = \begin{bmatrix} NDVI_1 \\ \vdots \\ NDVI_n \end{bmatrix}, \tag{9.7}$$

where fc_i is the FVC of the ith pixel that has in situ measurements. The least squares method (LSM) is then used to resolve Equation 9.7 and extract the parameters $NDVI_{veg}$ and $NDVI_{soil}$ for the DPMs. The retrieval parameters from the HJ-1/HSI and TM data are listed in Table 9.4 This method includes the in situ-measured FVC that can be seen as prior knowledge in the extraction of the parameters for the DPMs. Figure 9.3 illustrates the process for determining the parameters for this study's DPM.

Inserting the DPM parameters derived from spectral mixture decomposition into Equation 9.6, the study area's FVC—which has a value between

Table 9.4 Model parameters for each land cover type

Land cover types	HJ-1/HSI		Landsat TM	
	$NDVI_{veg}$	$NDVI_{soil}$	$NDVI_{veg}$	$NDVI_{soil}$
Farming lands	0.641	0.012	0.733	0.080
Grasslands	0.593	0.05	0.563	0.080
Forest lands	0.576	0.12	0.666	0.049

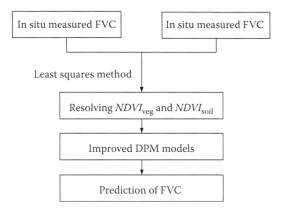

Figure 9.3. The framework for the extraction of the $NDVI_{veg}$ and $NDVI_{soil}$ parameters.

0 and 100, where 0 represents "pure" bare soil and 100 represents all or 100% vegetation coverage in the pixel—was then retrieved.

9.4 RESULTS AND DISCUSSION

9.4.1 Results

The DPM for the FVC retrieval as applied in Xinjiang's Shihezi area was implemented by programming in the ENVI/IDL6.5 environment. The FVC image as of July 24, 2009, of the study area was obtained using this model (see Figure 9.4a).

As shown in Figure 9.4a, the northern part of the study area is mainly sand lands with vegetation coverage of less than 5%. Some small patches surrounding the farming land have FVC values between 5% and 20%, but most have FVC values of less than 10%. The farming lands are located in the middle part of the area and have FVC values of more than 75%, indicating large patches of cotton and grape fields. In the same area, some pixels have FVC values between 50% and 75%, depending on the portion of the densely vegetated cotton and grape fields in a mixed pixel with 100 m resolution. In the wasteland surrounding the farmland, there is some medium vegetation of 20% to 50% coverage (either farmland or abandoned land) as well as some low-coverage sandy land. The steep areas in the piedmont hills of the Tianshan Mountain are sparsely vegetated with FVC values between 5% and 20% because of the lack of water content. The type of vegetation is mostly drought tolerant or ephemeral gramineae as well as alhagi pseudalhagi that the cattle and sheep would avoid. With the increase of altitude, some lands are medium-vegetated grass and have FVC between 20% and 50%. At an average altitude of 1,800 m in the Tianshan Mountains, there are distributions of Alpine meadows and Abies forests with the FVC

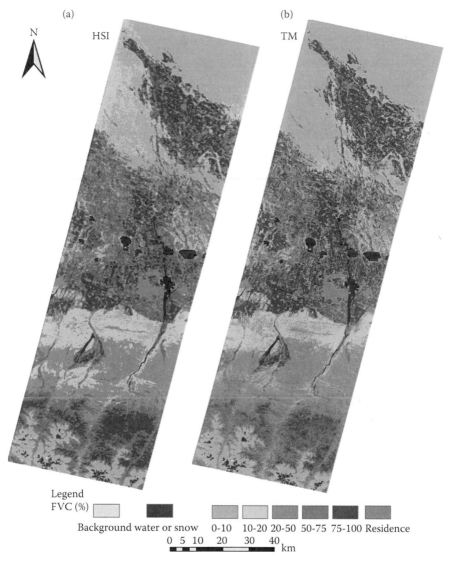

Legend
FVC (%)

Background water or snow   0-10   10-20 20-50 50-75 75-100 Residence
                            0 5 10   20   30   40
                            km

Figure 9.4 **(See color insert.)** The FVC maps retrieved from (a) the HJ-1/HSI data and (b) the Landsat TM data.

often being above 75% because of much better natural water supply (snow-melting water and larger rainfall) (see Figure 9.4a). In the skirting areas of the forest, high grasslands develop with FVC of more than 75%, which are densely vegetated high grassland. The averaged FVC values derived from the HJ-1/HSI data for farming land, grassland, forestland, and sandy land are 64.0%, 30.3%, 68.2%, and 4.9%, respectively.

In general, using the DPM improved by the narrow-band spectral indexes of the HJ-1/HSI data, the vegetation distribution of natural vegetation and farmland in the Shihezi area is quantitatively retrieved, thus offering an effective means for dynamic monitoring of vegetation.

9.4.2 Accuracy assessment and comparison

To assess the accuracy and viability of the HJ-1/HSI data in vegetation monitoring, a cross-check of ground quadratic data with the FVC results retrieved from the HJ-1/HSI data and from the Landsat TM data was made. Both to assess the FVC accuracy as retrieved from the HJ-1/HSI data, and to verify relative usability of the sensor's data in dynamic monitoring of vegetation, the Landsat TM data were first resampled into 100 m pixel size, as well as with the resolution of the HJ-1/HSI data. Figure 9.4b shows the FVC of the study area retrieved from one scene of the Landsat TM image acquired on July 23, 2009, using the aforementioned modeling method. By visually checking Figure 9.4a and Figure 9.4b, it is evident that there is great similarity between the FVC classes retrieved from the two data sources in terms of the spatial distribution patterns. The difference occurs in the desert area and the farmland in the plain. In the case of the farmland, the lower spatial resolution of the HJ-1/HSI data has resulted in a smaller area of high coverage of vegetation because of the marginal farmland and smaller patches of farmland in the mixed pixels. In the northwest corner of the FVC map, 2% to 3% higher FVC was retrieved from the HJ-1/HSI data than that from the TM data of the same area. This may be caused by the fact that the desert area has sparsely distributed desert vegetation Chenopodiaceae, which cannot be detected by the TM broadband sensor but can be recognized by the HSI. Comparing Figure 9.4a and Figure 9.4b, the difference shows that the HJ-1/HSI data can better differentiate those pixels of bare sand from those pixels of sparsely vegetated Chenopodiaceae. In the mountainous area, the HJ-1/HSI data resulted in more continuously distributed higher FVC values than the multispectral TM data. As a matter of fact, these subareas are densely vegetated forest and high grassland. Thus, the 100 m resolution hyperspectral data of the HJ-1/HSI sensor may be more suitable for FVC monitoring at a medium range, and its narrow-band-based vegetation index, in particular, can avoid the saturation problem in the multispectral broad-band-based vegetation index and the subjectivity in determining $NDVI_{veg}$ and $NDVI_{soil}$.

The averaged FVC values for farming land, grassland, forestland, and sand land derived from the Landsat TM and HJ-1/HSI data listed in pairs are 62.4% and 64.0%, 32.1% and 30.3%, 61.7% and 68.2%, and 3.1% and 4.9%, respectively. The two sets of averaged FVC values are very close and comparable, indicating the proposed model works well when applied to different remotely sensed data.

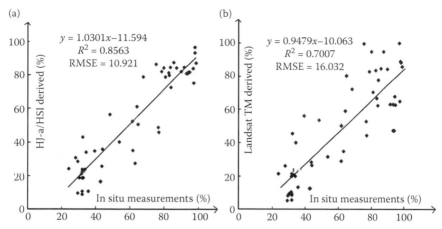

Figure 9.5 Comparison of (a) HJ-I/HSI-derived FVC and (b) Landsat TM-derived FVC using the ground quadratic data as the reference.

Figure 9.5 shows the results of cross-checking the FVC as retrieved from the HJ-1/HSI and Landsat TM data, respectively, against the 36 surface-measured quadrates. The vegetation coverage retrieved from the HJ-1/HSI data fits better with the surface quadrate results, with a correlation coefficient square of 0.86, while fitness between the TM data retrieved result and the surface quadrate result has a correlation coefficient square of 0.70. The average relative error of vegetation coverage values retrieved from HJ-1/HSI and TM data in relation to the measured values are 10.92% and 16.03%, respectively. Although ground survey also gives errors, generally speaking, the comparative study still indicates that the HJ-1/HSI data narrow-band-based vegetation index in the mixed pixel decomposition can better determine the parameters of the FVC retrieval model to improve the DPM to better depict the vegetation cover variations in the study area.

9.5 CONCLUSION

This chapter examines the mapping approaches and important spectroimaging systems for vegetation mapping and presents a case study of how the HJ-1 hyperspectral data were used to study the fractional vegetation covers in the arid area of Xinjiang, China. The preprocessing of the HJ-1/HSI data was first discussed, addressing the atmospheric radiation correction method on the basis of the MORDTRAN4 atmospheric transmission model and the determination of the parameters. Second, a strategy for using the MODIS LAI data in selecting the suitable hyperspectral narrow bands for vegetation index calculation was proposed to determine the parameters for the DPM and to assess the accuracy of the FVC retrieval. On such basis, the vegetation condition of the second half of July 2009 in Xinjiang's Shihezi area was retrieved using the proposed model established in this study. The retrieved FVC from the HJ-1/HSI data was verified with the in situ measurements, and the correlation coefficient square is 0.86 and the RMSE is 10.9%. The result showed that the FVC information can be estimated accurately by using the HJ-1/HSI data, indicating that the HJ-1/HSI data can satisfy the needs of dynamic and quantitative monitoring of vegetation coverage changes in a medium range.

REFERENCES

Ariana, P. D., R. Lu, and E. D. Guyer. 2006. Near-infrared hyperspectral reflectance imaging for detection of bruises on pickling cucumbers. *Computers and Electronics in Agriculture* 53: 60–70.

Blackburn, A. G. 1998. Quantifying chlorophylls and caroteniods at leaf and canopy scales: an evaluation of some hyperspectral approaches. *Remote Sensing of Environment* 66(3): 273–285.

Boardman, J. W., and F. A. Kruse. 1994. Automated spectral analysis: A geological example using AVIRIS data, North Grapevine Mountains, Nevada. In *Proceedings of the Tenth Thematic Conference on Geological Remote Sensing, Environmental Research Institute of Michigan, San Antonio, TX*, 407–418.

Broge, H. N., and E. Leblanc, 2001. Comparing prediction power and stability of broadband and hyperspectral vegetation indices for estimation of green leaf area index and canopy chlorophyll density. *Remote Sensing of Environment* 76(2): 156–172.

Cho, M.A., A. Skidmore, F. Corsi, S. E. van Wieren, and I. Sobhan. 2007. Estimation of green grass/herb biomass from airborne hyperspectral imagery using spectral indices and partial least squares regression. *International Journal of Applied Earth Observation and Geoinformation* 9: 375–391.

Darvishzadeh, R., A. Skidmore, C. Atzberger, and V. S. Wieren. 2008a. Estimation of vegetation LAI from hyperspectral reflectance data: effects data: effects of soil type and plant architecture. *International Journal of Applied Earth Observation and Geoinformation* 10: 358–373.

Darvishzadeh, R., A. Skidmore, M. Schlerf, C. Atzberger, F. Corsi, and M. Corsi. 2008b. LAI and chlorophyll estimation for a heterogeneous grassland using hyperspectral measurements. *Photogrammetry & Remote Sensing* 63(4): 409–426.

Du, Q., and C. Chang. 2004. Linear mixture analysis-based compression for hyperspectral image analysis. *IEEE Transactions on Geoscience and Remote Sensing* 42(4): 875–891.

Filippi, M. A., and R. J. Jensen. 2006. Fuzzy learning vector quantization for hyperspectral coastal vegetation classification. *Remote Sensing of Environment* 100(4): 512–530.

Gao, B., and A. F. H. Goetz. 1990. Column atmospheric water vapor and vegetation liquid water retrievals from airborne imaging spectrometer data. *Journal of Geophysical Research* 95(4): 3549–3564.

Goel, K. P., O. S. Prasher, M. R. Patel, A. J. Landry, R. B. Bonnell, and A. A. Viau. 2003. Classification of hyperspectral data by decision trees and artificial neural networks to identify weed stress and nitrogen status of corn. *Computers and Electronics in Agriculture* 39 (2): 67–93.

Green, A. A., M. Berman, P. Switzer, and M. D. Craig. 1988. A transformation for ordering multispectral data in terms of image quality with implications for noise removal. *IEEE Transactions on Geoscience and Remote Sensing* 26: 65–74.

Haboudane, D., R. J. Miller, E. Pattey, J. P. Zarco-Tejada, and B. I. Strachan. 2004. Hyperspectral vegetation indices and novel algorithms for predicting green LAI of crop canopies: modeling and validation in the context of precision agriculture. *Remote Sensing of Environment* 90(3): 337–352.

Haboudane, D., N. Tremblay, R. J. Miller, and P. Vigneault. 2008. Remote estimation of crop chlorophyll content using spectral indices derived from hyperspectral data. *IEEE Transactions on Geoscience and Remote Sensing* 46(2): 423–437.

Horler, D. N., M. Dockray, and J. Barber. 1983. The red edge of plant leaf reflectance. *International Journal of Remote Sensing* 4: 273–288.

Kruse, F. A. 2004. Comparison of ATREM, ACORN, and FLAASH atmospheric corrections using low-altitude AVIRIS data of Boulder, Colorado. In *Proceedings 13th JPL Airborne Geoscience Workshop, Jet Propulsion Laboratory, March 31–April 2, 2004, Pasadena, CA*, JPL publication 05-3, http://www.hgimaging. com/PDF/Kruse-JPL2004_ATM_Compare.pdf (last accessed on August 9, 2012).

Kruse, F. A., J. W. Boardman, and J. F. Huntington. 2003. Comparison of airborne hyperspectral data and EO-1 Hyperion for mineral mapping. *IEEE Transactions of Geoscience and Remote Sensing* 41(6): 1388–1400.

Landgrebe, D., S. B. Serpico, M. M. Crawford, and V. Singhroy. 2001. Introduction to the special issue on analysis of hyperspectral image data. *IEEE Transactions of Geoscience and Remote Sensing* 39(7): 1343–1345.

Lelong, C. D. C., C. P. Pinet, and H. Poilv. 1998. Hyperspectral imaging and stress mapping in agriculture: a case study on wheat in Beauce (France). *Remote Sensing of Environment* 66: 179–191.

Leprieur, C., M. M. Verstraete, and B. Pinty. 1994. Evaluation of the performance of various vegetation indices to retrieve cover from AVHRR data. *Remote Sensing Review* 10: 265–284.

Lewis, M., V. Jooste, and A. A. de Gaspairs. 2001. Discrimination of arid vegetation with airborne multispectral scanner hyperspectral imagery. *IEEE Transactions of Geoscience and Remote Sensing* 39(7): 1471–1479.

Li, G., K. Song, and S. Niu. 2007. Soybean LAI estimation with In-situ collected hyperspectral data based on BP-neural Networks. In *Proceedings of the 3rd International Conference on Recent Advances in Space Technologies (RAST '07), June 14–16, 2007, Istanbul, Turkey, 331–336*.

Liu, J., R. J. Miller, E. Pattey, and D. Haboudane, I. B. Strachan, and M. Hinther. 2004. Monitoring crop biomass accumulation using multi-temporal hyperspectral remote sensing data. *IEEE International, Geoscience and Remote Sensing Symposium* 3 1637–1640.

Liu, X., W. Fan, Q. Tian, and X. Xu. 2008. The LAI inversion based on directional second derivative using hyperspectral data. *IEEE International, Geoscience and Remote Sensing Symposium* 3: III-740–III-743.

Luo, B., and J. Chanussot. 2009. Unsupervised classification of hyperspectral images by using linear unmixing algorithm. *Image Processing, 2009 16th IEEE International Conference*: 2877–2880.

McGwire, K., T. Minor, and L. Fenstermaker. 2000. Hyperspectral mixture modeling for quantifying sparse vegetation cover in arid environments. *Remote Sensing of Environment* 72(3): 360–374.

Oldeland, J., W. Dorigo, and L. Lieckfeld. 2010. Combining vegetation indices, constrained ordination and fuzzy classification for mapping semi-natural vegetation units from hyperspectral imagery. *Remote Sensing of Environment* 114(6): 1155–1166.

Penuelas, F. J. 1994. The red edge position and shape as indicators of chlorophyll content, biomass and hydric status. *International Journal of Remote Sensing* 15(7): 1459–1470.

Qiao, J., M. O. Ngadi, and N. Wang, C. Gariépy, and S. O. Prasher. 2007. Pork quality and marbling level assessment using a hyperspectral imaging system. *Journal of Food Engineering* 83: 10–16.

Quamby, N. A., J. R. G. Townshend, and J. J. Settle. 1992. Linear mixture modeling applied to AHVRR data for crop area estimation. *International Journal of Remote Sensing* 13(3): 415–425.

Stagakis, S., N. Markos, O. Sykioti, and A. Kyparissis. 2010. Monitoring canopy biophysical and biochemical parameters in ecosystem scale using Satellite hyperspectral imagery: An application on a Phlomis fruticosa Mediterranean Ecosystem using multiangular CHRIS/PROBA observations. *Remote Sensing of Environment* 114(5): 977–994.

Tømmervik, H., J. A. Hogda, and I. Solheim. 2003. Monitoring vegetation changes in Pasvik (Norway) and Pechenga in Kola Peninsula (Russia) using multitemporal Landsat MSS/TM data. *Remote Sensing of Environment* 85(3): 370–388.

Tanvir, H. D., and D. S. Michael. 1990. High resolution derivative spectra in remote sensing. *Remote Sensing of Environment* 33: 55–64.

Tarabalka, Y., J. Chanussot, and A. J. Benediktsson. 2010. Segmentation and classification of hyperspectral images using watershed transformation. *Pattern Recognition* 43(7): 2367–2379.

Tong, Q., B. Zhang, and L. Zheng. 2004. Hyperspectral remote sensing technology and applications in China. In *Proceedings of the 2nd CHRIS/PROBA Workshop, ESA/ESRIN, Frascati, Italy, 28–30, ESA SP-578*.

Underwood, E., S. Ustin, and D. DiPietro. 2003. Mapping nonnative plants using hyperspectral imagery. *Remote Sensing of Environment* 86(2): 150–161.

Verstraete, M. M., C. LePrieur, S. De Brisis, and B. Pinty. 1993. GEMI: A new index to estimate the continental fractional vegetation cover. In *Proceedings of the 6th AVHRR Data User's Meeting, Belgirate, Italy, 29 June -2 July*, 143–149.

Vogt, F., and M. Tacke. 2001. Fast principal component analysis of large data sets. *Chemometrics and Intelligent Laboratory Systems* 59: 1–18.

Wang, J., and C. Chang. 2006. Independent component analysis-based dimensionality reduction with applications in hyperspectral image analysis. *IEEE Transactions on Geoscience and Remote Sensing* 44(6): 1586–1600.

White, M. A., R. G. Asner, P. R. Nemani, J. L. Rivette, and S. W. Running. 2000. Measuring fractional cover and leaf area index in arid ecosystem: Digital camera, radiation transmittance, and laser altimetry methods. *Remote Sensing of Environment* 74(l): 45–57.

Wu, C., Z. Niu, Q. Tang, and W. Huang. 2008. Estimating chlorophyll content from hyperspectral vegetation indices: modeling and validation. *Remote Sensing of Environment* 148(8-9) 1230–1241.

Yuan, J., and Z. Niu. 2007. Classification using EO-1 Hyperion hyperspectral and ETM+ data. In *Fuzzy Systems and Knowledge Discovery, 4th International Conference*, Vol. 3, 538–542. Washington, DC: IEEE Computer Society.

Zarco-Tejada, P. J., J. R. Miller, T. L. Noland, G. H. Mohammed, and P. H. Sampson. 2001. Scaling-up and model inversion methods with narrowband optical indices for chlorophyll content estimation in closed forest canopies with hyperspectral data. *IEEE Transactions on Geoscience and Remote Sensing* 39(7): 1491–1507.

Zare, A., J. Bolton, P. Gader, and M. Schatten. 2008. Vegetation mapping for land-mine detection using long-wave hyperspectral imagery. *Geoscience and Remote Sensing, IEEE Transactions*, 46(1): 172–178.

Zhang, M., Z. Qin, X. Liu, and S. Ustin. 2003. Detection of stress in tomatoes induced by late blight disease in California, USA, using hyperspectral remote sensing. *International Journal of Applied Earth Observation and Geoinformation* 4(4): 295–310.

Zhou, Q., and M. Robson. 2001. Automated rangeland vegetation cover and density estimation using ground digital images and a spectral-contextual classifier. *International Journal of Remote Sensing* 22(17): 3457–3470.

Zou, X., J. Zhao, M. Holmes, H. Mao, J. Shi, X. Yin, and Y. Li. 2010. Independent component analysis in information extraction from visible/near-infrared hyperspectral imaging data of cucumber leaves. *Chemometrics and Intelligent Laboratory Systems* 104(2): 265–270.

Zribi, M., S. Le Hégarat-Mascle, O. Taconet, V. Ciarletti, D. Vidal-Madjar, and M. R. Boussema. 2003. Derivation of wild vegetation cover density in semi-regions: ERS2/SAR evaluation. *International Journal of Remote Sensing* 24(6): 1335–1352.

(a)   (b)

(c)   (d)

Double mapping problems

(e)   (f)

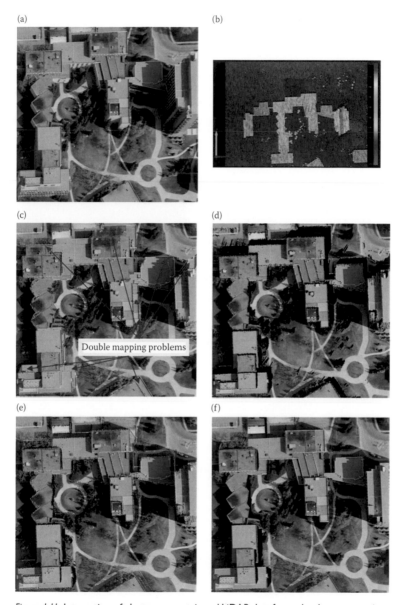

Figure 1.11 Integration of photogrammetric and LiDAR data for orthophoto generation. (a) Perspective image. (b) Corresponding LiDAR data. (c) Orthophoto with the double mapping problems. (d) True orthophoto after occlusion detection. (e) True orthophoto after occlusion filling. (f) True orthophoto with enhanced building boundaries.

Figure 1.12 3D visualization achieved by draping a true orthophoto on top of an enhanced LiDAR DSM with a DBM wireframe.

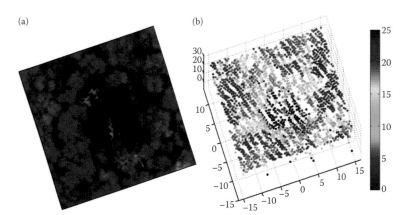

Figure 2.17 (a) Color-infrared aerial photograph and (b) first return LiDAR measurements for a gap and surrounding trees. The gap area is obvious in both the aerial photograph and LiDAR measurements because of differences in spectral reflectance and elevation. (After Zhang, K., "Identification of Gaps in Mangrove Forests with Airborne LIDAR," *Remote Sensing of Environment* 112: 2309–2325, 2008. With permission from Elsevier.)

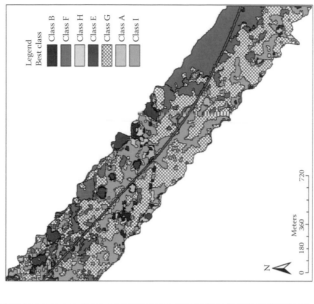

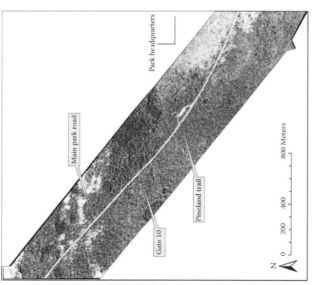

Figure 2.20 (a) Aerial photograph for an area in Everglades National Park. (b) Map showing the classification of the vegetation communities based on the height distributions of LiDAR measurements in a moving window. The classified vegetation classes include A: Hammock edge, B: Transition, E: Mature hammock center, F: Open ground with herbaceous plants, G: Pine trees with herbaceous understory, H: Pine forest edge, and I: Pine trees with shrub understory.

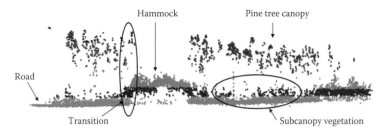

Figure 2.21 3D display of raw LiDAR points and vegetation classes. Colors indicate elevation ranges in meters (NAVD88): gray, 0.2–3.1 m; red, 3.2–5.4 m; green, 5.5–9.4 m; light blue, 9.5–14 m; dark blue, 14–18.8 m.

Figure 3.1 ETM+ image of the study area Near Littleport, England.

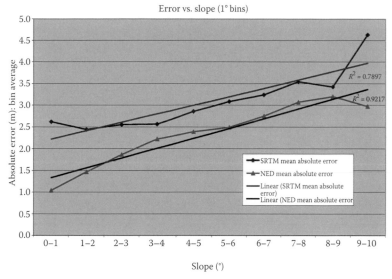

Error vs. slope (1° bins)

$R^2 = 0.7897$

$R^2 = 0.9217$

- ◆ SRTM mean absolute error
- ▲ NED mean absolute error
- —— Linear (SRTM mean absolute error)
- —— Linear (NED mean absolute error)

Slope (°)

Absolute error (m): bin average

Figure 4.1 Increasing elevation error with increasing slope for SRTM and the USGS National Elevation Dataset, as measured versus more than 13,000 geodetic control points in the conterminous United States.

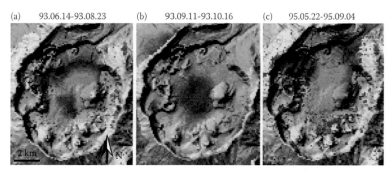

(a)   93.06.14-93.08.23   (b)   93.09.11-93.10.16   (c)   95.05.22-95.09.04

Figure 5.3 Deformation interferograms of Okmok volcano during three different time periods: (a) June 14 to August 23, 1993, with the perpendicular component of baseline, $B_\perp$, equal to 32 m; (b) September 11 to October 16, 1993, with $B_\perp = 25$ m; and (c) May 22 and September 4, 1995, with $B_\perp = 22$ m. The inflation was estimated and removed from those interferograms used for DEM generation (see Table 5.1). A full cycle of colors represents 28 mm surface deformation along the satellite look direction. Areas of coherence loss are uncolored.

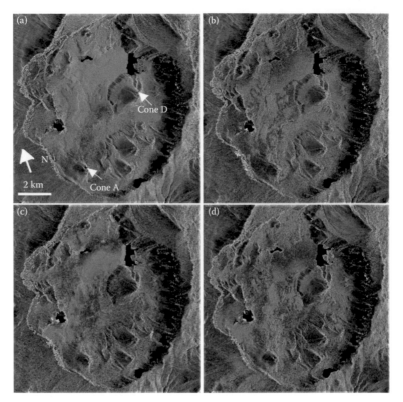

Figure 5.4 Residual interferograms produced by subtracting the topographic phase from the original interferograms (see Table 5.1). (a) The tandem interferogram with $B_\perp = 83$ m and the SRTM DEM was used to remove the topographic phase. (b) The interferogram with $B_\perp = 403$ m and the SRTM DEM was used to remove the topographic phase. (c) The interferogram with $B_\perp = 395$ m and the DEM produced from the interferogram with $B_\perp = 403$ m was used to remove the topographic phase. (d) The interferogram with $B_\perp = 690$ m and the DEM produced from the interferogram with $B_\wedge = 403$ m was used to remove the topographic phase. A full cycle of colors represents a phase change of 360 degrees.

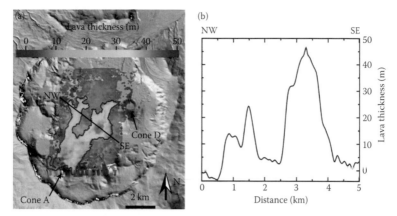

Figure 5.5 (a) Thickness of lava flows emplaced during the 1997 eruption at Okmok volcano, Alaska. Flow thickness was derived from the height difference between the posteruption SRTM DEM and a preeruption DEM constructed from multi-temporal interferograms. (b) Lava thickness along profile north-west-southeast, reaching nearly 50 m in the thickest part of the flow. The red line represents the lava perimeter based on field data collected in August 2001.

Figure 5.7 DEM height update from MTInSAR processing technique for Okmok volcano, Alaska. The height difference is with respect to an airborne DEM mosaic.

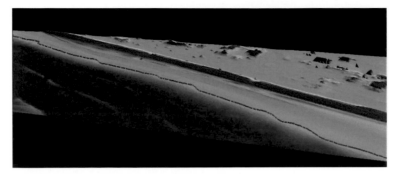

Figure 6.6 LiDAR-derived shoreline (magenta line).

Figure 6.7 (a) Thermal image of a stretch of shoreline in Monterey, California, acquired with an Itres TABI. (b) RGB imagery of same area acquired with an Applanix DSS digital aerial camera. These data were acquired in the late afternoon. (A seamline from the image mosaicing process is evident in the bottom image.)

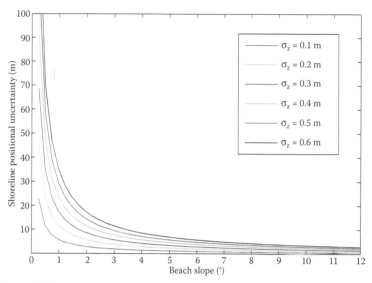

Figure 6.8 Plots of shoreline positional uncertainty as a function of beach slope and elevation uncertainty, σ_z.

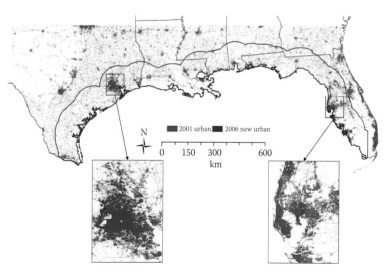

Figure 8.4 Distributions of 2001 and 2006 new urban land cover in the region. The lower left and right panels are for Houston, Texas, and Tampa Bay, Florida.

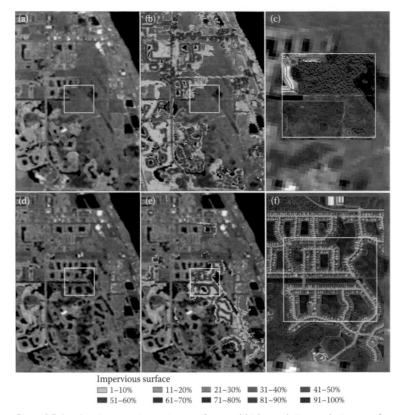

Impervious surface

◻ 1–10%   ▨ 11–20%   ▪ 21–30%   ▪ 31–40%   ▪ 41–50%
▪ 51–60%   ▪ 61–70%   ▪ 71–80%   ▪ 81–90%   ▪ 91–100%

Figure 8.7 Landsat imagery, impervious surface, and high-resolution orthoimagery for the site where the photos shown in Figure 8.6 were taken. (a) Landsat image in 2001, (b) 2001 impervious surface, and (c) 1 m orthoimagery in 2002 show the land cover in nonurban areas within the white rectangle. (d) Landsat image in 2006 shows the land cover has been transformed into urban, and (e) 2006 impervious surface captures these changes. (f) The orthoimagery in 2006 displays details of urban land cover growth in the area.

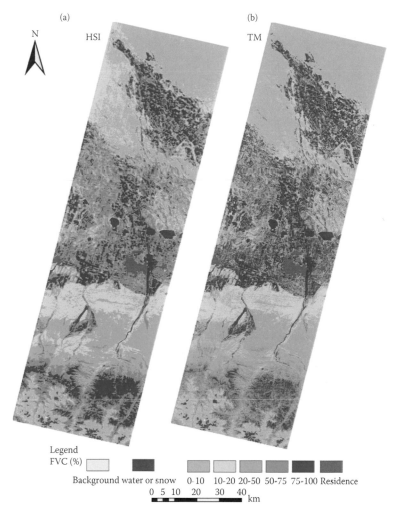

(a)

N

HSI

(b)

TM

Legend
FVC (%)

Background water or snow   0-10   10-20 20-50 50-75 75-100 Residence

0 5 10   20   30   40
km

Figure 9.4 The FVC maps retrieved from (a) the HJ-1/HSI data and (b) the Landsat TM data.

Figure 11.2 Landsat 30 m resolution data over Phoenix metropolitan area on April 19, 2000, displaying bands 5, 4, and 3 in red, green, and blue, respectively.

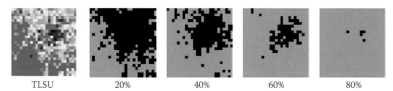

| TLSU | 20% | 40% | 60% | 80% |

Figure 13.2 Example showing the results of a threshold continuum reclassification. The tessellated spectral unmixing (TLSU) results are shown on the far left (red indicates high saltcedar proportions, green indicates low proportions). The four gray-scale images show threshold reclassifications at 20%, 40%, 60%, and 80% saltcedar land cover. For each threshold, all pixels equal to or greater than the threshold are in black. All other pixels are shown in gray.

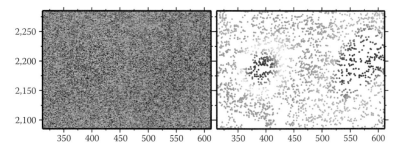

Figure 14.1 An example showing that the coherent points identified from (a) the highly noisy phases can be used to estimate (b) the deformation pattern.

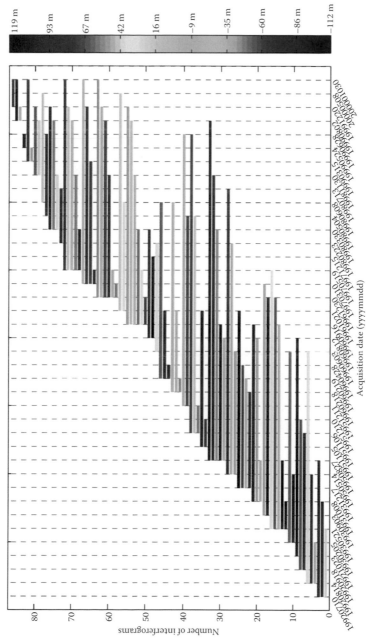

Figure 14.3 Spatiotemporal baselines of the 86 ERS-1/2 interferometric pairs used for PS solution.

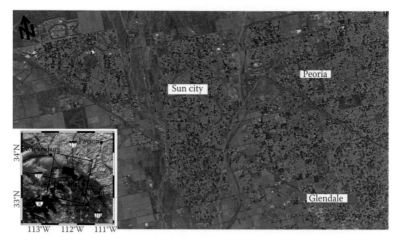

Figure 14.4 SAR amplitude image of the study area averaged from all the images. The PS candidates are superimposed as red points. The inset shows the geographic setting of the study area (tilted dark box) in Phoenix. The tilted rectangle identifies the SAR frame. (From Liu, G. X., S. M. Buckley, X. L. Ding, Q. Chen, and X. J. Luo, *IEEE Transactions on Geoscience and Remote Sensing* 47: 3209–3219, 2009. With permission.)

Figure 14.5 Distribution of linear subsidence rates in millimeters per year. The subsidence in the farmland is not available because of the lack of PS points in the area. P1 and P2 are two PS points to be analyzed subsequently. (From Liu, G. X., S. M. Buckley, X. L. Ding, Q. Chen, and X. J. Luo, *IEEE Transactions on Geoscience and Remote Sensing* 47: 3209–3219, 2009. With permission.)

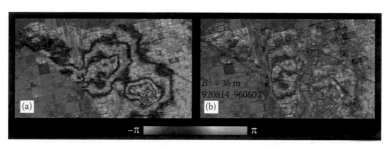

Figure 14.6 Comparison between (a) simulated and (b) observed differential interferograms with a time interval of about 4 years.

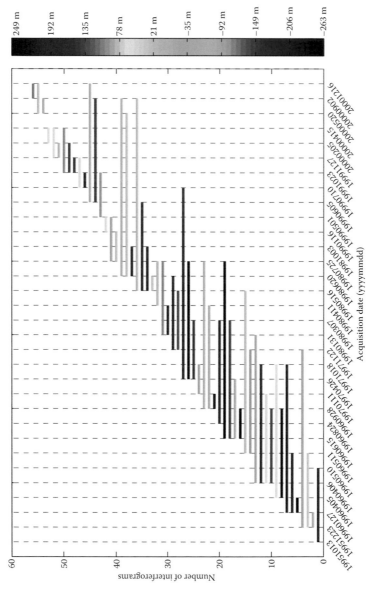

Figure 14.9 Spatiotemporal baselines of the 55 interferometric pairs used in the study.

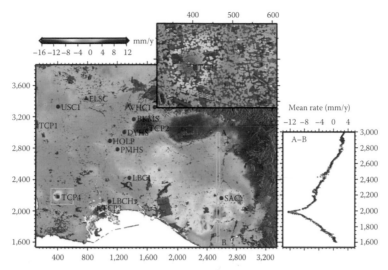

Figure 14.10 The distribution of deformation rates estimated by the TCP-based InSAR method over the Los Angeles basin.

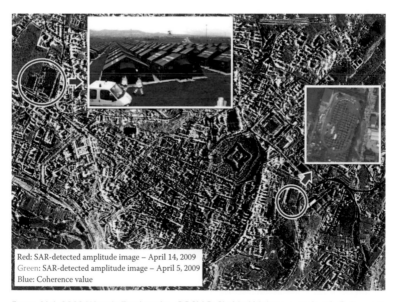

Figure 16.6 2009 L'Aquila Earthquake, COSMO-SkyMed Multitemporal with Coherence map for postearthquake change detection. The figure highlights new tent camps established to gather displaced persons. (From COSMO-SkyMed Product ©ASI Agenzia Spaziale Italiana, 2009. All rights reserved. Image processed by e-GEOS.)

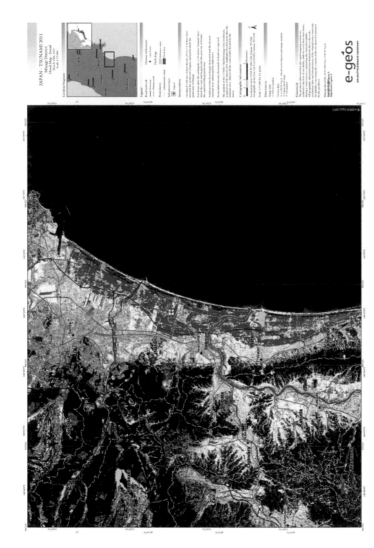

Figure 16.8 2011 Japan tsunami. Flood Map derived from the semiautomated analysis of COSMO-SkyMed images. (From COSMO-SkyMed Product ©ASI Agenzia Spaziale Italiana, 2011. All rights reserved. Image processed in very rush mode by e-GEOS Emergency Team.)

Chapter 10

Estimating and mapping forest leaf area index using satellite imagery

Ruiliang Pu

CONTENTS

Remote-sensing techniques, particularly using satellite remote-sensing techniques, offer a practical means to measure leaf area index (LAI) at a landscape or even global scale. This chapter provides an overview on using optical satellite remote-sensing techniques for estimating and mapping forest LAI. To estimate and map forest LAI, a wide range of analysis techniques and approaches that have been developed and demonstrated are reviewed. The spectral analysis techniques include spectral derivative analysis, vegetation index analysis, spectral position variable analysis, transformed feature extraction, spectral mixture analysis, and advanced statistical algorithms for mapping LAI, and two general categories of analysis methods: empirical and statistical methods and physically based modeling. Advantages and disadvantage for specific analysis techniques and approaches are discussed. An overview of estimating and mapping forest LAI with multisatellite sensors and systems, including moderate resolution, high resolution, coarse resolution, and hyperspectral resolution data, is conducted. The challenges and future directions for estimating and mapping forest LAI using various satellite imagery are discussed.

10.1 INTRODUCTION

Leaf area index (LAI), defined as one half of total leaf area per unit ground surface area (Lang, 1991; Chen and Black, 1992), is a structural parameter of vegetation that is important for quantitative analysis of many physical and biological processes related to vegetation dynamics (Chen et al., 2002), such as climate and ecological studies. LAI is an important driver to some ecosystem models applied at landscape to global scales, such as photosynthesis, respiration, transpiration, carbon and nutrient cycle, and rainfall interception (Running et al., 1989; Gong et al., 1995; Chen and Cihlar 1996; Fassnacht et al., 1997; White et al., 1997; Hu et al., 2000; Kalácska et al., 2004; Schlerf et al., 2005). Therefore, accurately mapping LAI spatial distribution is critical for improving the performance of such models over large areas. Although forest LAI can be directly (e.g., leaf harvest method and allometric estimate method) or indirectly (e.g., canopy gap fraction estimation method with optical instruments) measured by several approaches, the traditionally ground-based measurement of canopy LAI is labor intensive and, thus, is problematic over large areas (Gobron et al., 1997; Gower et al., 1999; White et al., 2000). Remote-sensing techniques, particularly satellite remote-sensing techniques, may offer a practical means to measure LAI at a landscape or even global scale. During the past three decades, numerous studies by statistical and empirical approaches and physically based modeling inversion techniques have shown that various satellite remote-sensing data are useful for retrieving and estimating forest LAI and mapping its spatial and temporal distribution, such as moderate-resolution data, including Landsat Thematic Mapper (TM) and SPOT high-resolution visible/infrared (HRVIR) data (Peterson et al., 1987; Spanner et al., 1994; Peddle et al., 1999; Cohen et al., 2003; Eklundh et al., 2003; Lu et al., 2005; Flores et al., 2006; Soudani et al., 2006; Stenberg et al., 2008); high-resolution data, including IKONOS and Quickbird data (Soudani et al., 2006; Song and Dickinson, 2008; Kovacs et al., 2009); coarse resolution data, including Moderate-Resolution Imaging Spectroradiometer (MODIS) data (Fang and Liang, 2005; Houborg et al., 2007; Sprintsin et al.,

2007, 2009; le Maire et al., 2011); and hyperspectral data, including Hyperion data (Gong et al., 2003; Pu et al., 2003, 2005, 2008).

Given a lot of currently available statistical and empirical approaches and physically based modeling inversion techniques that may be applied to estimating and mapping forest LAI with multiresolution satellite sensors and systems, it is necessary to conduct a relatively systematic review to these approaches and techniques. This chapter will provide an overview of optical satellite remote-sensing techniques used to estimate and map forest LAI. The main objectives of this chapter are as follows:

- Review suitable techniques and methods in estimating and mapping forest LAI with aerospace image data
- Conduct an overview of applying the techniques and methods with existing multisatellite sensors or systems data for estimating and mapping forest LAI
- Discuss challenges and identify future directions for estimating and mapping forest LAI using various satellite imagery

10.2 TECHNIQUES AND METHODS

In general, the techniques and methods used to estimate and map forest LAI from satellite image data can be classified into two categories: statistical and empirical methods and physically based modeling inversion techniques. To emphasize the applications of techniques and methods, the section reviews only those methods that are currently applied to various optical satellite images.

10.2.1 Statistical and empirical methods

The statistical and empirical methods establish linear or nonlinear relationships (with regression or other advanced algorithms, such as artificial neural networks [ANN]) between various spectral variables (e.g., original band reflectance, various vegetation indices [VIs], spectral position parameters, derivative spectra, and transformed features) and ground-based forest LAI measurement. Usually, in developing an empirical model, the various spectral variables are used as independent or explanatory variables and the ground-based LAI measurement is used as dependent or response variable. Therefore, different statistical and empirical methods mainly reflect the difference in selection and extraction of spectral variables and reflect whether a linear or nonlinear regression model or an advanced modeling algorithm is used. Each technique or method will be reviewed by (a) describing its characteristics, (b) briefly summarizing its advantages and disadvantages in estimating and mapping forest LAI with satellite data, (c) listing several major factors to be considered when applied in practice, and (d) presenting a few typical application cases. Furthermore, Table 10.1 summarizes a set

Table 10.1 Summary of statistical and empirical approaches suitable for estimating and mapping forest LAI by using optical satellite imagery

Approach	Characteristics and description	Advantages and disadvantages	Major factor	References
1. Derivative analysis	Normalized spectral difference of two continuous/neighbor narrow bands with their wavelength interval.	Remove or compress the effect of illumination variations with low frequency on target spectra but sensitive to the SNR of hyperspectral data and higher order spectral derivative processing is susceptible to the noise. Only suitable for hyperspectral data.	Spectral resolution <10 nm and also continuous, right threshold.	Pu et al., 2003; Asner et al., 2008
2. Vegetation index analysis	Calculate ratio of images of two bands or normalized difference of two or more than two bands.	Easy to use and reduce impact of Sun angle, atmosphere, shadow, topography. However, VI image usually is not normal.	Identify suitable bands to construct VIs.	See Table 10.2
3. Spectral position variable analysis	Wavelength positions rather than the amplitude of the spectral feature were investigated. The red edge, for example, located between 680 nm and 750 nm, shifts its spectral position according to chlorophyll content, its seasonal patterns, and LAI.	The spectral position variable is insensitive to changing of illumination conditions and sensitive to plant biophysical and biochemical variable. Only suitable for hyperspectral data or appropriate spectral setting for multispectral data.	Familiarizing modeling techniques for extracting the spectral position variables.	Pu et al., 2003, 2004

4. Transformed feature extraction	A linear or nonlinear combination of raw data to reduce dimensionality and preserve variance contained in raw data as much as possible in the first several component images.	Dimension reduction and usefully informative feature extraction; not easy to identify which are more signal components.	Identify informative features and components.	Pu and Gong, 2004; Darvishzadeh et al., 2008; Pu and Liu, 2011
5. Spectral mixture analysis	Spectral reflectances from different materials (>1 material) within a pixel are recorded as one spectral response (mixed spectrum). Using linear or nonlinear spectral mixture model to derive fraction (endmember) images from mixed pixels.	The fraction representing the areal proportion of each endmember within a mixed pixel, but don't know where the proportioned areas locate within the mixed pixel, and it is difficult to obtain endmember spectra and know spectral variation of varying LAI in mixed pixels.	Identify suitable and pure endmembers and extract their individual spectra for training and test purposes.	Chen et al., 2004; Somers et al., 2010
6. Advanced algorithms	Using advanced parameteric/nonparametric algorithms, for example, artificial neural networks, to estimate LAI value of one pixel or image-object. Usually dimension reduction and feature extraction are first considered.	There is a basis of statistic or probability or a rule for estimating LAI; usually it is difficult to obtain adequate training and test samples to supervise the method.	Identify and gather adequate training and test samples.	Jensen and Binford, 2004; Fang and Liang, 2005

of six general techniques and methods from their brief characteristics, advantages and disadvantages, major factors affecting estimating and mapping results, and some application examples. The application of multisatellite remote-sensing data in the implementation of techniques and methods to estimate and map forest LAI is reviewed in Section 10.3.

10.2.1.1 Derivative analysis

The first- and second-order derivative spectra are calculated from hyperspectral data, such as EO-1 hyperspectral sensor Hyperion and ESA-Proba CHRIS (Compact High Resolution Imaging Spectrometer). The derivative spectrum is the normalized spectral difference of two continuous or neighboring narrow bands with their wavelength interval. Spectral derivative analysis has been considered a desirable tool to remove or compress the effect of illumination variations with low frequency on target spectra, but it is sensitive to the signal-to-noise ratio (SNR) of hyperspectral data and higher order spectral derivative processing is susceptible to the noise (Cloutis, 1996; Pu, 2011b). In other words, lower order derivatives (e.g., the first-order derivative) are less sensitive to noise and hence are more effective in operational remote sensing. When implementing the spectral derivative analysis, a spectral resolution finer than 10 nanometer (nm) is required and spectral bands are continuous.

In spite of only a few cases of hyperspectral satellite data having been used to estimate and map forest LAI compared with other satellite imagery by spectral derivative analysis, existing studies have revealed herperspectral data's potential to retrieve and estimate forest LAI. For example, to extract red-edge optimal parameters from Hyperion data to estimate forest LAI, the polynomial fitting (Pu et al., 2003) and Lagrangian technique (Dawson and Curran, 1998) approaches need derivative spectra as input. Asner et al. (2008) conducted the spectral separability analysis between Hawaiian native and introduced (invasive) tree species with hyperspectral image data to detect and assess invasive species. They observed that the spectral differences (measured in reflectance, first and second derivative spectra) in canopy spectral signatures are linked to relative differences in leaf pigment, nutrient, and structural properties, as well as to canopy LAI. These relative differences associated with leaf and canopy properties of trees are helpful to separate invasive species from their background (native) species.

10.2.1.2 Vegetation index analysis

When developing statistical models (e.g., linear or nonlinear regression models) for estimating forest LAI, various VIs are frequently used for independent variables (Huete, 1988; Baret et al., 1989; Myneni et al., 1995).

Various VIs can be constructed with two or more bands from multi- and hyperspectral satellite data. Table 10.2 lists a set of 20 VIs that are developed from multi- and hyperspectral satellite image data and that are suitable for estimating and mapping forest LAI. These VIs have appeared in existing literature. To conveniently locate a VI for readers, the total 20 VIs are organized in an alphabetical order on the basis of the abbreviated name of the VI. The explicit advantages of VIs are that they are easy to use and can reduce the impact of Sun angle, atmosphere, shadow, and topography on target spectra. The probability distribution of VI images for different land cover types usually is not normal. A key factor to determine the usefulness of a VI depends on identifying suitable wavebands that are hopefully sensitive to the variation of the canopy LAI in forest areas.

Developing various VIs (see Table 10.2) from satellite data can help estimate and map forest LAI. For example, using the ground-measured LAI and VIs derived from Landsat 7 Enhanced Thematic Mapper Plus (ETM+) data, Lu et al. (2005) investigated relationships between the measured LAI and different VIs and found that the normalized difference vegetation index (NDVI) would be the most promising estimator to extract LAI in an alpine meadow in northwestern China. To investigate the use of TM data for estimating forest LAI in southern Sweden, Eklundh et al. (2003) used the individual TM bands and various VIs to estimate the forest LAI. They found that the best single band for deciduous stands was TM4 and the best VI was the simple ratio VI (TM4/TM3). Combining VIs (NDVI and enhanced vegetation index [EVI]) with model inversion techniques, Houborg et al. (2007) used Terra and Aqua MODIS reflectance data to extract and map forest LAI in the island of Zealand, Denmark. They remarked that the EVI outperformed NDVI for the retrieval of LAI because of the improved linearity of the EVI and its great ability to characterize a broad range in LAI. In addition, in producing MODIS LAI products, when the inversion of a three-dimensional radiative transfer (RT) model for MODIS LAI retrieval (Myneni et al., 2002) fails to optimize a solution, a backup algorithm based on a relationship between the NDVI and LAI is unitized (Myneni et al., 1995). In short, all examples reviewed here indicate that various VIs are important to extracting and mapping forest LAI.

10.2.1.3 Spectral position variable analysis

In practice, a weak correlation between VIs and forest LAI has been shown when canopy cover is low and there is a great spatial variation in understory reflectance (Spanner et al., 1990; Smith et al., 1991). Such a low correlation may be attributed to the sensitivity of VIs to changes of plant spectral characteristics, which are caused by various environmental factors (Guyot et al., 1992; Nemani et al., 1993), such as illustration condition changes.

Table 10.2 Summary of 20 vegetation spectral indices extracted from satellite imagery for estimating and mapping forest LAI

Spectral index	Characteristics and descriptions	Definition	References
ARVI, atmospherically resistant VI	Improve SAVI to make less sensitive to atmospheric effects by normalizing the radiance in the blue, red, and NIR bands.	$(R_{NIR} - R_{rb})/(R_{NIR} - R_{rb})$, where $R_{rb} = R_R - \gamma(R_B - R_R)$	Kaufman and Tanré, 1992
EVI, enhanced VI	Estimate vegetation LAI, biomass, and water content and improve sensitivity in high biomass region.	$2.5(R_{NIR} - R_{red})/(R_{NIR} + 6R_{red} - 7.5R_{blue} + 1)$	Huete et al., 2002
EVI2, two-band enhanced VI	Similar to EVI, but without blue band and good for atmospherically corrected data.	$2.5(R_{NIR} - R_{red})/(R_{NIR} + 2.4R_{red} + 1)$	Jiang et al., 2008
ISR, infrared simple ratio index	ISR is used to minimize uncertainties in atmospheric correction.	R_{554}/R_{677}	Fernandes et al., 2004
LAIDI, LAI determining index	Sensitive to LAI variation at canopy level with a saturation point > 8.	R_{1250}/R_{1050}	Delalieux et al., 2008
MSAVI, improved soil-adjusted vegetation index	A more sensitive indicator of vegetation amount than SAVI at canopy level.	$0.5[2R_{800} + 1 - ((2R_{800} + 1)2 - 8(R_{800} - R_{670}))^{1/2}]$	Qi et al., 1994
MSR, modified simple ratio	More linearly related to vegetation parameters than RDVI.	$(R_{800}/R_{670} - 1)/(R_{800}/R_{670} + 1)1/2$	Chen, 1996; Haboudane et al., 2004
NDMI, normalized difference moisture index	Sensitive to the reflectance of leaf chlorophyll content and to the absorbance of leaf moisture thus canopy LAI.	$(R_{NIR} - R_{SWIR})/(R_{NIR} + R_{SWIR})$	Hardisky et al., 1983
NDVI, normalized difference vegetation index	Responds to change in the amount of green biomass and more efficiently in vegetation with low to moderate density.	$(R_{NIR} - R_R)/(R_{NIR} + R_R)$	Rouse et al., 1973
NLI, nonlinear vegetation index	NLI linearizes relationships with surface parameters that tend to be nonlinear.	$(R_{NIR}^2 - R_R)/(R_{NIR}^2 + R_R)$	Gong et al., 2003; Goel and Qin, 1994

Index	Description	Formula	Reference
PVI, perpendicular VI	Eliminate differences in soil background and is most effective under conditions of low LAI, applicable for arid and semiarid regions.	$(R_{NIR} - aR_R - b)/(1 + a^2)^{1/2}$, where a = slope of the soil line, b = soil line intercept	Baret and Guyot, 1991; Huete et al., 1985
RDVI, renormalized difference VI	Suitable for low to high LAI values.		Reujean and Breon, 1995; Haboudane et al., 2004
RMPVI, SR multiplied by PVI	SR and PVI are combined as complementary vegetation indices in LAI estimation.	$R_{NIR}(R_{NIR} - aR_R - b)/ (R_R(1 + a^2)^{1/2}$, where a = slope of the soil line, b = soil line intercept	Wang et al., 2010
RSR, reduced simple ratio index	Unify deciduous and conifer species in LAI retrieval to show increased sensitivity to LAI.	$R_{NIR}(R_{SWIR-max} - R_{SWIR})/ [R_R(R_{SWIR-max} - R_{SWIRmin})]$	Brown et al., 2000
SAVI, soil-adjusted VI	L ranges from 0 for very high vegetation cover to 1 for very low vegetation cover; minimizes soil brightness-induced variations; L = 0.5 can reduce soil noise problems for a wide range of LAI.	$(R_{NIR} - R_R)(1 + L)/(R_{NIR} + R_R + L)$	Huete, 1988; van Leeuwen and Huete, 1996
sLAIDI, normalization or standard of the LAIDI	Sensitive to LAI variation at canopy level with a saturation point > 8.	$S(R_{1050} - R_{1250})/(R_{1050} + R_{1250})$, where $S = 5$	Delalieux et al., 2008
SPVI, spectral polygon vegetation index	Estimate LAI and canopy Chls.	$0.4[3.7(R_{800} - R_{670}) - 1.2 |R_{530} - R_{670}|]$	Vincini et al., 2006
SR, simple ratio	Same as NDVI.	R_{NIR}/R_R	Jordan, 1969
TSAVI, transformed soil-adjusted VI	Modify SAVI to compensate for soil variability due to changes in solar elevation and canopy structure.	$a(R_{NIR} - aR_R - b)/[(aR_{NIR} + R_R - ab + X(1 + a^2)]$, where X = adjustment factor to minimize soil noise, a and b = slope and intercept of the soil line	Baret and Guyot, 1991
WDRVI, wide dynamic range VI	Estimate LAI, vegetation cover, biomass; Better than NDVI	$(0.1R_{NIR} - R_{red})/(0.1R_{NIR} + R_{red})$	Gitelson, 2004

A wavelength position variable is insensitive to the changing illumination conditions and is sensitive to plant biophysical and biochemical variables (e.g., chlorophyll content and LAI). For example, red-edge parameters, compared with VIs, are relatively insensitive to changes of bioenvironmental factors, such as soil cover percentage and optical properties, canopy structure and leaf optical properties, atmospheric effects, and irradiance and solar zenith angle (Horler et al., 1983; Baret et al., 1987; Leprieur, 1989; Curran et al., 1995). To extract the spectral position parameters with the corresponding extraction technique, however, the hyperspectral data or appropriate band setting for multispectral (MS) data are required, and for a specific application purpose, some modeling techniques (Pu et al., 2003) for extracting the spectral position variables are also required.

There are two primary red-edge optical parameters, red-edge position (REP) and red-well position (RWP). REP, located between 680 nm and 750 nm, is defined as the wavelength of the inflection point of the reflectance slope at the red edge. RWP is the wavelength position corresponding to a plant's minimum reflectance in red (maximum chlorophyll absorption), which also functions as REP (Belanger et al., 1995). Experimental and theoretical studies show that REP shifts according to changes of chlorophyll content (Munden et al., 1994; Belanger et al., 1995), LAI (Danson and Plummer, 1995), plant health levels (Vane and Goetz, 1988) and seasonal patterns (Miller et al., 1991). When a plant is healthy with high chlorophyll content and high LAI, the REP shifts toward the longer wavelengths; when it suffers from disease or chlorosis and low LAI, it shifts toward the shorter wavelengths. Researchers have demonstrated this behavior of the red edge using spectral reflectance measurements obtained either in the laboratory or from airborne sensors. Airborne and spaceborne imaging spectrometers now have the potential to determine the red-edge optical parameters of vegetation canopies at regional scale (Dawson and Curran, 1998). For example, Pu et al. (2003) used the first satellite hyperspectral data Hyperion to test the four techniques of extracting red-edge optical parameters to estimate and map forest LAI in the Rio Negro province in the Patagonia region of southern Argentina. The four extraction techniques include four-point interpolation (Guyot et al., 1992), Polynomial fitting (Pu et al., 2003), Lagrangia technique (Dawson and Curran, 1998), and inverted-Gaussian modeling (Miller et al., 1990).

10.2.1.4 Transformed feature extraction

Transformed feature extraction is a linear or nonlinear combination of relatively high-dimension data used to reduce dimensionality and preserve spectral variance as much as possible in the first several components, images or transformed spectral features. The most popular transformation techniques are principal component analysis (PCA) and its modified version, maximum noise fraction (MNF). The PCA technique has been applied to reduce the data dimension and feature extraction from multi- and hyperspectral data to assess leaf or canopy parameters (e.g., Gong et al., 2002; Walsh et al., 2008). Canonical discriminant analysis (CDA) is a dimension-reduction technique equivalent to canonical correlation analysis (CCA) used for dimensional reduction and feature extraction (e.g., Zhao and Maclean, 2000; van Aardt and Wynne, 2001, 2007). Although CDA produces canonical variables, given a classification variable and several quantitative variables, linear combinations of these quantitative variables summarize between-class variation much in the same way that PCA summarizes most of the variation in the first several principal components (Zhao and Maclean, 2000). Differing from PCA, the CDA involves human effort and knowledge derived from training samples, whereas PCA performs a relatively automatic data transformation and tries to concentrate the majority of data variance in the first several PCAs. Wavelet transform (WT) is a relatively new signal-processing tool that provides a systematic means for analyzing signals at various scales or resolutions and shifts. In the past two decades, WT has been successfully applied to image processing, data compression, pattern recognition (Mallat, 1998), image texture feature analysis (e.g., Fukuda and Hirosawa, 1999), and feature extraction (e.g., Simhadri et al., 1998; Pittner and Kamarthi, 1999). Wavelets have proven to be quite powerful in these remote-sensing application areas. This is attributed to the fact that WT can decompose a spectral signal into a series of shifted and scaled versions of the mother wavelet function and that the local energy variation (represented as "peaks and valleys") of a spectral signal in different bands at each scale can be detected automatically and provides some useful information for further analysis of hyperspectral data (Pu and Gong, 2004). WT can decompose signals over dilated (scaled) and translated (shifted) wavelets (Mallat, 1989; Rioul and Vetterli, 1991). In addition, other transformed feature extraction methods are still used to transform and extract spectral features for estimating and mapping forest LAI, such as partial least square (PLS) regression, stepwise discriminant analysis (SDA), and stepwise multiple linear regression (SMLR).

Many previous studies have tested the abilities of various data transformation and feature extraction methods. CCA has been demonstrated to outperform PCA, including MNF (e.g., Zhao and Maclean, 2000; Pu and Liu, 2011), and PLS regression outperforms stepwise discriminant analysis (SDA) and stepwise multiple linear regression (SMLR) (e.g., Darvishzadeh et al., 2008). For example, when conducting a comparative analysis between CDA and PCA for spectral transformation for forest-type delineation, Zhao and Maclean (2000) reported Kappa accuracies using CDA images were significantly higher than those derived using PCA at $\alpha = 0.05$ and suggested that CDA transformation was superior to PCA transformation in improving the overall classification accuracy. Pu and Liu (2011) also demonstrated that CDA outperforms PCA and SDA methods in identifying 13 urban tree species in Tampa, Florida, with in situ hyperspectral measurements. By using hyperspectral measurements and spectral indices extracted from hyperspectral data, Darvishzadeh et al. (2008) compared the accuracies of estimating canopy LAI and chlorophyll content between PLS and SMLR and concluded that PLS using the entire reflectance spectra increased all R^2 values with an independent test set by 0.1 to 0.14 compared with the use of SMLR. When applying WT techniques in data transformation, many different types of mother wavelets and wavelet bases can be selected for use. In practice, researchers need to test most wavelet families to find the most useful wavelet family for a particular project. After a set of WT coefficients for each level or scale of a pixel-based spectrum is calculated, the energy feature of the wavelet decomposition coefficients is computed at each scale for both approximation and details and is used to form an energy feature vector (Pittner and Kamarthi, 1999; Bruce et al., 2001; Li et al., 2001; Pu and Gong, 2004). This can become a feature extraction through a dimension reduction. With hyperspectral data of vegetation and the wavelet transform technique, several studies have demonstrated the benefits of wavelet analysis. For example, Pu and Gong (2004) used mother wavelet db3 in MATLAB® (Misiti et al., 1996) to transfer Hyperion data (167 available bands in their analysis) to extract features through a dimension reduction for mapping forest LAI and crown closure (CC).

10.2.1.5 Spectral mixture analysis

Spectral mixture analysis (SMA) provides an estimate of the physically interpretable canopy cover, which is related to the canopy LAI. A large portion of remotely sensed data is spectrally mixed because the spatial resolution (pixel size) of image data cannot resolve individual materials. To identify various "pure materials" and to determine their spatial proportions from remotely sensed data, the spectral mixing process has to be properly modeled. Then the model can be inverted to derive the spatial proportions and spectral properties of those "pure materials." There are two types of spectral mixing models, linear spectral mixing model and nonlinear spectral mixing model. Linear spectral mixing modeling and its inversion have been widely used since the late 1980s. Nonlinear spectral mixture models can be found in Sasaki et al. (1984) and Zhang et al. (1998). In a linear spectral mixture model analysis, there are two solutions: a linear least square solution and a nonlinear solution (e.g., a neural network-based, nonlinear, subpixel classifier by Walsh et al., 2008). At present, because the SMA method is easy to use, it has been widely and successfully applied for mapping the abundance of canopy cover with multi- and hyperspectral satellite data. For example, in a study by Chen et al. (2004), the linear SMA analyses of IKONOS and ETM+ data were used to isolate spectral members (bare soil, understory grass, and tree and shade) and to calculate their subpixel fractional coverages. They then compared these endmember cover estimates to similar cover estimates derived from light detection and ranging (LiDAR) data and field measures. The IKONOS-derived tree or shade fraction was significantly correlated with the field-measured canopy effective LAI ($R^2 = 0.55$, $p < 0.001$) and with the LiDAR-derived estimate of tree occurrence ($R^2 = 0.79$, $p < 0.001$). In the SMA analysis, the fraction represents the areal proportion of each endmember within a mixed pixel, but one does not know where the proportioned areas are located within the mixed pixel, and it is difficult to identify the spatial variation of LAI in mixed pixels, although people can estimate LAI in the mixed pixel. In addition, a successful SMA analysis requires adequately identifying suitable endmembers and correctly extracting their corresponding individual spectra for training and test purposes.

This simple mixture model (i.e., SMA) has the advantage that it is relatively simple and provides a physically meaningful measure of abundance in mixed pixels. There are a number of limitations to the simple mixing concept, however. For instance, the maximum number of components a linear SMA can map is limited by the number of bands in the image data (Li and Mustard, 2003). Therefore, Roberts et al. (1998) introduced a multiple endmember spectral mixture analysis (MESMA), a technique used to identify materials in a hyperspectral image using endmembers from a spectral library. According to Roberts et al. (1998), the MESMA overcomes these limitations of the simple mixing model. Using the MESMA, the number of endmembers and types are allowed to vary for each pixel in the image. The general procedure of the MESMA approach is to start with a series of candidate two-endmember models, evaluate each model based on selection criteria, and then, if required, construct candidate models that incorporate more endmembers (Roberts et al., 1998). For example, Somers et al. (2010) applied the novel spectral unmixing technique (MESMA) to Landsat-5 TM and EO-1 Hyperion data acquired over a Eucalyptus globulus (*Labill.*) plantation in southern Australia. This technique combines an iterative mixture analysis cycle allowing endmembers to vary on a per-pixel basis (MESMA) and a weighting algorithm that prioritizes wavebands based on their robustness against endmember variability.

10.2.1.6 Advanced statistical algorithms

Theoretically, after spectral features and component images in a relatively lower dimension are produced by running appropriate extraction and transformation algorithms, some advanced algorithms, such as artificial neural networks (ANN) and projection pursuit regression (PPR), are used to estimate and map forest LAI. Basically, the nonparametric ANN method establishes a mapping function between spectral features or component images and a set of biophysical variables of interest such as LAI. The PPR is another nonparametric multiregression method (Fang and Liang, 2005). For example, Jensen and Binford (2004) compared simple, multiple regression and ANN methods to estimate and map forest LAI with forest LAI ground measurement and spectral variables extracted from Landsat TM imagery (e.g., various VIs). They concluded that the most accurate way to estimate regional-scale forest LAI is to train an ANN using in situ LAI data and remote-sensing brightness values. With nonparametric regression methods (ANN and PPR), Fang and Liang (2005) proposed an alternative to increase inversion accuracy of forest LAI from MODIS data.

In practice, to estimate forest LAI with satellite imagery, sometimes it is difficult to obtain adequate training samples for developing a reliable LAI estimate model with advanced algorithms because a large amount of fieldwork is required. The major factors for estimating and mapping forest LAI with advanced algorithms are to understand the algorithms' principles and significance in mathematics and application and to gather adequate training and test samples.

The six statistical and empirical methods reviewed are relatively simple and easy to implement, but the derived empirical relationships between spectral variables and LAI are recognized as being sensor specific and site specific and dependent on sampling conditions, and thus they are expected to change in space and time (Baret and Guyot, 1991; Delalieux et al., 2008; Gonsamo, 2010). Because of this, physically based modeling techniques have attracted the attention of a lot of researchers who retrieved biophysical and biochemical parameters, such as forest LAI, by inversing various physically based models from simulated spectra or real satellite image data.

10.2.2 Physically based modeling inversion techniques

The physically based models, including RT models and geometric-optical (GO) models, consider the underlying physics and complexity of the canopy internal structure and are therefore robust and have the potential to replace the statistically based approaches (Zhang et al., 2008a, 2008b). In the context of the remote sensing of biophysical and biochemical parameters, such models have been used in the forward mode to calculate leaf or canopy reflectance and transmittance and in inversion to estimate leaf or canopy chemical and physical properties.

Physically based models must be inverted to retrieve vegetation characteristics (e.g., LAI) from observed reflectance data. A successful inversion of physically based model requires considering three aspects: a good model, an appropriate inversion procedure, and a set of calibrated reflectances and bioparameters (Jacquemound et al., 2000; Schlerf and Atzberger, 2006). So far, three general inversion techniques are used to retrieve forest LAI from simulated or real image data with canopy optical property models. The traditional iterative optimization methods (Goel and Thompson, 1984; Liang and Strahler, 1993; Jacquemoud et al., 1995; Jacquemoud et al., 2000; Meroni et al., 2004) are used to estimate the model parameters by minimizing a merit function (Goel, 1988). The iterative process is necessary to find the optimal estimates of these parameters (Schlerf and Atzberger, 2006). The traditional iterative method, however, is time consuming and often requires a simplification of the models when processing large data sets. It is also difficult to achieve globally optimal and stable results. This may result in a decrease in inversion accuracy and makes the retrieval of biophysical and biochemical variables unfeasible for large geographic areas (Houborg et al., 2007). As an alternative to the numerical optimization techniques, look-up table (LUT) approaches (Knyazikhin et al., 1998; Weiss et al., 2000; Combal et al., 2002, 2003; Gastellu-Etchegorry et al., 2003) have been proposed and applied in inversion process of physically based models. LUT methods can partially overcome the drawback of the optimization techniques. They operate through a database of simulated canopy reflectance variable in structural and radiometric properties. LUT creation, however, can be complicated and requires an extensive set of reliable field measurements. To overcome the limitations from traditional optimal methods and LUT approaches, ANN techniques (Gong et al., 1999; Weiss and Baret, 1999; Walthall et al., 2004; Schlerf and Atzberger, 2006) have been employed. ANN techniques are expected to reduce the complexity of inversion. The major advantages of ANN techniques are that they are computationally very simple and fast and that they can establish a mapping function (nonlinear) between input and output variables, provided there are enough neurons in the hidden layer. The major problem with the ANN techniques is related to the often time-consuming training phase and the unpredictable behavior of ANN when the spectral characteristics of targets are not well represented by the modeled spectral or real imaging data (Liang, 2004; Schlerf and Atzberger, 2006).

Using either simulated spectra or real satellite image data, many researchers have employed physically based models at canopy level to retrieve and map forest LAI. For example, le Maire et al. (2011) used MODIS reflectance time-series data and physically based model inversion techniques to estimate canopy LAI during full rotations of a eucalyptus plantation in Brazil. The inversion models included models of leaf reflectance and transmittance, soil reflectance, and canopy RT. The inversion results of canopy LAI from MODIS reflectance time-series data indicated that a good fit between ground-measured and estimated LAI was proven. Houborg et al. (2007) explored the benefits of combining vegetation index and physically based approaches for the spatial and temporal mapping of green LAI, total chlorophyll content, and total vegetation water content. A numerical optimization method was employed for the inversion of a canopy reflectance model using Terra and Aqua MODIS multispectral, multitemporal, and multiangle reflectance observations to aid the determination of vegetation-specific physiological and structural canopy parameters (e.g., canopy LAI). Their impressive performance results were reported for LAI retrieved for barley, wheat, and deciduous forest sites given that no site-specific in situ measurements were used to calibrate the model.

In short, the physically based model inversion methods are founded on inversion of physically based canopy reflectance models by determining a set of canopy biophysical variables, and thus the main advantage of model inversion is its general applicability to different sites and sampling conditions (Schlerf and Atzberger, 2006; Houborg et al., 2007). Application of those model inversions, however, is often limited by the requirements of considerably large homogeneous areas or additional unknown input data (Fang et al., 2003), and a drawback to the use of physically based models is the ill-posed nature of model inversion (Combal et al., 2002; Houborg and Boegh, 2008).

10.3 ESTIMATING AND MAPPING LAI WITH MULTISATELLITE SENSORS AND SYSTEMS

The multisatellite sensors and systems' data include satellite moderate-resolution, high-resolution, and coarse-resolution data and hyperspectral data. Table 10.3 presents a list of optical satellite remote-sensing sensors and systems frequently used for estimating and mapping forest LAI. The table summarizes satellite sensors and systems from satellite platforms, band setting (bands suitable for estimating LAI), corresponding spatial resolution, and sensors and systems' revisiting periods. This section provides an overview of the application of various satellite image data to estimating and mapping canopy LAI of different forest types. The section discusses image characteristics, most frequently used methods and techniques, LAI products, and examples of typical applications.

10.3.1 With moderate resolution data

The typical moderate-resolution satellite data are sensors onboard Landsat series (Landsat 1-7) and SPOT satellite series (SPOT 1-5). Indian Remote-Sensing (IRS) satellite series and EO-1 ALI also belong to this category. The representative MS data include blue, green, red, near infrared (NIR), and shortwave infrared (SWIR) as well as an additional panchromatic (Pan) band (see Table 10.3) for some sensors and systems. Their spatial resolutions range from 10 m to 100 m for MS data and from 2.5 m to 15 m for Pan band. The moderation-resolution data include the most important bands—for example, red (absorbed by plants' pigments) and NIR band (multireflected by the plant leaf internal structure)—for modeling and

Table 10.3 A list of optical satellite remote-sensing sensors and systems frequently used to estimate and map forest LAI

Sensor/system	Platform	Revisited period	No. of bands	Spectral range/band	Spatial resolution MSS	Spatial resolution Pan.
Moderate resolution						
MSS, multispectral scanner	Landsat 1–5	16	4	B, G, R, NIR	79 m	
TM, thematic mapper	Landsat 4, 5	16	6	B, G, R, NIR, SWIR	30 m	
ETM+, enhanced thematic mapper plus	Landsat 7	16	7	B, G, R, NIR, SWIR, G-R-NIR	30 m	15 m
HVR, high-resolution visibility	SPOT* 1, 2, 3	26	4	G, R, NIR, VNIR	20 m	10 m
HRVIR, high-resolution visible and infrared	SPOT 4	26	5	G, R, NIR, SWIR, VNIR	20 m	10 m
HRG, high-resolution geometrical	SPOT 5	26	5	G, R, NIR, SWIR, VNIR		
ASTER, advanced spaceborne thermal emission and reflectance radiometer	Terra		10	G, R, NIRn, NIRb, 6-SWIR	15 m (G, R, NIR); 30 m (SWIR	
IRS, Indian remot-sensing satellite, IC, ID	Indian Remote-Sensing Satellite		4	G, R, NIR, SWIR	23 m	5.8 m
ALI, advanced land imager	EO-1		10	2-B, G, R, 2-NIR, 3-SWIR, V	30 m	15 m
High resolution						
Quickbird		3–4	5	B, G, R, NIR, VNIR	2.4 m	0.6 m
IKONOS		3	5	B, G, R, NIR, VNIR	4 m	1 m
WorldView-2		1.1–3.7	9	2_b, G, Y, R, RE, 2-NIR, VNIR	2 m	0.5 m
GeoEye-1		2–3	5	B, G, R, NIR, VNIR	1.65 m	0.41

Continued

Table 10.3 A list of optical satellite remote-sensing sensors and systems frequently used to estimate and map forest LAI (Continued)

Sensor/system	Platform	Revisited period	No. of bands	Spectral range/band	Spatial resolution	
					MSS	Pan.
Coarse resolution						
AVHRR, advanced very high-resolution radiometer	NOAA satellite	0.5	2	V, NIR	1.1 km	
MODIS, moderate-resolution imaging spectroradiometer	Terra/Aqua	1–2	19	R, NIR; B, G, 3-SWIR; 11-VNIR	1–2, 250 m; 3–7, 500 m; 8–19, 1 km	
VEGETATION	SPOT 4, 5	1–2	4	G, R, NIR, SWIR	1.165 km	
Hyperspectral						
Hyperion, hyperspectral imager	EO-1		220	VNIR-SWIR	30 m	
CHRIS, compact high-resolution imaging spectrometer	ESA-Proba satellite	7	19	VNIR	18 m	

SPOT = Satellite Pour l'Observation de la Terre; EO-1 = Earth Observing-1; B = blue; G = green; R = red; RE = red edge; NIR = near infrared; SWIR = shortwave infrared; V = visible; VNIR = visible and near infrared; Pan. = pancromatic.

mapping forest LAI. The data sources are mostly easy and free (or cheap if charging) to access.

Table 10.2 lists the most frequently used methods and techniques for each category of data using various VIs, extracted from the different MS data as predictors and from ground-based forest LAI measurement as dependent variables, to develop univariate or multivariate regression models to estimate and map forest LAI. The single-band reflectance and transformed spectral features are often employed as predictors to develop the statistical and empirical modes. Sometimes, linear and nonlinear SMA techniques are applied to retrieve and estimate forest LAI information. Retrieving LAI from various VIs calculated from different satellite image data has a phenomenon of LAI saturation, which means that the VIs and LAI do not increase linearly, and the complicated structure of forest canopy and effects of understory and soil background will affect the reflected radiance and need to be considered when modeling the canopy LAI with the satellite image data (Zheng and Moskal, 2009).

Among numerous studies for estimating and mapping forest LAI, many researchers have focused on using moderate-resolution satellite imagery in temperate, boreal forest areas and forest plantations in tropical and subtropical areas. For example, to assess the potential of moderate-resolution data to estimate LAI in East Africa rainforest ecosystems, Kraus et al. (2009) employed the spectral VIs and texture measures calculated from ASTER and SPOT-4 HRVIR reflectance data to predict effective LAI with regression models. Their experimental results showed that ASTER data generally were better suited, with simple-ratio VI performing best at early and intermediate forest stages and texture measures (gray-level cooccurrence matrix variance) derived from SWIR information rendering superior results at later forest stages. They have achieved an LAI mapping root mean square error (RMSE) of 0.39. To compare the potential use of IKONOS, ETM+, and SPOT HRVIR sensors' data for LAI estimation in temperate coniferous and deciduous forest stands, Soudani et al. (2006) reported that the three sensor types showed the similar estimate ability for stand LAIs, with an average RMSE of about 1.0 for LAI between 0.5 and 6.9. By using multidate ETM+ data and multitype regression models to develop an improved strategy, Cohen et al. (2003) first extracted different spectral VIs from the multidate ETM+ image data and then applied a CCA to integrate these multiple indices into a single index that represents a significant strategic improvement over existing uses of regression analysis in remote sensing. Among the three regression models—traditional (Y on X) ordinary least square (OLS) regression, inverse (X on Y) OLS regression, and an orthogonal regression method called reduced major axis (RMA)—the RMA has provided an intermediate set of predictions in terms of the RMSE, but the variance in the observations was preserved in the prediction. In a study using ETM+ imagery to estimate leaf area and response to silvicultural treatments in loblolly pine stands, Flores et al. (2006) reported that stand LAI of loblolly pine plantations could be accurately estimated from existing available remote-sensing data. Furthermore, Jensen and Binford (2004) developed and verified several models to estimate LAI using in situ field measurements, Landsat TM imagery, VIs, simple and multiple regression, and ANNs in forested ecosystem in the southeastern U.S. coastal plain. Their results showed that although multiple-band regression and regression with individual VIs could be used to estimate forest stand LAI, the most accurate way to estimate LAI at a regional scale was to train an ANN using in situ LAI data and remote-sensing brightness values.

10.3.2 With high resolution data

The most important and accessible high-resolution sensors include IKONOS, QuickBird, WorldView-2, and GeoEye-1. Most of their MS bands cover visible and NIR spectral ranges and have an additional Pan band (see Table 10.3) to cover visible and NIR (VNIR). Their band setting and band wavelength ranges are the same as or similar to those multispectral bands with Landsat series. The spatial resolutions range from 1.5 m to 4.0 m for MS data and from 0.4 m to 1.0 m for Pan band. For modeling and estimating forest LAI, this category of data also includes critical bands, such as red and NIR bands. The high-resolution data are not frequently used and are relatively expensive, and thus the utilization of the data is limited in relatively small areas, although the data sources are relatively easy to access.

The most frequently used methods and techniques for high-resolution data are similar to those for moderate-resolution multispectral data. Therefore, general methods include using various VIs (see Table 10.2), single-band reflectance, texture measures, and image spatial information as spectral predictors and ground-based LAI measurement as the dependent variable to develop statistical and empirical models. Because the data sources are usually available with Pan band, the pan-sharpening (PS) high-resolution image usually needs to be produced before extracting spectral variables (e.g., VIs and texture measures). The VIs calculated from PS MS bands might be slightly different from those derived from original MS bands, however, because the spectral property of PS MS bands has been modified by the Pan spectrum in the PS process. In addition, the high-resolution data are often used for training and validation data sets for linear and nonlinear SMA techniques applied to moderate or coarse resolution data for retrieving and estimating forest canopy information. For example, Kovacs et al. (2004, 2009) used the high-resolution satellite image data—IKONOS and Quickbird—to extract NDVI and simple ratio (SR) VI as spectral predictors and ground LAI measurement as the dependent variable to develop univariate regression models to map a devaluating mangrove forest LAI in the Mexican Pacific. Their regression analyses of the in situ LAI measurement and image-derived VIs revealed significant positive relationships for both sensors' data (R^2 values ranging from 0.63 to 0.73). To efficiently utilize the spatial information from IKONOS data, Song and Dickinson (2008) extracted image spatial information, defined as semivariance, from the high-resolution data. They also constructed NDVI and SR VIs from the same high-resolution image and calculated the canopy LAI of conifer- and hardwood-dominated stands from the DBH (diameter at breast height) and used a species-specific allometric equation to estimate forest canopy structures (crown size and LAI). Their experimental results indicate that the tree crown size is more sensitive to the image spatial information than LAI, and the image variance information is more useful in estimating LAI than NDVI and SR VIs. They suggested that combining both spatial (semivariance) and spectral information (VIs) could provide an improvement in estimating LAI compared with using spatial information alone.

10.3.3 With coarse resolution data

The coarse resolution data sources for retrieving and mapping forest LAI mainly include MODIS (onboard EOS-Terra/Aqua), AVHRR (onboard NOAA satellite series), and VEGETATION (onboard SPOT 4 & 5). Their typical spatial resolution at nadir is about 1 km (see Table 10.3), and their most frequently used multispectral bands cover VNIR, NIR, and SWIR bands. For mapping forest LAI or creating LAI products, those sensors and systems all include visible (or red) and NIR bands. Distinct characteristics of the data sources are its spacious coverage and high frequency revisiting of the same place globally. These sensors' data usually cover the whole global in just half to a couple of days. The data sources and LAI products are free and easy to access. Therefore, they are the most frequently used global application data sources. The coarse resolution LAI products are important in climate, weather, land use and land cover, and ecological studies at a global scale, and they have been routinely estimated and produced from the coarse resolution satellite sensors.

Currently, the most frequently used coarse LAI products are MODIS LAI products. The basic algorithms and methods used to create the LAI products with MODIS data with both Terra and Aqua platforms are based on three-dimensional RT theory, and LAI is retrieved by comparing the observed and modeled bidirectional reflectance factor (BRF) (Sprintsin et al. (2007). The latter utilizes a soil reflectance model (Jacquemound et al., 1992) for each biome for varying sun-view geometry and canopy-soil patterns. Under optimal circumstances, an LUT method is used to achieve inversion of the three-dimensional RT model for MODIS LAI retrieval (Myneni et al., 2002). When this method fails to optimize a solution, a back-up algorithm based on a relationship between the NDVI and LAI is utilized (Myneni et al., 1995). The 1 km global LAI product (MCD15A2, Collection 4) is updated every 8 days to eliminate the contamination from cloud cover (Knyazikhin et al., 1999). This product's algorithm uses six major biomes: cereal crops, shrubs, broad-leaf crops, savannas, broad-leaf forest, and needle-leaf forest to constrain the vegetation structural and optical parameter space (Fang and Liang, 2005). LAI estimates from MODIS data have become widely used since their release in 2000 and subsequently have received considerable validation for the previous product. Collection 4 (c4) was released in 2003 and the current version, Collection 5 (c5), was released in 2007. The c4 product was generally found to produce a significant overestimate compared with in situ observations. The latest version of the MODIS LAI c5 (MCD15A2) attempts to increase LAI estimate accuracy. The LUTs of RT model simulations used to invert biophysical estimates from observations were also recalibrated using a new stochastic RT model (Shabanov et al., 2005) to better depict three-dimensional effects. In the new product, the eight biomes are used by splitting up evergreen and deciduous forests types from the six biomes in c4 to improve retrievals in forested areas (De Kauwe et al., 2011). Recently, in contrast to the combined Terra and Aqua MODIS (MCD15A2) versions of the 8-day LAI and FPAR products, the level-4 combined (Terra and Aqua) MODIS global LAI and FPAR product is composited every 4 days at 1 km resolution. The increased temporal frequency of the 4 day product helps to monitor phenology and associated rapid changes (that occur within less than a week) especially during transition periods (green-up and senescence) (see http://edcdaac. usgs.gov/modis/dataproducts.asp).

To create coarse resolution LAI products from NOAA/AVHRR and SPOT/VEGETATION sensors' data, most methods are applied by first calculating NDVI or SR VIs with AVHRR channels 1 and 2 and VEGETATION red and NIR bands, and then fitting a linear relationship with ground-measured LAI or estimated LAI from moderate- or high-resolution image data (Duchemin et al., 1999; Chen et al., 2002; Wang et al., 2005; Peckham et al., 2008). For example, Canada-wide LAI maps are now being produced using cloud-free AVHRR imagery every 10 days at 1 km resolution. The archive of these products began in 1993. LAI maps at the same resolution are also being produced with images from the SPOT VEGETATION sensor (Chen et al., 2002). Both LAI products were produced with a relationship between AVHRR and VEGETATION-derived NDVI or SR VIs and ground-measured or high-resolution estimated LAI. In studying the NDVI–LAI relationship from 1996 to 2001 at a deciduous forest site, Wang et al. (2005) used NDVI time-series derived from various remote-sensing sources, namely NOAA AVHRR (1996, 1997, 1998, 2000), SPOT VEGETATION (1998–2001), and Terra MODIS (2001) to explore the seasonal and annual variability of this relationship. The results suggested that the NDVI–LAI relationship could vary both seasonally and interannually in tune with the variations in phenological development of the trees and in response to temporal variations of environmental conditions.

10.3.4 With hyperspectral data

At present, only a few of satellite hyperspectral sensors' data (e.g., EO-1/ Hyperion and ESA-Proba/CHRIS) may be used to estimate and map forest LAI. Their hyperspectral bands cover VNIR and SWIR (Hyperion) and VNIR only (CHRIS), and their moderate spatial resolution ranges from 18 m to 30 m. Because they are hyperspectral sensors, their images have the characteristics of hyperspectral data, such as available spectral diagnostic features extraction (Pu and Gong, 2011). The category of data has been demonstrated to have high potential for extracting and mapping forest LAI. The data, however, are relatively expensive and not available for most areas in the world and usually the data sources are difficult to access.

On the basis of the characteristics of hyperspectral data, all methods and techniques summarized in Table 10.1, such as spectral derivative and spectral position variables, can be used to retrieve and map forest LAI. Because many narrow and continuous bands in wavelength are available for the hyperspectral data, some important spectral features need to be extracted first, including spectral position variables and data transformation, and then the extracted spectral features and variables can be used as independent variables coupled with ground-based LAI measurement as the dependent variable to develop statistical and empirical models. We can inverse physically based RT models with hyperspectral reflectance data as the input for retrieving forest physical variables, including LAI. When using hyperspectral data to construct spectral VIs, the hyperspectral remote sensing provides more chances and flexibility to choose applicable spectral bands. For instance, with hyperspectral data, one can choose many of such red and NIR narrow-band combinations (Gong et al., 2003) to construct VIs.

Although hyperspectral satellite data are not popularly applied to estimating and mapping forest LAI as other categories of satellite imagery, a few studies have revealed the potential of the data resources to retrieve and estimate forest LAI. For example, by using reflectance images retrieved from Hyperion data to estimate forest LAI, Pu et al. (2003) tested and compared four approaches for extracting red-edge optical parameters used as spectral predictors to develop linear regression models for estimating forest LAI. The four approaches are four-point interpolation (Guyot et al., 1992), polynomial fitting (Pu et al., 2003), Lagrangian technique (Dawson and Curran, 1998), and inverted-Gaussian modeling (Miller et al., 1990). Experimental results indicate that the four-point approach is the most practical and suitable method for extracting the two red-edge parameters from Hyperion data. The polynomial fitting approach is a direct method and thus has practical value if hyperspectral data are available. When

Twele et al. (2008) evaluated the utility of narrow band of Hyperion and broadband ETM+ remote-sensing data to estimate LAI in a tropical environment in Sulawesi, Indonesia, with regression models, they found that the predictive power of most regression models was notably higher when employing narrow-band data instead of broadband data. Highly significant relationships between LAI and spectral reflectance were observed near the red-edge region and in most SWIR bands. They also found that differing from most NIR narrow bands, the correlation between SWIR reflectance and LAI was not confounded when including multivegetation types and did not suffer from saturation. Their results demonstrate that the canopy LAI under a tropical environment can be estimated with satisfactory accuracy from hyperspectral remote-sensing data. For mapping conifer forest CC and LAI in California, with EO-1 Hyperion data, Pu and Gong (2004) compared the performance of three feature extraction methods: band selection (SB), PCA, and WT. The experimental results indicate that the energy features extracted using the WT method are the most effective for mapping forest CC and LAI (mapped accuracy [MA] for CC = 84.90%, LAI MA = 75.39%), followed by the PCA method (CC MA = 77.42%, LAI MA = 52.36%). The SB method performed the worst (CC MA = 57.77%, LAI MA = 50.87%).

10.4 CHALLENGES AND FUTURE DIRECTIONS

Following this relatively extensive review of estimating and mapping forest LAI using multisensor and system satellite imagery, and to further improve methods and techniques of mapping LAI and increase the accuracy of estimating LAI with the methods and techniques, this section briefly addresses some of the challenges we are facing and will face and several future directions we should consider.

10.4.1 Challenges

Limited by spectral characteristics of remotely sensed data recorded by current sensors and systems themselves, we are encountering and will face several challenges when analyzing the spectral data for estimating and mapping forest LAI, including the following: saturation problems associated with various VIs correlated with LAI in estimating high-forest LAI, seasonal change of LAI resulting from the phenology change of plants versus remote-sensing temporal and frequent resolution, shadow problems of high-resolution data in retrieving forest canopy LAI, and effects of understory and background on spectral estimation of canopy LAI.

Some of the standard methods and techniques used to estimate LAI from different satellite remote-sensing data are inadequate across the range of

LAI, especially in high-forest LAI. For example, one major issue associated with retrieving LAI from various VIs calculated from the different band combinations of multi- and hyperspectral sensors is the saturation of LAI, which means that as LAI increases, reflected energy in the red and NIR wavelengths reaches a maximum, resulting in a threshold or saturation value for LAI (Carlson and Ripley, 1997; Lawrence and Ripple, 1998; Luo et al., 2002; Zheng and Moskal, 2009). Therefore, some new VIs need to be developed to increase the saturation point of canopy LAI. In this case, adding more spectral bands in a red-NIR band-based VI may improve the estimation of LAI saturation. For instance, EVI is a modified NDVI with a soil adjustment factor, L, and two coefficients, $C1$ and $C2$, and includes a blue band in correction of the red band for atmospheric aerosol scattering. This algorithm has improved sensitivity to high biomass regions and improved vegetation monitoring through a decoupling of the canopy background signal and a reduction in atmospheric influences (Huete et al., 2002). The complicated structure of forest canopy such as the angular distribution of foliage element will also affect the reflected radiances. To map maximum forest LAI, because of the seasonal change of LAI associated with plant phenology change, it may not be possible to use current optical remote-sensing techniques because maximum LAI mostly appears in rainy and cloudy seasons (summer) (Luo et al., 2002). This issue is also related to the temporal or frequent resolution of remote-sensing systems in addition to considering their spatial resolution. The temporal resolution challenges related to satellite sensors and systems are not easy to solve because they are related to technical and economic issues, as well as to huge data volume storage and processing issues. In addition, when mapping forest canopy LAI, shadow may be problematic. According to Seed and King (2003), the shadow fraction remains sensitive to increases in LAI in addition to considering the phenomenon of saturation of spectral VIs. One way to solve the problem may be to first use a spectral unmixing method to extract the shadow fraction (e.g., Aragão et al., 2005). Then spectral unmixing can be applied to extract the shadow fraction for tropical rainforest in Brazil from Landsat ETM+ imagery. Finally, a threshold of SR or NDVI can be used to separate tree canopy shadow from non-tree-canopy shadow (Pu, 2011a). Furthermore, the effects of understory and soil background need to be considered when mapping forest canopy LAI (Iiames et al., 2008), especially the effect of understory for those forest stands with a low CC.

10.4.2 Future directions

The integrative application of multisensor data, which can synergize high spatial resolution (usually low temporal resolution, e.g., IKONOS data) and high temporal resolution (usually low spatial resolution, e.g., MODIS data), provides a potential way to estimate and map high heterogeneous forest canopy LAI through integration of different spatial and spectral characteristics of sensor data. The disadvantage of using multisensor data for estimating and mapping LAI is the difficulty associated with various image acquisition and processing and the use of appropriate retrieving and mapping techniques. Application of multisensor data will become increasingly important in future study of estimating and mapping LAI in high heterogeneous areas, and thus more advanced image processing and canopy LAI estimating and mapping techniques are needed. Accurately retrieving and mapping forest LAI with multisensor data remains an active research topic, and new techniques continue to be developed. An advanced LAI estimating and mapping technique must be easy to use and must be able to provide accurate retrieving and mapping results.

With the development of remote-sensing technology, accurate, timely, and dynamic acquisition of LAI from a plot to a landscape level requires us to develop new algorithms that can efficiently utilize the spectral and spatial characteristics of the remote sensors to estimate and map forest LAI. In terms of LAI estimation, the most important issue of estimating LAI using multisensor data is that the LAI estimation result remains consistent, while the spatial scale changes. According to Zheng and Moskal (2008), the two ways to scale transferring of LAI, according to the direction of scaling, are up-scaling and down-scaling. Various LAI products may be generated from remote-sensing sensors' data with different spatial resolutions. When the LAI products with high resolution are transferred to low resolution by an aggregation scheme, up-scaling can be applicable to this method. For example, the LAI map created from Landsat TM with 30 m resolution can be up-scaled to 1 km images by executing an aggregation function to the 30 m resolution LAI map, and the up-scaled 1 km LAI can then be compared with the 1 km resolution LAI product created with MODIS data. Similarly, a down-scaling method can be used when the LAI products with low resolution are transferred to high resolution by a "splitting scheme." This kind of practical application with the down-scaling method may be useful when we consider using time-series of LAI products created from high-frequency remote-sensing data (e.g., MODIS and AVHRR sensors) to a required high-resolution application. Here, the heterogeneity of forest stands at canopy level is a key concept and is the major source of error when considering the application of an up- or down-scaling method.

As the availability of the high-resolution satellite sensors (e.g., World-View-2) increases, how to efficiently utilize the spatial information from the high spatial resolution data is another issue for improving the accuracy of estimating and mapping canopy LAI. Extraction of texture information and development of texture indices are a direction to efficiently utilize the high-resolution image. For instance, Colombo et al. (2003) found that the inclusion of texture indices strongly increased the linear relationship between NDVI derived from IKONOS data and LAI (increase in R^2 from

0.19 to 0.73, mean LAI values from 3 to 6), and Wulder et al. (1998) also found that texture information significantly improved the prediction of LAI from NDVI.

ACKNOWLEDGMENTS

The anonymous reviewers' comments and suggestions were greatly valuable to improve this chapter. The author sincerely appreciates their efforts.

REFERENCES

Aragão, L. E. O. C., Y. E. Shimabukuro, F. D. B. Espírito Santo, and M. Williams. 2005. Landscape pattern and spatial variability of leaf area index in Eastern Amazonia. *Forest Ecology and Management* 211: 240–256.

Asner, G. P., M. O. Jones, R. E. Martin, D. E. Knapp, and R. F. Hughes. 2008. Remote sensing of native and invasive species in Hawaiian forests. *Remote Sensing of Environment* 112: 1912–1926.

Baret, F., I. Champion, G. Guyot, and A. Podaire. 1987. Monitoring wheat canopies with a high spectral resolution radiometer. *Remote Sensing of Environment* 22: 367–378.

Baret, F., and G. Guyot. 1991. Potentials and limits of vegetation indices for LAI and APAR assessment. *Remote Sensing of Environment* 35(2–3): 161–173.

Baret, F., G. Guyot, and D. Major. 1989. TSAVI: a vegetation index which minimizes soil brightness effects on LAI and APAR estimation. In *12th Canadian Symposium on Remote Sensing and IGARSS'90, Vancouver, Canada, July 10–14, Vol. 3*, 1355–1358. Washington, DC: IEEE.

Belanger, M. J., J. R. Miller, and M. G. Boyer. 1995. Comparative relationships between some red edge parameters and seasonal leaf chlorophyll concentrations. *Canadian Journal of Remote Sensing* 21(1): 16–21.

Brown, L., J. M. Chen, S. G. Leblanc, and J. Cihlar. 2000. A shortwave infrared modification to the simple ratio for LAI retrieval in boreal forests: an image and model analysis. *Remote Sensing of Environment* 71: 16–25.

Bruce, L. M., C. Morgan, and S. Larsen. 2001. Automated detection of subpixel hyperspectral targets with continuous and discrete wavelet. *IEEE Transactions on Geoscience Remote Sensing* 39: 2217–2226.

Carlson, T. N., and D. A. Ripley. 1997. On the relation between NDVI, fractional vegetation cover, and leaf area index. *Remote Sensing of Environment* 62: 241–252.

Chen, J., and J. Cihlar. 1996. Retrieving leaf area index of boreal conifer forests using Landsat TM images. *Remote Sensing of Environment* 55: 153–162.

Chen, J. M. 1996. Evaluation of vegetation indices and a modified simple ratio for boreal applications. *Canadian Journal of Remote Sensing* 22: 229–242.

Chen, J. M., and T. A. Black. 1992. Defining leaf area index for non-flat leaves. *Plant Cell Environment* 15: 421–429.

Chen, J. M., G. Pavlic, L. Brown, J. Cihlar, S. G. Leblanc, H. P. White, R. J. Hall, et al. 2002. Derivation and validation of Canada-wide coarse-resolution leaf area index maps using high-resolution satellite imagery and ground measurements. *Remote Sensing of Environment* 80: 165–184.

Chen, X., L. Vierling, E. Rowell, and T. DeFelice. 2004. Using lidar and effective LAI data to evaluate IKONOS and Landsat 7 ETM+ vegetation cover estimates in a ponderosa pine forest. *Remote Sensing of Environment* 91: 14–26.

Cloutis, E.A. 1996. Hyperspectral geological remote sensing: evaluation of analytical techniques. *International Journal of Remote Sensing* 17(12): 2215–2242.

Cohen, W. B., T. K. Maiersperger, S. T. Gower, and D. P. Turner. 2003. An improved strategy for regression of biophysical variables and Landsat ETM+ data. *Remote Sensing of Environment* 84: 561–571.

Colombo, R., D. Bellingeri, D. Fasolini, and C. Marino. 2003. Retrieval of leaf area index in different vegetation types using high resolution satellite data. *Remote Sensing of Environment* 86: 120–131.

Combal, B., F. Baret, and M. Weiss. 2002. Improving canopy variables estimation from remote sensing data by exploiting ancillary information: case study on sugar beet canopies. *Agronomie* 22(2): 205–215.

Combal, B., F. Baret, M. Weiss, A. Trubuil, D. Mace, A. Pragnere, R. Myneni, et al. 2003. Retrieval of canopy biophysical variables from bidirectional reflectance: using prior information to solve the ill-posed inverse problem. *Remote Sensing of Environment* 84(1): 1–15.

Curran, P. J., W. R. Windham, and H. L. Gholz. 1995. Exploring the relationship between reflectance red edge and chlorophyll concentration in slash pine leaves. *Tree Physiology* 15: 203–206.

Danson, F. M., and S. E. Plummer. 1995. Red-edge response to forest leaf area index. *International Journal of Remote Sensing* 16(1): 183–188.

Darvishzadeh, R., A. Skidmore, M. Schlerf, C. Atzberger, F. Corsi, and M. Cho. 2008. LAI and chlorophyll estimation for a heterogeneous grassland using hyperspectral measurements. *ISPRS Journal of Photogrammetry and Remote Sensing* 63: 409–426.

Dawson, T. P., and P. J. Curran. 1998. A new technique for interpolating the reflectance red edge position. *International Journal of Remote Sensing* 19: 2133–2139.

De Kauwe, M. G., M. I. Disney, T. Quaife, P. Lewis, and M. Williams. 2011. An assessment of the MODIS collection 5 leaf area index product for a region of mixed coniferous forest. *Remote Sensing of Environment* 115: 767–780.

Delalieux, S., B. Somers, S. Hereijgers, W. W. Verstraeten, W. Keulemans, and P. Coppin. 2008. A near-infrared narrow-waveband ratio to determine Leaf Area Index in orchards. *Remote Sensing of Environment* 112: 3762–3772.

Duchemin, B., J. Goubier, and G. Courrier. 1999. Monitoring phenological key stages and cycle duration of temperate deciduous forest ecosystems with NOAA/AVHRR data. *Remote Sensing of Environment* 67: 68–82.

Eklundh, L., K. Hall, H. Eriksson, J. Ardö, and P. Pilesjö. 2003. Investigating the use of Landsat thematic mapper data for estimation of forest leaf area index in southern Sweden. *Canadian Journal of Remote Sensing* 29(3): 349–362.

Fang, H., and S. Liang. 2005. A hybrid inversion method for mapping leaf area index from MODIS data: experiments and application to broadleaf and needleleaf canopies. *Remote Sensing of Environment* 94: 405–424.

Fang, H., S. Liang, and A. Kuusk. 2003. Retrieving leaf area index using a genetic algorithm with a canopy radiative transfer model. *Remote Sensing of Environment* 85: 257–270.

Fassnacht, K. S., S. T. Gower, M. D. MacKenzie, E. V. Nordheim, and T. M. Lillesand. 1997. Estimating the leaf area index of north central Wisconsin forests using the Landsat Thematic Mapper. *Remote Sensing of Environment* 61: 229–245.

Fernandes, R., J. R. Miller, J. M. Chen, and I. G. Rubinstein. 2004. Evaluating image-based estimates of leaf area index in boreal conifer stands over a range of scale using high resolution CASI imagery. *Remote Sensing of Environment* 89: 200–216.

Flores, F. J., H. L. Allen, H. M. Cheshire, J. M. Davis, M. Fuentes, and D. Kelting. 2006. Using multispectral satellite imagery to estimate leaf area and response to silvicultural treatments in loblolly pine stands. *Canadian Journal of Forest Research* 36: 1587–1596.

Fukuda, S., and H. Hirosawa. 1999. A wavelet-based texture feature set applied to classification of multifrequency polarimetric SAR images. *IEEE Transactions on Geoscience and Remote Sensing* 37: 2282–2286.

Gastellu-Etchegorry, J. P., F. Gascon, and P. Esteve. 2003. An interpolation procedure for generalizing a look-up table inversion method. *Remote Sensing of Environment* 87(1): 55–71.

Gitelson, A. A. 2004. Wide Dynamic Range Vegetation Index for remote quantification of crop biophysical characteristics. *Journal of Plant Physiology* 161: 165–173.

Gobron, N., B. Pinty, and M. M. Verstraete. 1997. Theoretical limits to the estimation of the leaf area index on the basis of visible and near-infrared remote sensing data. *IEEE Transactions on Geoscience and Remote Sensing* 35: 1438–1445.

Goel, N. S. 1988. Models of vegetation canopy reflectance and their use in estimation of biophysical parameters from reflectance data. *Remote Sensing Reviews* 4: 1–212.

Goel, N. S., and W. Qi. 1994. Influences of canopy architecture on relationships between various vegetation indices and LAI and FPAR: a computer simulation. *Remote Sensing Reviews* 10: 309–347.

Goel, N. S., and R. L. Thompson. 1984. Inversion of vegetation canopy reflectance models for estimating agronomic variables: Estimation of leaf area index and average leaf inclination angle using measured canopy reflectance. *Remote Sensing of Environment* 15: 69–85.

Gong, P., R. Pu, G. S. Biging, and M. Larrieu. 2003. Estimation of forest leaf area index using vegetation indices derived from Hyperion hyperspectral data. *IEEE Transactions on Geoscience and Remote Sensing* 41(6): 1355–1362.

Gong, P., R. Pu, and R. C. Heald. 2002. Analysis of in situ hyperspectral data for nutrient estimation of giant sequoia. *International Journal of Remote Sensing* 23(9): 1827–1850, 2002.

Gong, P., R. Pu, and J. R. Miller. 1995. Coniferous forest leaf area index estimation along the Oregon transect using compact airborne spectrographic imager data. *Photogrammetric Engineering and Remote Sensing* 61: 1107–1117.

Gong, P., D. Wang, and S. Liang. 1999. Inverting a canopy reflectance model using an artificial neural network. *International Journal of Remote Sensing* 20(1): 111–122.

Gonsamo, A. 2010. Leaf area index retrieval using gap fractions obtained from high resolution satellite data: comparisons of approaches, scales and atmospheric effects. *International Journal of Applied Earth Observation and Geoinformation* 12: 233–248.

Gower, S. T., C. T. Kucharil, and J. M. Norman. 1999. Direct and indirect estimation of leaf area index, fapar, and net primary production of terrestrial ecosystems. *Remote Sensing of Environment* 70: 29–51.

Guyot, G., F. Baret, and S. Jacquemoud. 1992. Imaging spectroscopy for vegetation studies. In *Imaging Spectroscopy: Fundamentals and Prospective Application*, Vol. 2, edited by Toselli, F., and J. Bodechtel, 145–165. Norwell, MA: Kluwer.

Haboudane, D., J. R. Miller, E. Pattery, P. J. Zarco-Tejad, and I. B. Strachan. 2004. Hyperspectral vegetation indices and novel algorithms for predicting green LAI of crop canopies: modeling and validation in the context of precision agriculture. *Remote Sensing of Environment* 90: 337–352.

Hardinsky, M. A., V. Lemas, and R. M. Smart. 1983. The influence of soil salinity, growth form, and leaf moisture on the spectral reflectance of Spartina alternifolia canopies. *Photogrammetric Engineering and Remote Sensing* 49: 77–83.

Horler, N. N. H., M. Dockray, and J. Barber. 1983. The red edge of plant leaf reflectance. *International Journal of Remote Sensing* 4(2): 273–288.

Houborg, R., and E. Boegh. 2008. Mapping leaf chlorophyll and leaf area index using inverse and forward canopy reflectance modeling and SPOT reflectance data. *Remote Sensing of Environment* 112(1): 186–202.

Houborg, R., H. Soegaard, and E. Boegh. 2007. Combining vegetation index and model inversion methods for the extraction of key vegetation biophysical parameters using Terra and Aqua MODIS reflectance 909 data. *Remote Sensing of Environment* 106(1): 39–58.

Hu, B., K. Inannen, and J. R. Miller. 2000. Retrieval of leaf area index and canopy closure from CASI data over the BOREAS flux tower sites. *Remote Sensing of Environment* 74: 255–274.

Huete, A. 1988. A soil adjusted vegetation index (SAVI). *Remote Sensing of Environment* 25: 295–309.

Huete, A., K. Didan, T. Miura, E. P. Rodriguez, X. Gao, and L. G. Ferreira. 2002. Overview of the radiometric and biophysical performance of the MODIS vegetation indices. *Remote Sensing of Environment* 83: 195–213.

Huete, A. R., R. D. Jackson, and D. F. Post. 1985. Spectral response of a plant canopy with different soil backgrounds. *Remote Sensing of Environment* 17: 37–53.

Iiames, J. S., R. G. Congalton, A. N. Pilant, and T. E. Lewis. 2008. Leaf area index (LAI) change detection analysis on loblolly pine (pinus taeda) following complete understory removal. *Photogrammetric Engineering & Remote Sensing* 74(11): 1389–1400.

Jacquemoud, S., C. Bacour, H. Poilve, and J.-P. Frangi. 2000. Comparison of four radiative transfer models to simulate plant canopies reflectance: direct and inverse mode. *Remote Sensing of Environment* 74(3): 471–481.

Jacquemoud, S., F. Baret, B. Andrieu, F. M. Danson, and K. Jaggard. 1995. Extraction of vegetation biophysical parameters by inversion of the PROSPECT + SAIL models on sugar beet canopy reflectance data: application to TM and AVIRIS sensors. *Remote Sensing of Environment* 52(3): 163–172.

Jacquemound, S., F. Baret, and J. F. Hanocq. 1992. Modeling spectral and bidirectional soil reflectance. *Remote Sensing of Environment* 41(2-3): 123–132.

Jensen, R. R., and M. W. Binford. 2004. Measurement and comparison of Leaf Area Index estimators derived from satellite remote sensing techniques. *International Journal of Remote Sensing* 25(20): 4251–4265.

Jiang, Z., A. R. Huete, K. Didan, and T. Miura. 2008. Development of a two-band enhanced vegetation index without a blue band. *Remote Sensing of Environment* 112: 3833–3845.

Jordan, C. F. 1969. Derivation of leaf area index from quality of light on the forest floor. *Ecology* 50: 663–666.

Kalácska, M., G. A. Sánchez-Azofeifa, B. Rivard, J. C. Calvo-Alvarado, A. R. P. Journet, J. P. Arroyo-Mora, and D. Ortiz-Ortiz. 2004. Leaf area index measurements in a tropical moist forest: a case study from Costa Rica. *Remote Sensing of Environment* 91: 134–152.

Kaufman, Y. J., and D. Tanré. 1992. Atmospherically Resistant Vegetation Index (ARVI) for EOS-MODIS. *IEEE Transactions on Geoscience and Remote Sensing* 30: 261–270.

Knyazikhin, Y., J. Glassy, J. L. Privette, Y. Tian, A. Lotsch, and Y. Zhang. 1999. *MODIS Leaf Area Index (LAI) and Fraction of Photosynthetically Active Radiation Absorbed by Vegetation (FPAR) Product (MOD 15) Algorithm, Theoretical Basis Document*. Greenbelt, MD: NASA Goddard Space Flight Center.

Knyazikhin, Y., J. V. Martonchik, D. Diner, R. B. Myneni, M. M. Verstraete, B. Pinty, and N. Gobron. 1998. Estimation of vegetation canopy leaf area index and fraction of absorbed photosynthetically active radiation from atmosphere-corrected MISR data. *Journal of Geophysical Research* 103(D24): 32239–32256.

Kovacs, J. M., F. Flores-Verdugo, J. Wang, and L. P. Aspden, 2004. Estimating leaf area index of a degraded mangrove forest using high spatial resolution satellite data. *Aquatic Botany* 80: 13–22.

Kovacs, J. M., J. M. L. King, F. Flores de Santiago, and F. Flores-Verdugo. 2009. Evaluating the condition of a mangrove forest of the Mexican Pacific based on an estimated leaf area index mapping approach. *Environmental Monitoring and Assessment* 157: 137–149.

Kraus, T., M. Schmid, S. W. Dech, and C. Samimi. 2009. The potential of optical high resolution data for the assessment of leaf area index in East African rainforest ecosystems. *International Journal of Remote Sensing* 30(19): 5039–5059.

Lang, A. R. G. 1991. Application of some of Cauchy's theorems to estimation of surface areas of leaves, needles and branches of plants, and light transmittance. *Agricultural and Forest Meteorology* 55: 191–212.

Lawrence, R. L., and W. J. Ripple. 1998. Comparisons among vegetation indices and bandwise regression in a highly disturbed, heterogeneous landscape: Mount St. Helens, Washington. *Remote Sensing of Environment* 64: 91–102.

le Maire, G., C. Marsden, W. Verhoef, F. J. Ponzoni, D. L. Seen, A. Bégué, J.-L. Stape, et al. 2011. Leaf area index estimation with MODIS reflectance time series and model inversion during full rotations of Eucalyptus plantations. *Remote Sensing of Environment* 115: 586–599.

Leprieur, C. E. 1989. Preliminary evaluation of AVIRIS airborne measurements for vegetation. In *Proceedings of the 9th EARSel Symposium, Espoo, Finland, June 27–July 1*, 524–530.

Li, J., L. M. Bruce, J. Byrd, and J. Barnett. 2001. Automated detection of pueraria Montana (Kudzu) through Haar analysis of hyperspectral reflectance data. In *IEEE International Geoscience and Remote Sensing Symposium, Sydney, Australia, July 9–13, 2001*. Washington, DC: IEEE.

Li, L., and J. F. Mustard. 2003. Highland contamination in lunar mare soils: improved mapping with multiple end-member spectral mixture analysis (MESMA). *Journal of Geophysical Research* 108(E6): 5053. doi:10.1029/2002 JE001917.

Liang, S. 2004. *Quantitative Remote Sensing of Land Surfaces*. Wiley Praxis Series in Remote Sensing. Hoboken, NJ: Wiley & Sons.

Liang, S, and A. H. Strahler. 1993. Calculation of the angular radiance distribution for a coupled atmosphere and leaf canopy. *IEEE Transactions on Geoscience and Remote Sensing* GE31: 1081.

Lu, L., X. Li, C. L. Huang, M. G. Ma, T. Che, J. Bogaert, F. Veroustraete, et al. 2005. Investigating the relationship between ground-measured LAI and vegetation indices in an alpine meadow, north-west China. *International Journal of Remote Sensing* 26: 4471–4484.

Luo, T., R. P. Neilson, H. Tian, C. J. Vörösmarty, H. Zhu, and S. Liu. 2002. A model for seasonality and distribution of leaf area index of forests and its application to China. *Journal of Vegetation Science* 13(6): 817–830.

Mallat, S. G. 1989. A theory for multiresolution signal decomposition: The wavelet representation. *IEEE Transactions on Pattern Analysis and Machine Intelligence* 11: 674–693.

Mallat, S. G. 1998. *A Wavelet Tour of Signal Processing*. San Diego: Academic Press.

Meroni, M., R. Colombo, and C. Panigada. 2004. Inversion of a radiative transfer model with hyperspectral observations for LAI mapping in poplar plantations. *Remote Sensing of Environment* 92(2): 195–206.

Miller, J. R., E. W. Hare, and J. Wu. 1990. Quantitative characterization of the vegetation red edge reflectance: an inverted-Gaussian reflectance model. *International Journal of Remote Sensing* 11: 1775–1795.

Miller, J. R., J. Wu, M. G. Boyer, M. Belanger, and E. W. Hare. 1991. Season patterns in leaf reflectance red edge characteristics. *International Journal of Remote Sensing* 12(7): 1509–1523.

Misiti, M., Y. Misiti, G. Oppenheim, and J. M. Poggi. 1996. *Wavelet Toolbox User's Guide*. Natick, MA: Mathworks.

Munden, R., P. J. Curran, and J. A. Catt. 1994. The relationship between red edge and chlorophyll concentration in Broadbalk winter wheat experiment at Rothamsted. *International Journal of Remote Sensing* 15(3): 705–709.

Myneni, R. B., F. G. Hall, P. J. Sellers, and A. L. Marshak. 1995. The interpretation of spectral vegetation indexes. *IEEE Transactions on Geoscience and Remote Sensing* 33: 481–486.

Myneni, R. B., S. Hoffman, Y. Knyazikhin, J. L. Privette, J. Glassy, Y. Tian, Y. Wang, et al. 2002. Global products of vegetation leaf area and fraction absorbed PAR from year one of MODIS data. *Remote Sensing of Environment* 83: 214–231.

Nemani, R. R., L. L. Pierce, S. W. Running, and L. E. Band. 1993. Forest ecosystem processes at the watershed scale: sensitivity to remotely-sensed leaf area index estimates. *International Journal of Remote Sensing* 14(13): 2519–2534.

Peckham, S. D., D. E. Ahl, S. P. Serbin, and S. T. Gower. 2008. Fire-induced changes in green-up and leaf maturity of the Canadian boreal forest. *Remote Sensing of Environment* 112: 3594–3603.

Peddle, D. R., F. G. Hall, and E. F. Ledrew. 1999. Spectral mixture analysis and geometric optical reflectance modeling of boreal forest biophysical structure. *Remote Sensing of Environment* 67: 288–297.

Peterson, D., M. Spanner, S. Running, and K. Teuber. 1987. Relationship of Thematic Mapper data to leaf area index of temperate coniferous forest. *Remote Sensing of Environment* 22: 323–341.

Pittner, S., and S. V. Kamarthi. 1999. Feature extraction from wavelet coefficients for pattern recognition tasks. *IEEE Transactions on Pattern Analysis and Machine Intelligence* 21: 83–88.

Pu, R. 2011a. Mapping urban forest tree species using IKONOS imagery: preliminary results. *Environmental Monitoring and Assessment* 172: 199–214.

Pu, R. 2011b. Detecting and mapping invasive plant species by using hyperspectral data. In *Hyperspectral Remote Sensing of Vegetation*, edited by P. S. Thenkabail, 447–467. Boca Raton, FL: CRC Press.

Pu, R., L. Foschi, and P. Gong. 2004. Spectral feature analysis for assessment of water status and health level of coast live oak (Quercus Agrifolia) leaves. *International Journal of Remote Sensing* 25(20): 4267–4286.

Pu, R., and P. Gong. 2004. Wavelet transform applied to EO-1 hyperspectral data for forest LAI and crown closure mapping. *Remote Sensing of Environment* 91: 212–224.

Pu, R., and P. Gong. 2011. Hyperspectral remote sensing of vegetation bioparameters. *Advances in Environmental Remote Sensing: Sensors, Algorithms, and Applications*, edited by Q. Weng, 101–142. Boca Raton, FL: CRC Press.

Pu, R., P. Gong, G. S. Biging, and M. R. Larrieu. 2003. Extraction of red edge optical parameters from Hyperion data for estimation of forest leaf area index. *IEEE Transactions on Geoscience and Remote Sensing* 41(4): 916–921.

Pu, R., P. Gong, and Q. Yu. 2008. Comparative analysis of EO-1 ALI and Hyperion, and Landsat ETM+ data for mapping forest crown closure and leaf area index. *Sensors* 8: 3744–3766. doi:10.3390/s8063744.

Pu, R., and D. Liu. 2011. Segmented canonical discriminant analysis of in situ hyperspectral data for identifying thirteen urban tree species. *International Journal of Remote Sensing* 32(8): 2207–2226.

Pu, R., Q. Yu, P. Gong, and G. S. Biging. 2005. EO-1 Hyperion, ALI and Landsat 7 ETM+ data comparison for estimating forest crown closure and leaf area index. *International Journal of Remote Sensing* 26(3): 457–474.

Qi, J., A. Chehbouni, A. R. Huete, Y. H. Kerr, and S. Sorooshian. 1994. A modified soil adjusted vegetation index. *Remote Sensing of Environment* 48: 119–126.

Reujean, J., and F. Breon. 1995. Estimating PAR absorbed by vegetation from bidirectional reflectance measurements. *Remote Sensing of Environment* 51: 375–384.

Rioul, O., and M. Vetterli. 1991. Wavelet and signal processing. *IEEE Signal Processing Magazine* 8: 14–38.

Roberts, D. A., M. Gardner, R. Church, S. Ustin, G. Scheer, and R. O. Green. 1998. Mapping Chaparral in the Santa Monica Mountains using multiple endmember spectral mixture models. *Remote Sensing of Environment* 65: 267–279.

Rouse, J. W., R. H. Haas, J. A. Schell, and D. W. Deering. 1973. Monitoring vegetation systems in the Great Plains with ERTS. In *Proceedings, Third ERTS Symposium*, 309–317. Washington, DC: NASA Goddard Space Flight Center.

Running, S. W., R. R. Nemani, D. L. Peterson, L. E. Band, D. F. Potts, L. L. Pierce, and M. A. Spanner. 1989. Mapping regional forest evapotranspiration and photosynthesis by coupling satellite data with ecosystem simulation. *Ecology* 70: 1090–1101.

Sasaki, K., S. Kawata, and S. Minami. 1984. Estimation of component spectral curves from unknown mixture spectra. *Applied Optics* 23: 1955–1959.

Schlerf, M., and C. Atzberger. 2006. Inversion of a forest reflectance model to estimate structural canopy variables from hyperspectral remote sensing data. *Remote Sensing of Environment* 100: 281–294.

Schlerf, M., C. Atzberger, and J. Hill. 2005. Remote sensing of forest biophysical variables using HyMap imaging spectrometer data. *Remote Sensing of Environment* 95: 177–194.

Seed, E. D., and D. J. King. 2003. Shadow brightness and shadow fraction relations with effective leaf area index: importance of canopy closure and view angle in mixed wood boreal forest. *Canadian Journal of Remote Sensing* 29: 324–335.

Shabanov, N. V., D. Huang, W. Yang, B. Tan, Y. Knyazikhin, R. B. Myneni, D. E. Ahl, et al. 2005. Analysis and optimization of the MODIS leaf area index algorithm retrievals over broadleaf forests. *IEEE Transactions on Geoscience and Remote Sensing* 43: 1855–1865.

Simhadri, K. K., S. S. Iyengar, R. J. Holyer, M. Lybanon, and J. M. Zachary. 1998. Wavelet-based feature extraction from oceanographic images. *IEEE Transactions on Geoscience and Remote Sensing* 36: 767–778.

Smith, N. J., G. A. Borstad, D. A. Hill, and R. C. Kerr. 1991. Using high-resolution airborne spectral data to estimate forest leaf area and stand structure. *Canadian Journal of Forest Research* 21: 1127–1132.

Somers, B., J. Verbesselt, E. M. Ampe, N. Sims, W. W. Verstraeten, and P. Coppin. 2010. Spectral mixture analysis to monitor defoliation in mixed-aged Eucalyptus globulus Labill plantations in southern Australia using Landsat 5-TM and EO-1 Hyperion data. *International Journal of Applied Earth Observation and Geoinformation* 12: 270–277.

Song, C., and M. B. Dickinson. 2008. Extracting forest canopy structure from spatial information of high resolution optical imagery: tree crown size versus leaf area index. *International Journal of Remote Sensing* 29(19): 5605–5622.

Soudani, K., C. François, G. le Maire, V. L. Dantec, and E. Dufrêne. 2006. Comparative analysis of IKONOS, SPOT, and ETM+ data for leaf area index estimation in temperate coniferous and deciduous forest stands. *Remote Sensing of Environment* 102: 161–175.

Spanner, M. A., L. Johnson, J. R. McCreight, J. Freemantle, J. Runyon, and P. Gong. 1994. Remote sensing of seasonal leaf area index across the Oregon transect. *Ecological Applications* 4: 258–271.

Spanner, M. A., L. L. Pierce, D. L. Peterson, and S. W. Running. 1990. Remote sensing of temperate coniferous forest leaf area index: the influence of canopy closure, understory vegetation and background reflectance. *International Journal of Remote Sensing* 11: 95–111.

Sprintsin, M., A. Karnieli, P. Berliner, E. Rotenberg, D. Yakir, and S. Cohen. 2007. The effect of spatial resolution on the accuracy of leaf area index estimation for a forest planted in the desert transition zone. *Remote Sensing of Environment* 109: 416–428.

Sprintsin, M., A. Karnieli, P. Berliner, E. Rotenberg, D. Yakir, and S. Cohen. 2009. Evaluating the performance of the MODIS Leaf Area Index (LAI) product over a Mediterranean dryland planted forest. *International Journal of Remote Sensing* 30(19): 5061–5069.

Stenberg, P., M. Rautiainen, T. Manninen, P. Voipio, and M. Mottus. 2008. Boreal forest leaf area index from optical satellite images: model simulations and empirical analyses using data from central Finland. *Boreal Environment Research* 13(5): 433–443.

Twele, A, S. Erasmi, and M. Kappas. 2008 Spatially explicit estimation of leaf area index using EO-1 Hyperion and Landsat ETM+ data: implications of spectral bandwidth and shortwave infrared data on prediction accuracy in a tropical montane environment. *Giscience and Remote Sensing* 45(2): 229–248.

van Aardt, J. A. N., and R. H. Wynne. 2001. Spectral separability among six southern tree species. *Photogrammetric Engineering and Remote Sensing* 67(12): 1367–1375.

van Aardt, J. A. N., and R. H. Wynne. 2007. Examining pine spectral separability using hyperspectal data from an airborne sensor: an extension of field-based results. *International Journal of Remote Sensing* 28(2): 431–436.

van Leeuwen, W. J. D., and A. R. Huete. 1996. Effects of standing litter on the bio-physical interpretation of plant canopies with spectral indices. *Remote Sensing of Environment* 55(2): 123–138.

Vane G., and A. F. H. Goetz. 1988. Terrestrial imaging spectroscopy. *Remote Sensing of Environment* 24: 1–29.

Vincini, M., E. Frazzi, and P. D'Alessio. 2006. Angular dependence of maize and sugar beet VIs from directional CHRIS/PROBA data. In *Proceedings of the 4th ESA CHRIS PROBA Workshop 2006, ESRIN, Frascati, Italy, September 19–21*, http://earth.esa.int/workshops/4th_chris_proba/posters/p09_Vincini. pdf (last accessed on August 6, 2012).

Walsh, S. J., A. L. McCleary, C. F. Mena, Y. Shao, J. P. Tuttle, A. González, and R. Atkinson. 2008. QuickBird and Hyperion data analysis of an invasive plant species in the Galapagos Islands of Ecuador: implications for control and land use management. *Remote Sensing of Environment* 112: 1927–1941.

Walthall, C., W. Dulaney, M. Anderson, J. Norman, H. Fang, and S. Liang. 2004. A comparison of empirical and neural network approaches for estimating corn and soybean leaf area index from Landsat ETM+ imagery. *Remote Sensing of Environment* 92(4): 465–474.

Wang, F., J. Huang, and L. Chen. 2010. Development of a vegetation index for estimation of leaf area index based on simulation modeling. *Journal of Plant Nutrition* 33: 328–338.

Wang, Q., S. Adiku, J. Tenhunen, and A. Granier. 2005. On the relationship of NDVI with leaf area index in a deciduous forest site. *Remote Sensing of Environment* 94: 244–255.

Weiss, M., and F. Baret. 1999. Evaluation of canopy biophysical variable retrieval performances from the accumulation of large swath satellite data. *Remote Sensing of Environment* 70(3): 293–306.

Weiss, M., F. Baret, R. B. Myneni, A. Pragnere, and Y. Knyazikhin. 2000. Investigation of a model inversion technique to estimate canopy biophysical variables from spectral and directional reflectance data. *Agronomie* 20(1): 3–22.

White, J. D., S. W. Running, R. Nemani, R. E. Keane, and K. C. Ryan. 1997. Measurement and remote sensing of LAI in Rocky Mountain Montane ecosystems. *Canadian Journal Forest Research* 27: 1714–1727.

White, M. A., G. P. Asner, R. R. Nemani, J. L. Privette, and S. W. Running. 2000. Measuring fractional cover and leaf area index in arid ecosystems: digital camera, radiation transmittance, and altimetry methods. *Remote Sensing of Environment* 74: 45–75.

Wulder, M. A., E. F. Ledrew, S. E. Franklin, and M. B. Lavigne. 1998. Aerial image texture information in the estimation of northern deciduous and mixed wood forest leaf area index (LAI). *Remote Sensing of the Environment* 64: 64–76.

Zhang, L., D. Li, Q. Tong, and L. Zheng. 1998. Study of the spectral mixture model of soil and vegetation in Poyang Lake area, China. *International Journal of Remote Sensing* 19: 2077–2084.

Zhang, Y., J. M. Chen, J. R. Miller, and T. L. Noland. 2008a. Leaf chlorophyll content retrieved from airborne hyperspectral remote sensing imagery. *Remote Sensing of Environment* 112: 3234–3247.

Zhang, Y., J. M. Chen, J. R. Miller, and T. L. Noland. 2008b. Retrieving chlorophyll content in conifer needles from hyperspectral measurements. *Canadian Journal of Remote Sensing* 34(3): 296–310.

Zhao, G., and A. L. Maclean. 2000. A comparison of canonical discriminant analysis and principal component analysis for spectral transformation. *Photogrammetric Engineering & Remote Sensing* 66(7): 841–847.

Zheng, G., and L. M. Moskal. 2009. Retrieving leaf area index (LAI) using remote sensing: theories, methods and sensors. *Sensors* 9: 2719–2745.

Chapter 11

Effects of the spatial pattern of vegetation cover on urban warming in a desert city

Soe W. Myint

CONTENTS

Vegetation cover lowers air and surface temperatures in urban areas. What is less known, however, is the spatial pattern of vegetation and its variable influence on urban warming. Hence, we combine remote-sensing techniques with climate data to address the role of spatial patterns of vegetation in relation to air temperatures. Landsat Enhanced Thematic Mapper Plus image at 30 meter (m) resolution over the Phoenix metropolitan area acquired on April 19, 2000, was used. Multiple endmember spectral mixture analysis, an extension of the spectral mixture analysis approach, was used to quantify vegetation fractions at subpixel level. The Getis statistic was used to determine the spatial pattern of vegetation fractions (clustered, random, dispersed) at different levels of spatial scales (i.e., 11 × 11, 17 × 17, 23 × 23, 29 × 29, 35 × 35 window size). Results from this study suggest that spatial arrangements of vegetation play an important role in lowering air temperatures (i.e., maximum, minimum, mean). It was found that clustered vegetation can lower air temperatures more effectively than dispersed vegetation. The spatial pattern of clustered vegetation can lower minimum air temperatures (nighttime) more effectively than maximum air temperatures (daytime). The temperature difference of the lowering between dispersed and clustered patterns of vegetation cover for minimum air temperature is 7.44°C.

II.I INTRODUCTION

It has been proven that the urban heat island phenomena is strongly influenced by the amount and percent distribution of green vegetation biomass in urban areas (Oke, 1982; Huang et al., 1987; Sailor, 1995; Spronken-Smith and Oke, 1998; Carlson and Arthur, 2000; Bonan, 2002; Weng et al., 2004; Grossman-Clarke et al., 2005; Jenerette et al., 2007). The reflection and absorption of electromagnetic radiation and the process of evapotranspiration from vegetation biomass alter urban environmental conditions and energy fluxes (Gallo et al., 1993). The growth of population pressure and the extension of built-up areas are often associated with warmer climate and are more polluted than their surrounding rural environments (Streutker, 2002; Arnfield, 2003; Brazel et al., 2007). Trees in urban areas can mitigate the urban heat island effect as they provide shade and evaporation. The urban forest also functions as a major absorber of carbon dioxide and buffer of air pollutant loads. The modification of an urban landscape influences local climate (micro climate), especially the urban heat island effect that consequently could lead to regional and global climate change (Lo et al., 1997; Brazel et al., 2000; Quattrochi et al., 2000; Lo and Quattrochi, 2003; Voogt and Oke, 2003). Brazel et al. (2000) demonstrated that the temperature trend in Phoenix has demonstrated a 5.5°C increase in the minimum temperatures from the late 1940s. Myint et al. (2010) examined the role of percent cover of impervious surfaces and vegetation in relation to air temperatures at different spatial scales and developed a multiple regression model to predict maximum air temperatures in the city of Phoenix. Their study demonstrated that vegetation cover mitigates the urban heat island effect in this rapidly urbanizing desert city. Therefore, identification of percent distribution and growth of vegetation biomass and vegetation cover types in an urban-suburban environment and their relationships to air temperatures is an important step toward understanding how vegetation effectively mitigates urban warming effect in a desert city for effective decision making.

Remotely sensed images, including Landsat 5, 7, Enhanced Thematic Mapper Plus (ETM+), Moderate-Resolution Imaging Spectroradiometer (MODIS), and NOAA's Advanced Very High Resolution Radiometer (AVHRR), have provided an indirect approach to demonstrate the urban heat island and correlate temperatures to land cover information (Carlson et al., 1977; Wilson et al., 2003; Weng et al., 2004; Hung et al., 2006; Xian and Crane, 2006; Myint et al., 2010). Theoretically, the air temperature of the atmospheric boundary layer is modified by land surface temperature

(Voogt and Oke, 2003). Kawashima et al. (1999) demonstrated that surface temperature alone explained more than 80% of the observed variation in air temperature and accurately predicted air temperature by multiple regression using spatially averaged surface temperature and normalized vegetation index. Myint et al. (2010) suggested that a higher abundance of vegetation at reasonably large scales (i.e., 210 m × 210 m and 270 m × 270 m) can effectively decrease maximum air temperature. The same study reported that a small amount of impervious cover can still increase maximum air temperature, even though there is a large amount of vegetation cover. Although the data sources and the magnitude of the surface and air temperatures vary, a common conclusion among these studies is a negative relationship between vegetation cover and both types of temperatures. However, relationships of how spatial arrangements of vegetation influence urban warming are unknown. It is important to explore the spatial pattern of vegetation cover to understand how the spatial distribution of vegetation influences surface temperatures. To better understand the nature of spatial arrangements of vegetation cover, hypothetical images of a land cover in different neighborhoods with the same fraction (or same coverage in percent) that show different spatial patterns were created (see Figure 11.1). Hence, the goal of this study is to examine the role of spatial arrangements or patterns (e.g., clustered, dispersed) of vegetation cover at different spatial scales (local windows) in controlling air temperatures in an urban environment. The study area selected is the Phoenix metropolitan area in Arizona, located in the Sonoran Desert. We combine remote-sensing techniques with climate data to address the role of spatial patterns of vegetation in relation to air temperatures in this rapidly urbanizing landscape.

11.2 DATA AND STUDY AREA

A Landsat ETM+ image (L1G product of path 37 and row 37) at 30 meter (m) spatial resolution with six channels ranging from blue to short-wave infrared portions of the spectrum was used to quantify varying amounts and the distribution of vegetation in urban and suburban areas. The image data were acquired for the Phoenix metropolitan area under cloud-free conditions on April 19, 2000. The original image was subset to extract the Phoenix metropolitan area (upper left longitude W 112°47'10.96" and latitude N 33°49'59.62", lower right longitude W 111°34'18.56" and latitude N 33°12'09.81").

The Phoenix metropolitan area selected for the study is shown in Figure 11.2 by displaying Landsat ETM+ channel 4 (0.750–0.900 μm), channel 3 (0.630–0.690 μm), and channel 2 (0.525–0.605 μm) in red, green, and blue, respectively. The study covers common urban-suburban land use and land cover classes: high-density residential, low-density

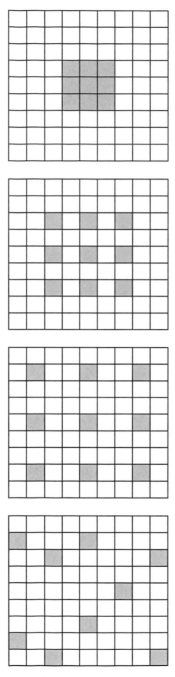

Figure 11.1 Hypothetical images showing spatial patterns of a vegetation cover in a neighborhood with same percent of distribution (Percent vegetation cover = 9/81 = 11.11%).

Figure 11.2 **(See color insert.)** Landsat 30 m resolution data over Phoenix metropolitan area on April 19, 2000, displaying bands 5, 4, and 3 in red, green, and blue, respectively.

residential, commercial, wild grass, woodlands, manmade grass, riparian vegetation, agriculture, cement roads, tar roads, cement and tar parking, rivers, lakes, sandbars, and exposed soil.

We used daily maximum, minimum, and mean temperatures that were compiled by the Central Arizona–Phoenix Long-Term Ecological Research project from major national, state, and local meteorological networks with weather stations distributed evenly over the Phoenix metropolitan area. Only air temperature data were used for the study. In particular, we used daily air temperature measured within the urban canopy layer (ca. 2 m height) by four networks: (a) Phoenix Realtime Instrumentation for Surface Meteorological Studies (Pon et al., 1998), (b) Arizona Meteorological Network (see http://ag.arizona.edu/azmet), (c) Maricopa County Flood Control District (see http://fcd.maricopa.gov), and (d) co-op stations obtained from NOAA's National Climatic Data Center (data available from http://www.ncdc.noaa.gov and summarized monthly at http://www.wrcc.dri.edu). Locations of weather stations used in our study are presented in Figure 11.3. We extracted air temperatures recorded at the above stations on April 19, 2000. The air temperatures recorded as minimum, maximum, and mean air temperatures at different weather stations are presented in Table 11.1.

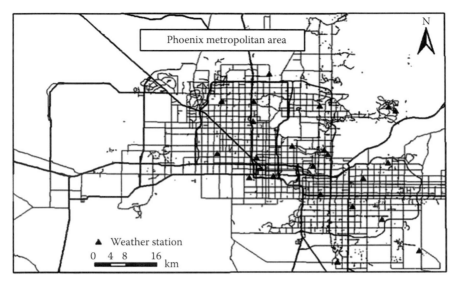

Figure 11.3 Weather stations.

11.3 METHOD

11.3.1 Subpixel analysis to quantify vegetation fractions

The subpixel approach that we employed in this study is based on a model proposed by Ridd (1995) that land cover in an urban environment is a linear combination of three land cover types (i.e., impervious, soil, vegetation). We used multiple endmember spectral mixture analysis (MESMA; Roberts et al., 1998), an extension of the spectral mixture analysis (SMA) approach, which allows the number and type of endmembers to vary for each pixel within an image.

We employed 28 endmembers and 544 endmember models using MESMA to identify fractions of soil, impervious surface, vegetation, and shade in the Phoenix metropolitan area. The detailed methodology employed in the study and validation of output fraction images were presented in Myint and Okin (2009). The mean RMS error for the selected land use–land cover classes range from 0.003 to 0.018. The Pearson correlation between the fraction outputs from MESMA and reference data from Quickbird 60 centimeter (cm) resolution data for vegetation was 0.8032. We also conducted several field visits for validation. Results from the study demonstrated that the selected models were reliable, and the MESMA algorithm quantified the selected urban land covers accurately (Myint and Okin, 2009). Figure 11.4 shows the vegetation distribution for the Phoenix metropolitan area. White portions of the image represent 100% distribution of a particular land

Table 11.1 Air temperatures recorded at different weather stations

Latitude (N)	Longitude (W)	Max. air temp. (°C)	Min. air temp. (°C)	Mean air temp. (°C)
33.48	−112.10	26.30	12.80	19.00
33.62	−112.11	25.40	10.50	18.70
33.39	−111.92	27.00	13.00	20.00
33.51	−112.00	26.00	12.00	19.00
33.36	−111.83	26.00	12.00	19.00
33.47	−111.73	25.00	10.00	17.50
33.60	−111.71	25.00	8.00	16.50
33.46	−111.94	25.00	11.00	10.00
33.57	−112.11	26.00	13.00	19.50
33.25	−111.64	26.00	7.00	16.50
33.49	−112.21	26.00	11.00	18.50
33.43	−111.80	30.00	13.00	21.50
33.22	−111.87	26.00	11.00	18.50
33.62	−112.20	27.22	10.00	18.61
33.43	−112.12	28.33	12.78	20.56
33.61	−111.73	26.67	9.44	18.06
33.50	−111.91	27.22	10.56	18.89
33.33	−111.75	27.78	10.56	19.17
33.61	−111.92	26.67	10.56	18.61
33.62	−112.20	27.22	10.00	18.61
33.49	−111.90	27.22	10.56	18.89
33.43	−112.05	26.67	14.44	20.56
33.69	−112.07	25.00	11.11	18.06

cover and black portions represent 0% cover. The study demonstrated that selected subpixel models were reliable, and the MESMA algorithm quantified the selected endmembers accurately.

11.3.2 Spatial pattern analysis

The two commonly applied spatial autocorrelation techniques are Moran's I and Geary's C (Cliff and Ord, 1973; Lee and Wong, 2001). Both assess how dispersed, uniformly distributed, or clustered points (weighted by their attributes) are in space and whether these patterns occur by chance (Goodchild, 1986). The general Getis statistic (Gi) was introduced as a technique to characterize the presence of hot spots (high clustered values) and cold spots (low clustered values) over an entire area (Getis and Ord, 1992; Ord and Getis, 1995). One key difference between Gi and spatial autocorrelation measures is that the computation of the Gi method takes the form of multiplying attribute values of neighborhood polygons or pixels, whereas the spatial autocorrelation techniques consider differences

Figure 11.4 Vegetation fractions.

between these attribute values (Myint et al., 2007). Hence, we employed the Getis statistic to determine the spatial pattern of vegetation fractions at different levels of spatial scales or local window sizes (i.e., 11×11, 17×17, 23×23, 29×29, 35×35).

The Getis statistic (Getis and Ord, 1992; Ord and Getis, 1995) is computed as follows:

$$Gi(d) = \frac{\sum_{i}^{n} \sum_{j}^{n} w_{ij}(d) z_i z_j}{\sum_{i}^{n} \sum_{j}^{n} z_i z_j}, \text{ for } i \neq j. \tag{11.1}$$

The Gi is defined by a distance, d, within which areal units can be regarded as neighbors of i. The weight $w_{ij}(d)$ is 1 if areal unit j is within d and is 0 otherwise. The relationship among the neighboring points is determined by a distance threshold, d. The value of d needs to be defined before computing the Gi. A high Getis value represents a clustered pattern of a particular spatial feature, and a low Getis value implies a dispersed distribution in a neighborhood. To demonstrate this, two hypothetical layers with attribute values that represent clustered and dispersed patterns were created (see Figure 11.5). From Figure 11.5 it can be observed that

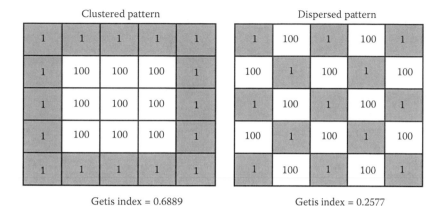

Figure 11.5 Clustered and dispersed patterns with Getis index values.

the computed value for a clustered pattern is higher (0.6889) than for a dispersed pattern (0.2577).

11.3.3 Regression analysis

We extracted the percent distribution of vegetation from the original image (30 m × 30 m) and computed Getis index values with the use of the selected window sizes. The extracted Getis-transformed images based on vegetation fractions were later integrated with air temperatures measured at the weather stations in the Phoenix metro area. A linear regression analysis was performed to determine the correlation between air temperatures (i.e., maximum, minimum, mean) and the Gi indices of vegetation fractions at different spatial scales (i.e., 330 m to 1,050 m). The research design is presented in Figure 11.6.

11.4 RESULTS AND DISCUSSION

Table 11.2 shows the correlations between air temperatures and Gi indices based on vegetation fractions extracted in local windows (i.e., 11 × 11, 17 × 17, 23 × 23, 29 × 29, 35 × 35). It can be observed that spatial patterns of vegetation fractions are negatively correlated to air temperatures. This result suggests that clustered vegetation patterns can lower air temperatures more effectively than dispersed vegetation patterns. However, no strong relations are found between spatial patterns (Gi indices) of vegetation fractions and maximum air temperatures at different spatial scales, except the largest window (35 × 35). The maximum air temperature represents the highest air temperature during daytime. The air temperature in Phoenix normally reaches the highest point of the day between 3:00 P.M. and 4:00 P.M. in the

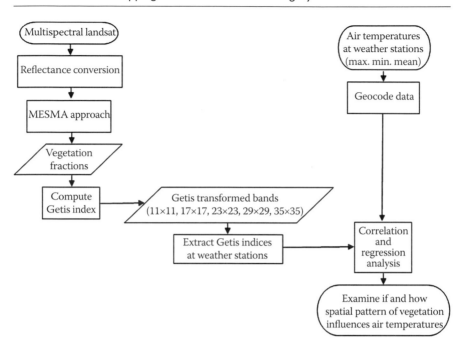

Figure 11.6 Research design.

afternoon. The low correlations between spatial patterns of vegetation and maximum air temperatures could be due to the fact that we need a dense enough vegetation cover that occupies a reasonably large area to lower the maximum temperature in a desert environment. This finding is consistent with the previous study that suggests that we need a dense enough vegetation cover that occupies a large area (i.e., equal to or larger than 270 m × 270 m) to lower the maximum temperature in Phoenix (Myint et al., 2010). The same study reports that a small amount of impervious cover has a higher impact on maximum air temperature than a large amount of vegetation cover. The relations between spatial patterns of vegetation and maximum air temperature are evident, however, when using larger window sizes even though they are low. There are statistically significant relations between

Table 11.2 Correlations between spatial patterns of vegetation fractions and air temperatures

Air	Getis index of vegetation fractions				
Temperature	*11 × 11*	*17 × 17*	*23 × 23*	*29 × 29*	*35 × 35*
Maximum	−0.278	−0.392	−0.399	−0.407	−0.416*
Minimum	−0.653**	−0.634**	−0.627**	−0.652**	−0.625**
Mean	−0.601**	−0.653**	−0.649**	−0.675**	−0.654**

* Correlation significant at 0.05; ** correlation significant at 0.01.

Gi indices generated at different spatial scales and minimum and mean air temperatures. Spatial patterns (clustered or dispersed) of vegetation fractions are negatively correlated to all types of air temperatures. The image and air temperatures were acquired in summertime. The air temperature usually reaches its maximum (hottest time) in the mid-afternoon (2:00 P.M. to 4:00 P.M.) and its minimum (coolest) during mid-night. From Figures 11.7, 11.8, and 11.9, it can be observed that the coefficients of determination are

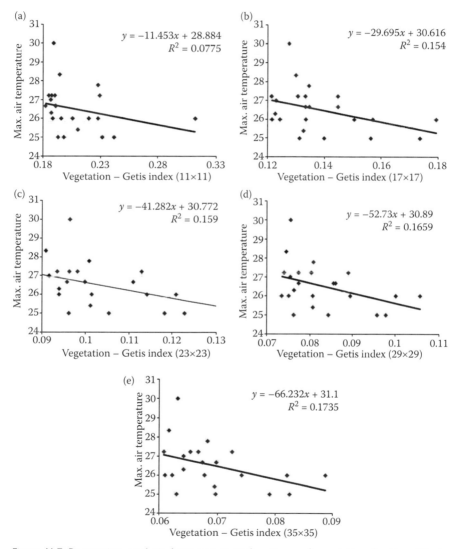

Figure 11.7 Regression analysis between spatial pattern of vegetation fraction (Getis index) and maximum air temperatures: (a) 11 × 11, (b) 17 × 17, (c) 23 × 23, (d) 29 × 29, and (e) 35 × 35.

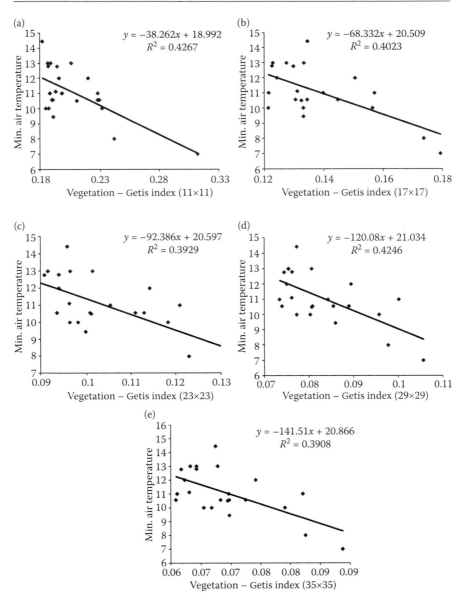

Figure 11.8 Regression analysis between spatial pattern of vegetation fraction (Getis index) and minimum air temperatures: (a) 11 × 11, (b) 17 × 17, (c) 23 × 23, (d) 29 × 29, and (e) 35 × 35.

stronger and the slope becomes steeper as the local window gets larger. This result implies that spatial arrangements of the vegetation pattern have a more effective influence on air temperatures when considering their spatial pattern in a larger neighborhood (e.g., 35 × 35 or 1,050 m × 1,050 m).

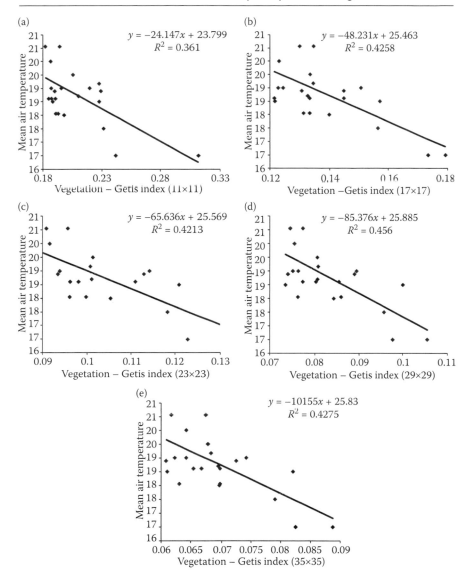

Figure 11.9 Regression analysis between spatial pattern of vegetation fraction (Getis index) and mean air temperatures: (a) 11 × 11, (b) 17 × 17, (c) 23 × 23, (d) 29 × 29, and (e) 35 × 35.

In other words, the spatial pattern of vegetation fractions in a small neighborhood (e.g., 11 × 11 or 330 m × 330 m) may not have a significant impact on air temperatures. It seems that correlation coefficients for the minimum and mean air temperatures dropped for areas larger than the 29 × 29 window size (see Table 11.2). This trend suggests that spatial arrangements of vegetation pattern (clustered or dispersed) in a neighborhood larger than 870 m × 870 m (29 × 29 local window) may have less impact on air temperature. Regression slopes for minimum air temperatures are steeper than for maximum air temperatures, suggesting that spatial arrangements of vegetation fractions influence nighttime temperatures more effectively. This finding is favorable because nighttime warming in urban areas is considered more important than daytime warming because the temperature difference between urban areas and surrounding rural areas usually is larger at night than during the day. The same trend of slopes was observed for minimum air temperatures when dealing with larger window sizes. In general, coefficients of determinations are high, correlations are statically significant, and slopes are steep for mean air temperatures. The relationships between vegetation patterns and mean air temperatures are higher in comparison to both minimum and maximum air temperatures. This relationship implies that spatial arrangements of vegetation have a significant impact on air temperatures throughout the day. Regression equations and slopes reflect the effect of spatial patterns of vegetation on air temperatures, suggesting that more clustered vegetation patterns effectively lower air temperatures. A dispersed vegetation distribution is not as effective in lowering air temperatures, and hence, clustered vegetation patterns may be more favorable for combating local and regional or global warming impacts. The temperature differences of the lowering between dispersed and clustered patterns of vegetation for minimum air temperatures and maximum air temperatures are 7.44°C and 5°C, respectively.

11.5 CONCLUSION

Results from this study suggest that spatial arrangements of vegetation play an important role in lowering air temperatures (i.e., maximum, minimum, mean) in urban areas. It was found that spatial patterns of vegetation (through analysis of the fraction patterns) are negatively correlated with air temperatures. This implies that clustered vegetation can lower air temperatures more effectively than dispersed vegetation patterns. For example, planting trees closer to each other in a neighborhood is more effective than planting trees far apart or arranging trees evenly distributed over the same neighborhood to mitigate the urban heat island effect. In general, regression models with bigger window sizes have steeper slopes, implying that spatial patterns of vegetation cover in a large spatial coverage have higher impact. The spatial pattern of vegetation in a neighborhood larger than 870 m × 870 m has a little lower coefficient of determination for all types of temperatures, implying that spatial arrangements of vegetation in a large area have less impact on air temperatures. The same situation may also be true for areas smaller than 510 m × 510 m. The spatial pattern of clustered vegetation can lower minimum air temperatures (nighttime) more effectively than maximum air temperatures (daytime). The temperature difference of the lowering between dispersed and clustered patterns of vegetation cover for nighttime air temperature is 7.44°C. In general, it can be concluded that clustered patterns of vegetation have a significant impact on air temperatures throughout the day. Further studies should focus on the spatial pattern of different vegetation types and other land cover categories in relation to air temperatures and should investigate what degree of clustered vegetation patterns we need to effectively lower air temperatures.

ACKNOWLEDGMENTS

The author thanks Anthony Brazel, School of Geographical Sciences and Urban Planning, Arizona State University, and Ariane Middel, Decision Center for a Desert City, Arizona State University, for their constructive comments and suggestions on the draft of this manuscript.

REFERENCES

Arnfield, A. J. 2003. Two decades of urban climate research: a review of turbulence, exchanges of energy and water, and the urban heat island. *International Journal of Climatology* 23: 1–26.

Bonan, G. B. 2002. *Ecological Climatology: Concepts and Applications*. Cambridge: Cambridge University Press.

Brazel, A. J., P. Gober, S. J. Lee, S. Grossman-Clarke, J. Zehnder, B. Hedquist, and E. Comparri. 2007. Determinants of changes in the regional urban heat island in metropolitan Phoenix (Arizona, USA) between 1990 and 2004. *Climate Research* 33: 171–182.

Brazel, A. J., N. Selover, R. Vose, and G. Heisler. 2000. The tale of two climates: Baltimore and Phoenix LTER sites. *Climate Research* 15: 123–135.

Carlson, T. N., and S. T. Arthur. 2000. The impact of land use–land cover changes due to urbanization on surface microclimate and hydrology: a satellite perspective. *Global Planet Change* 25(1): 49–65.

Carlson, T. N., J. A. Augustine, and F. E. Boland. 1977. Potential application of satellite temperature measurements in the analysis of land use over urban areas. *Bulletin of the American Meteorological Society* 58: 1301–1303.

Cliff, A. D., and J. K. Ord. 1973. *Spatial Autocorrelation*. London: Pion Press.

Gallo, K. P., A. L. McNab, T. R. Karl, J. F. Brown, J. J. Hood, and J. D. Tarpley. 1993. The use of a vegetation index for assessment of the urban heat island effect. *International Journal of Remote Sensing* 14(11): 2223–2230.

Getis, A., and J. K. Ord. 1992. The analysis of spatial association by use of distance statistics. *Geographical Analysis* 24(3): 1269–1277.

Goodchild, M. F. 1986. *Spatial Autocorrelation. Concepts and Techniques in Modern Geography, Catmog 47.* Norwich: Geo Books. http://qmrg.org.uk/files/2008/11/47-spatial-aurocorrelation.pdf (accessed October 31, 2011).

Grossman-Clarke, S., J. A. Zehnder, W. L. Stefanov, Y. B. Liu, and M. A. Zoldak. 2005. Urban modifications in a mesoscale meteorological model and the effects on near-surface variables in an arid metropolitan region. *Journal of Applied Meteorology* 44(9): 1281–1297.

Huang, Y. J., H. Akbari, H. Taha, and A. H. Rosenfeld. 1987. The potential of vegetation in reducing summer cooling loads in residential buildings. *Journal of Climatology and Applied Meteorology* 26: 1103–1116.

Hung, T., D. Uchihama, S. Ochi, and Y. Yasuoka. 2006. Assessment with satellite data of the urban heat island effects in Asian mega cities. *International Journal of Applied Earth Observation* 8(1): 34–48.

Jenerette, G. D., S. L. Harlan, A. J. Brazel, N. Jones, L. Larsen, and W. Stefanov. 2007. Regional relationships between surface temperature, vegetation, and human settlement in a rapidly urbanizing ecosystem. *Landscape Ecology* 22(3): 353–365.

Kawashima, S., T. Ishida, M. Minomura, and T. Miwa. 1999. Relations between surface temperature and air temperature on a local scale during winter nights. *Journal of Applied Meteorology* 39(9): 1570–1579.

Lee, J., and D. Wong. 2001. *Statistical Analysis with ArcView GIS.* New York: Wiley & Sons.

Lo, C. P., and D. A. Quattrochi. 2003. Land-use and land-cover change, urban heat island phenomenon, and health implications: a remote sensing approach. *Photogrammetric Engineering and Remote Sensing* 69(9): 1053–1063.

Lo, C. P., D. A. Quattrochi, and J. C. Luvall. 1997. Application of high-resolution thermal infrared remote sensing and GIS to assess the urban heat island effect. *International Journal of Remote Sensing* 18(2): 287–304.

Myint, S. W., A. Brazel, G. Okin, and A. Buyantuyev. 2010. An interactive function of impervious and vegetation covers in relation to the urban heat island effect in a rapidly urbanizing desert city. *GIScience and Remote Sensing* 47(3): 301–320.

Myint, S. W., and G. S. Okin. 2009. Modelling land-cover types using multiple endmember spectral mixture analysis in a desert city. *International Journal of Remote Sensing* 30(9): 2237–2257.

Myint, S. W., E. Wentz, and S. Purkis. 2007. Employing spatial metrics in urban land use/land cover mapping: comparing the Getis and Geary indices. *Photogrammetric Engineering and Remote Sensing* 73(21): 1403–1415.

Oke, T. R. 1982. The energetic basis of the urban heat island. *Quarterly Journal of the Royal Meteorology Society* 108(455): 1–24.

Ord, J., and A. Getis. 1995. Local spatial autocorrelation statistics: distributional issues and an application. *Geographical Analysis* 27(4): 286–306.

Quattrochi, D. A., J. C. Luvall, D. L. Rickman, M. G. Estes, Jr., C. A. Laymon, and B. F. Howell. 2000. A decision support information system for urban landscape management using thermal infrared data *Photogrammetric Engineering and Remote Sensing* 66(10): 1195–1207.

Ridd, M. 1995. Exploring a V-I-S (vegetation-impervious surface-soil) model for urban ecosystem analysis through remote sensing: comparative anatomy of cities. *International Journal of Remote Sensing* 16(12): 2165–2185.

Roberts, D. A., M. Gardner, R. Church, S. Ustin, G. Scheer, and R. O. Green. 1998. Mapping Chaparral in the Santa Monica Mountains using multiple endmember spectral mixture models. *Remote Sensing of Environment* 65(3): 267–279.

Sailor, D. J. 1995. Simulated urban climate response to modifications in surface albedo and vegetative cover. *Journal of Applied Meteorology* 34(7): 1694–1704.

Spronken-Smith, R. A., and T. R. Oke. 1998. The thermal regime of urban parks in two cities with different summer climates. *International Journal of Remote Sensing* 19(11): 2085–2104.

Streutker, D. R. 2002. A remote sensing study of the urban heat island of Houston, Texas. *International Journal of Remote Sensing* 23(13): 2595–2608.

Voogt, J. A., and T. R. Oke. 2003. Thermal remote sensing of urban climates. *Remote Sensing of Environment* 86(3): 370–384.

Weng, Q., D. Lu, and J. Schubring. 2004. Estimation of land surface temperature–vegetation abundance relationship for urban heat island studies. *Remote Sensing of Environment* 89(4): 467–483.

Wilson, J. S., M. Clay, E. Martin, D. Struckey, and K. Vedder-Risch. 2003. Evaluating environmental influences of zoning in urban ecosystems with remote sensing. *Remote Sensing of Environment* 86(3): 303–321.

Xian, G., and M. Crane. 2006. An analysis of urban thermal characteristics and associated land cover in Tampa Bay and Las Vegas using satellite data. *Remote Sensing of Environment* 104(2): 147–156.

Chapter 12

Remote sensing of algal blooms in inland waters using the matrix inversion method and semiempirical algorithms

Glenn Campbell

CONTENTS

Water resource managers have the responsibility to deliver water of sufficient quality to urban, agricultural, and industrial users as well as to maintain the recreational and ecological amenity of the inland water bodies under their control. Remote sensing is a useful tool to allow managers to monitor the quality of water economically.

This chapter gives an overview of methods used to retrieve the chlorophyll *a* concentration in inland waters. It utilizes in situ radiometric observations and Medium Resolution Imaging Spectrometer (MERIS) images of a tropical inland water impoundment, Burdekin Falls Dam, Australia, in a case study of the performance of a semianalytical and four semiempirical algorithms. It finds that all the semiempirical algorithms could be successfully applied to in situ radiometric observations, but two fail when applied to simulated MERIS bands and MERIS images. The other two semiempirical approaches were successfully applied to the one MERIS image but failed when applied to another MERIS image. Although the semianalytical approach resulted in a less accurate retrieval of chlorophyll *a* from the first image, it managed to successfully invert both images successfully. The case study highlights the need to consider the implicit and explicit assumptions of any approach when applying it to a new environment.

12.1 INTRODUCTION

Water resource managers have the responsibility to deliver water of sufficient quality to urban, agricultural, and industrial users as well as to maintain the recreational and ecological amenity of the inland water bodies under their control. To deliver these objectives, it is critical that these managers monitor and maintain the quality of the water in their storage reservoirs. Two important qualities of water that are relevant to the managers' objectives are the turbidity of the water body and the level of algal activity within it. The turbidity, or clarity of the water, which is a major influence on the ecology of aquatic systems, is determined by the absorption and scattering processes that take place within the water column. Three water quality parameters—algal cells, tripton (the nonalgal particles of the suspended particulate matter), and colored dissolved organic matter (CDOM)—represent the major absorption and scattering agents within the water.

Algal blooms, especially cyanobacterial (blue-green algae) blooms, in water can cause adverse health effects in humans and animals, ranging from skin irritations to permanent organ damage and death (Australian State of the Environment Committee, 2001; Chorus et al., 2003). Almost all water bodies in the state of Queensland, Australia, including those used for human consumption, feature cyanobacteria blooms at some time (Garnett et al., 2003), and approximately 80% of the water used in Australia is extracted from surface waters (Australian State of the Environment Committee, 2001). In Queensland, it is predicted by global climate change models that average daily minimum temperatures may rise more quickly than maximum temperatures. The increased growth rate of cyanobacteria that results from higher nighttime temperatures will allow hazardous blooms to develop faster (Garnett et al., 2003). The problem of algal blooms in the water supply is serious, ubiquitous, and has the potential to get worse. Hence, the monitoring of water quality parameters in general, and algal blooms in particular, is critical to maintain usable water resources.

With regards to regular environmental monitoring, single-point sampling may be of limited utility because the horizontal spatial distribution of suspended sediment and phytoplankton concentrations in water bodies is highly variable (Jupp, Kirk, and Harris, 1994; Hotzel and Croome, 1999; Vos et al., 2003; Kutser, 2004). The variability of the phytoplankton concentration can be addressed by a more intensive point-sampling routine, but the taking of water samples is labor intensive so the cost per sample point is a limit. In practice, the number of sampling sites and samples taken is a function of the aims of the monitoring, the morphology of the water body, and the financial resources available (Hotzel and Croome, 1999). Satellite remote sensing has been used for the simultaneous measurement of water quality parameters in each pixel of an image, meaning that the marginal cost for each measurement is small. It is not possible, however, to make a generic assessment of remote sensing's cost effectiveness, as each application will have its own trade-offs between the expediency of point sampling and the usefulness of a synoptic view, and each algorithm will require different amounts of fieldwork and laboratory processing to deliver a water quality parameter map. The area monitored by a point sample will change depending on the spatial variability of the physical quantity being measured. It is perhaps useful to compare the alternatives using a "cost per monitored area" measure (Bukata, 2005). For example, if for a large lake one sample per 5,000 ha was adequate to describe the spatial variability, then the cost of remote sensing would not be justified; however, if one sample per 10 ha was required, then it most probably would. Nevertheless, it is unlikely that remote sensing will ever evolve to a stage at which it can measure, at the same levels of accuracy and precision, all of the quality parameters that in situ sampling can.

Morel and Prieur (1977) classified water into two types: case-I, in which the optical properties are determined only by phytoplankton and the water itself; and case-II, in which contributions to the optical properties are made by colored dissolved organic matter (CDOM) and suspended inorganic particles as well. When optical remote sensing has been used to retrieve water quality parameters, the primary focus of work to date has been on oceanic (case-I) and coastal waters (case-II). The application of remote sensing to case-II, inland waters, has received less attention from researchers, perhaps because of the complexities introduced by high CDOM and tripton concentrations. This may change, as recent studies have shown that inland water bodies have a disproportionate effect on the global carbon cycle, with water bodies that are supersaturated with carbon dioxide consequently emitting it to the atmosphere (Cole et al., 2007). Higher carbon dioxide concentrations in warm lakes have been attributed to higher rates of respiration (Kosten et al., 2010). Remote sensing is an essential tool to understand the spatial distribution of the factors involved in the ecology of aquatic systems, but its application to inland impoundments has been limited.

12.1.1 Remote sensing of inland water bodies

Two dominant approaches exist to optical remote sensing of water quality: the empirical approach and the semianalytical approach. The empirical approach seeks to find correlations between the desired water quality parameter and the reflectance value of specific bands or band ratios. One of the influential drivers of this type of algorithm development has been a desire to develop a method that is mathematically simple and requires unsophisticated processing. In contrast, the semianalytical approach relies on modeling the interaction of the light field with the optical properties of the water. This approach is not totally analytical as it uses empiricism to parameterize several of the terms in the model (Rijkeboer, Dekker, and Gons, 1997; O'Reilly et al., 1998).

Within the empirical algorithms, the simplest approach is to examine the relationship between the concentration of chlorophyll *a* and the measured reflectance or radiance at a particular wavelength (Giardino, Candiani, and Zilioli, 2005; Matthews, Bernard, and Winter, 2010) or the wavelength of particular spectral maxima (Schalles et al., 1998). Rather than rely solely on statistical relationships between the in situ and remotely sensed measurements, semiempirical algorithms attempt to find relationships between the in situ measurements and selected wavelength subsets or wavelength combinations. To normalize the effect of the absorption and scattering of other water quality parameters, a number of researchers have used ratios of spectrally close wavelengths (Mittenzwey, Gitelson, and Kondratiev, 1992; Dekker, 1993; Koponen et al., 2001). In inland water, most of these approaches use ratios of near infrared (NIR) or red wavelengths, but success has been reported with blue and green ratios for water that is low in CDOM and tripton concentrations (George and Malthus, 2001; Giardino, Candiani, and Zilioli, 2005). The next stage of complexity, the fluorescent or reflectance line height algorithm, corrects the height of the reflectance or radiance maximum often found around 700 nm by subtracting a baseline under the peak that is interpolated by unaffected neighboring wavelengths (Yacobi, Gitelson, and Mayo, 1995; Schalles et al., 1998; Gower, Doerffer, and Borstad, 1999). In a more sophisticated attempt to allow for the scattering and absorption of other water quality parameters, three and four wavelength algorithms can be used. An NIR or red ratio with the value for backscattering derived from a 776 nm band was developed, but the author found that the result was improved when backscattering was not presumed to be constant (Gons, 1999). Likewise, Dall'Olmo et al. (2003)

adapted a terrestrial vegetation chlorophyll a method and applied it to productive inland waters. This method was modified in subsequent work (Le et al., 2009; Yang et al., 2010). Two of these methods will be applied in the case study in Section 12.2 and will be explained in more detail later. More complex still, Fraser (1998) established relationships between chlorophyll a concentrations and the first derivative of the reflectance spectrum.

There are three general types of semianalytical algorithms: the look-up table (LUT) approach, which matches measured spectra to large number of previously calculated spectra (Keller, 2001; Matarrese et al., 2004; Mobley et al., 2005); the neural network (NN) approach, which uses a large set of training data to relate the measured spectra to the parameters used to create the training set (Schaale, Fischer, and Olbert, 1998; Baruah et al., 2001; Su et al., 2006; Doerffer and Schiller, 2007); and the inversion-optimization algorithms (Lee, Carder, and Arnone, 2002; Maritorena, Siegel, and Peterson, 2002; Santini et al., 2010). In the inversion-optimization approach, a forward model is used to simulate the spectra from a number of parameters, and the set of parameters that minimizes a selected cost function is selected as the solution. If the forward model is linear and the cost function is the sum of the squares of the residuals, then this reduces to the linear matrix inversion method (Hoge and Lyon, 1996). The first inland water study using this method, a study of the Dutch Lake Braassem, refers to this model as the matrix inversion method (MIM) (Hoogenboom, Dekker, and de Haan, 1998) and this term and approach is used in this chapter.

12.2 CASE STUDY: BURDEKIN FALLS DAM, AUSTRALIA

Burdekin Falls Dam, Australia, has been used previously as a study site to investigate the use of the MIM algorithm with differentially weighted, overdetermined systems of equations. To parameterize and apply the method, it was first necessary to characterize the inherent optical properties (IOPs) of the water body, take in situ measurements of water quality parameters, and develop and validate a site-specific atmospheric correction. This paper will give a brief overview of this work that is reported in more depth in Campbell et al. (2011a,b). This previously reported work will be compared with the results that can be obtained with three semiempirical methods applied to the same field and image data.

12.2.1 Study site

The Burdekin Falls Dam (20°37′ S, 147°0′ E) receives inputs from four major subcatchments that cover a total area of 114,000 km². From the north, the Burdekin River has its origin in tropical rainforest, but it primarily flows through tropical savannah. From the west, the lake is fed from the Cape River, which rises in reasonably steep sedimentary country and then flows through flat less erodible areas. The Belyando and Suttor Rivers meet just beyond the inundated area and feed the lake from the south. The Belyando River and Suttor River suffer from persistent turbidity. The impoundment is split into an upper and lower basin by a narrow neck of land. Water released from the dam enters the Burdekin River and discharges into the Great Barrier Reef lagoon approximately 200 km downstream.

12.2.2 Water quality parameter concentration measurements

In situ water quality parameter measurements and near-coincident MERIS images were obtained as part of two field campaigns in October 2008 and August 2009.

12.2.2.1 October 2008 measurements

During October 2008, the water quality parameters of the storage were measured at 11 stations (see Figure 12.1). Water samples were taken from approximately 0.3 m below the surface and kept cool for later laboratory measurement of tripton (*TR*), chlorophyll *a* (*CHL*) concentration, and CDOM spectral absorption. A detailed description of the of the water quality parameter measurements and subsequent processing is provided in Campbell et al. (2011a).

12.2.2.2 August 2009 measurements

A second field campaign was conducted in August 2009 to obtain a larger validation data set for chlorophyll *a* that was independent of the

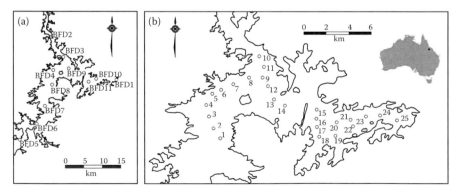

Figure 12.1 Location of the validation sample sites for the October 2008 (a) and the August 2009 (b) fieldwork activities.

measurements used to parameterize the algorithm. Water samples were taken from approximately 0.3 m below the surface at 25 observation stations (see Figure 12.1).

Two replicates were prepared for each water sample by filtering through a 47 mm diameter GF/F glass-fiber filter (Whatman, nominal pore size; 0.7 μm) and then freezing the filter. The pigments were measured using the U.S. Environmental Protection Agency method 445.0 (Arar and Collins, 1997). This method measures the combined concentrations of chlorophyll *a* and pheophytin *a*. With the aid of high-performance liquid chromatography analysis, no pheophytin *a* was detected in the October 2008 samples, and so it is assumed that the measured concentration is only that of chlorophyll *a*.

12.2.3 The semianalytical algorithm

The water quality parameter concentrations and the reflectance spectrum are linked by the IOPs of the water. These IOPs have magnitudes that are independent of the geometric structure of the light field. The absorption coefficient *a* describes the chances of a photon being absorbed, the scattering coefficient *b* describes the chances of a photon being scattered, and the volume scattering function (*VSF*) $\beta(\theta)$ describes the probability of a scattered photon being scattered in a particular direction. For practical purposes, the latter two IOPs are combined into the backscattering parameter (b_b), which describes the probability of a photon being scattered more than 90° from its original path. A version of the Gordon et al. (1975) subsurface irradiance reflectance ($R(0^-)$) model, with the higher order terms neglected, was used:

$$R(0^-, \lambda_i) = f(\omega_b, \mu_0, \lambda_i) \frac{b_b(\lambda_i)}{a(\lambda_i) + b_b(\lambda_i)}, \qquad (12.1)$$

where *a* is the linear sum of the absorption of the main color-producing agents in the water: CDOM, tripton, phytoplankton, and the water itself. It was assumed that CDOM is not a scattering agent, so the backscattering is made up of only three parts. The values for absorption of pure water were obtained from Pope and Fry (1997) and Smith and Baker (1981), and the backscattering coefficient for pure water was obtained from Morel (1974), assuming a scattering-to-backscattering ratio of 2:1. The proportionality factor (*f*) was modeled as a cubic function of the subsurface reflectance, and the Sun zenith angle was calculated from Hydrolight® simulations (Campbell and Phinn, 2010). Within the model, the absorption and backscattering resulting from the water quality constituents are related to their respective constituents by previously measured specific inherent optical properties (SIOPs). The specific absorption spectra were sourced from the field measurements previously described. Notable intraimpoundment variation in the specific absorption and specific scattering of phytoplankton and tripton was found, and as a result, the water in Burdekin Falls Dam was characterized by two SIOP sets, the upper basin and the lower basin (Campbell, Phinn, and Daniel, 2011). The two SIOP sets are shown in Figure 12.2.

Utilizing the SIOPs, Equation 12.1 can be rearranged and put into matrix form for all wavelengths of the spectra (Hoogenboom, Dekker, and de Haan, 1998), as follow:

$$y = Ax, \tag{12.2}$$

where A is a $3 \times N$ dimension matrix, with N being the number of wavelengths utilized by the inversion, y is a $1 \times N$ dimension matrix of constant values, and x is a column vector of the unknown water quality parameter concentrations. The minimum number of wavelengths required to solve this equation is three, but a more robust solution can be obtained by using more and differentially weighting the wavelengths. The solution of this

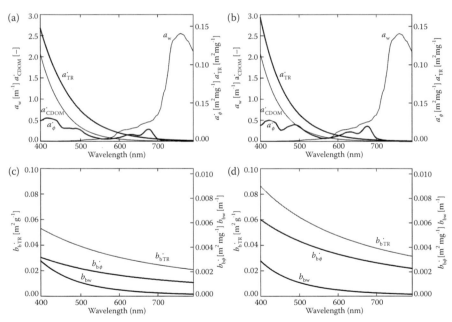

Figure 12.2 SIOP sets upper basin (a, c) and lower basin (b, d) for Burdekin Falls Dam measured during the October 2008 fieldwork. Graphs a and b show the spectral absorption of water (w) and the chlorophyll a–specific absorption spectra of phytoplankton (φ), tripton (TR), and colored dissolved organic matter (CDOM). Graphs c and d show the spectral backscattering of water (w) and the specific backscattering spectra of chlorophyll a(φ) and tripton (TR). (From Campbell, G., S. R. Phinn, and P. Daniel, Hydrobiologia 658, 245, 2010. With permission from Springer Science+Business Media.)

equation uses a square $(N \times N)$ weight matrix, which is a diagonal matrix (W), where W_{ii} = relative weight of wavelength i. The solution then becomes

$$x = [A^T W A]^{-1} A^T W y. \tag{12.3}$$

For each image pixel, the optimal SIOP is set selected using a measure based on the optical closure between the measured spectrum and a forward modeled spectrum.

In this case study, the best performed weighting scheme from Campbell et al. (2011b) was used and is shown in Figure 12.3.

12.2.4 Semiempirical algorithms

Dall'Olmo et al. (2003) recognized the similarity between the interaction of light with green vegetation with the case of inland water remote sensing. For vegetation it was found that:

$$Chl\ a \propto [R^{-1}(\lambda_1) - R^{-1}(\lambda_2)]\, R(\lambda_3), \tag{12.4}$$

where $R(\lambda_1)$ is the reflectance in a spectral region where the absorption due to the pigment is a maximum, and $R(\lambda_2)$ is the reflectance in an adjacent spectral region where the absorption due to the pigment is minimal, but the absorption due to other constituents is approximately equal to what it was in the λ_1 region. In the vegetation case, this difference was still affected by the variability of the leaf structure and thickness. Dall'Olmo et al. (2003) supposed that this was much like the effect of backscattering in the case of water. They incorporated $R(\lambda_3)$, which is the reflectance in a spectral region where the absorption is minimal, but it can be used to account for the backscattering differences between the samples. Le et al. (2009) explained that for many environments, this assumption is valid; however, in the case of highly turbid waters, it is not necessarily true that the absorption and backscattering due to tripton is negligible in the NIR. They introduced a fourth wavelength region close to λ_3 to mitigate the effect of tripton absorption. Because of the difficulty in obtaining four narrow wavelength bands within the range of 660–760 nm, Yang et al. (2010) proposed an alternative formulation:

$$Chl\ a \propto [R^{-1}(\lambda_1) - R^{-1}(\lambda_2)][R^{-1}(\lambda_3) - R^{-1}(\lambda_2)]. \tag{12.5}$$

The assumption is made that the backscattering is spectrally invariant $(b_b(\lambda_1) = b_b(\lambda_2) = b_b(\lambda_3))$ and the wavelength regions are chosen so that $a_{TR}(\lambda_1) + a_{CDOM}(\lambda_1) \sim a_{TR}(\lambda_2) + a_{CDOM}(\lambda_2)$ and $a_\phi(\lambda_1) \gg a_\phi(\lambda_2)$. The region λ_3 is chosen such that the absorption of all water quality parameters is negligible so $R(\lambda_3) \propto \dfrac{b_b(\lambda_3)}{a_w(\lambda_3)}$.

12.2.5 Atmospheric correction

The c-WOMBAT-c software (Brando and Dekker, 2003) was used, but it was modified to use the 6S radiative transfer model and the Thuillier et al. (2003) reference Sun irradiance spectrum (Campbell et al., 2011b). The c-WOMBAT-c approach corrects for the adjacency effect by applying an $n \times n$ low pass filter to the image to supply an average radiance ($L_{rs,b}$) image. A value of $n = 9$ was adopted, resulting in a 2.7 km × 2.7 km adjacency window.

The aerosol optical thickness (AOT) was estimated by taking advantage of the homogeneity of aerosols over small spatial scales of 50 to 100 km (Vidot and Santer, 2005) to calculate the AOT over adjoining dense dark vegetation (DDV) and then applying this value to the water body. The Vidot and Santer approach assumes that the reflectance value of the DDV in the blue and red regions is known and uses these values to identify the aerosol type and retrieve the AOT. In this work, the aerosol type was selected based on the water-body location and the prevailing wind conditions before the image acquisition. The reference DDV values for three bands (412 nm, 443 nm, and 665 nm), corrected for the bidirectional reflectance distribution function effects using the Leroy et al. (1998) model, were extracted from the MERIS auxiliary files. Image pixels were designated as DDV pixels if their atmospherically resistant vegetation index (ARVI) (Kaufman and Tanre, 1992) was above a given threshold.

A simple ARVI threshold is prone to select normal vegetation that is shadowed by cloud at the time of the image acquisition. The reflectance value in the 865 nm band was used to separate cloud shadow from DDV using a minimum reflectance of 17%, and a subset of image pixels that represented the highest 0.5% of ARVI values was selected. The DDV pixel subset was averaged to get the DDV spectrum for that AOT value. The AOT value was iterated until the image DDV value matched the reference DDV.

12.2.6 In situ spectroradiometric data

Two RAMSES spectroradiometers were mounted in a cage. One spectroradiometer was fitted with a cosine collector and was orientated upward in the cage to measure the downwelling irradiance, and one radiance collector was orientated downward to measure the upwelling radiance. The cage was lowered on the unshaded side of the vessel to minimize the shading effects. For each station of the 2008 field campaign, the simultaneous measurements of downwelling irradiance and upwelling radiance were combined to calculate the above-surface reflectance. The observed reflectance spectra were convolved with the MERIS band response functions to produce simulated in situ spectra. The observations for stations 9–11 were made within 90 minutes of the acquisition of the October 15, 2008, MERIS image.

12.3 RESULTS

12.3.1 In situ measurements

The range, mean, and standard deviation of the laboratory measurements of the water quality parameters from the two field campaigns are shown in Table 12.1.

Two replicates for each station in the August 2009 field campaign were created by first dividing the water sample in two before filtering each half onto separate filters. The differences between the replicates had a mean of 0.3 µgl^{-1} (8%) and a standard deviation of 0.35 µgl^{-1} (9%), and the maximum difference was 1.36 µgl^{-1} (33%).

12.3.1.1 Atmospheric correction

The 2008 image was corrected with an AOT at 550 nm value of 0.15 was compared to the in situ measurements for stations 9–11. The comparison is shown in Figure 12.4.

12.3.1.2 Semiempirical approaches

Following the approach of Dall'Olmo et al. (2003), the three semiempirical algorithms were tuned to select wavelengths that provided the minimum standard error when the ratio was plotted against the measured chlorophyll *a* concentrations. The MERIS bands closest to the optimized wavelength values were used to calculate the ratios for the simulated MERIS and MERIS image spectra. Figure 12.5 shows the result of changes in the available data on the validity of the relationship between then the semiempirical ratio and the chlorophyll *a* concentration. It would be preferable to split the data into a model parameterization set and an independent validation set. Unfortunately, the 2008 data set was too small to do this without the risk of overfitting the data in the algorithm development.

The Dall'Olmo et al. (2003) semiempirical relationships proved to be the least robust as the suitability of the data was reduced. In the first instance, the Dall'Olmo et al. approach showed an acceptable correlation when the

Table 12.1 Range (mean and standard deviation) of chlorophyll *a*, tripton, a_{CDOM}(440), and secchi depth for each field campaign

Field campaign	Measurement stations	Chlorophyll a (µgl^{-1})	Tripton (mgl^{-1})	a_{CDOM}(440) (m^{-1})	Secchi depth (m)
October 2008	11	2.8 – 7.7 (4.74 ± 1.54)	5.6 – 10.3 (6.78 ± 1.31)	0.88 – 1.27 (1.05 ± 0.11)	0.9 – 1.3 (1.04 ± 0.13)
August 2009	25	1.8 – 6.9 (1.72 ± 1.15)	0.8 – 7.7 (3.20 ± 1.63)	n.d.	n.d.

n.d. = no data.

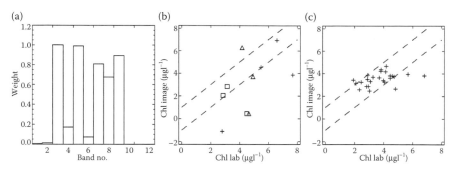

Figure 12.3 The (a) MER_BU_RAN2 weighting scheme and the laboratory chlorophyll *a* concentrations versus image-retrieved chlorophyll *a* concentrations for semianalytical approach for the (b) 2008 and (c) 2009 images. The dashed lines show the bounds of 1 μgl^{-1} for chlorophyll *a*. In plot (b), the points marked with a cross and a triangle had in situ samples taken two days before and one day before the satellite overpass respectively, and the points marked with a square had in situ samples taken on the day of the satellite overpass.

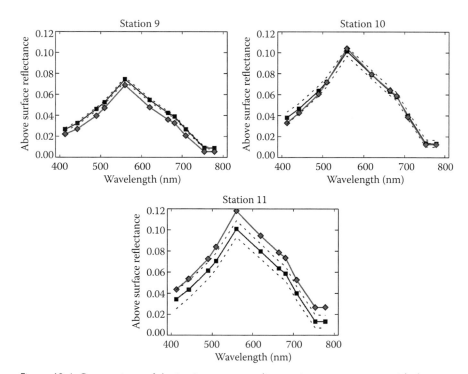

Figure 12.4 Comparison of the in situ spectroradiometric measurements with the corrected image for October 15, 2008 at Burdekin Falls Dam. The in situ measurements are shown by squares; the corrected image spectra are shown by diamonds. The dotted lines represent one standard deviation either side of the mean for the in situ measurements.

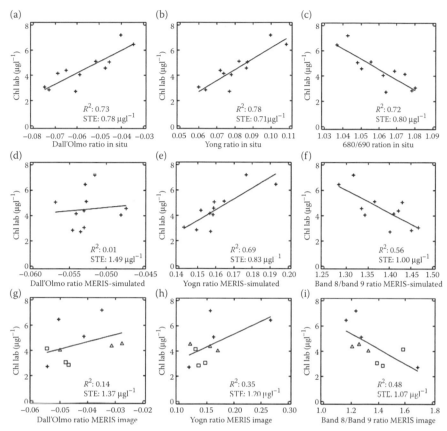

Figure 12.5 The ratio versus in situ chlorophyll *a* measurements for the three semiempirical algorithms for three data types. For each algorithm, a plot is provided showing the result using the in situ spectroradiometric data [(a)–(c)],
the in situ spectroradiometric data convolved with the MERIS band spectral
response functions [(d)–(f)], and the MERIS images [(g)–(i)]. In plots
(g) through (i), points marked with a cross and a triangle had in situ samples
taken two days before and on the day before the satellite overpass, respectively, and the points marked with a square had in situ samples taken on the
day of the satellite overpass.

optimized wavelengths of $\lambda_1 = 680$ nm, $\lambda_2 = 691$ nm, and $\lambda_3 = 696$ nm
were used. These wavelengths are very close together because of the need
to comply with the assumptions inherent in the relationship. When the in
situ spectra were convolved onto the MERIS bands, it was necessary to use
the closest bands to these values, Band 7 (665 nm nominal center), Band 8
(681.25 nm nominal center), and Band 9 (708.75 nm nominal center).
Figure 12.5d shows that the ratio and the in situ–measured chlorophyll *a*
concentrations show no discernable correlation. When the same bands are

used and applied to the image, there would appear to be more evidence of correlation, but it is at such a low level that the outcome might be as easily ascribed to chance. The band combination used by Gitelson et al. (2009) (Bands 7, 9, and 10; 753.75 nm nominal center) was trialed, but it showed even less correlation. As before, in the first instance, the Yang approach showed an acceptable correlation when the optimized wavelengths of $\lambda_1 = 669$ nm, $\lambda_2 = 663$ nm, and $\lambda_3 = 709$ nm were used. The MERIS band combination that was closest to the optimized band positions was Band 8 (681.25 nm nominal center), Band 7 (665 nm nominal center), and Band 9 (708.75 nm nominal center). The plot of the ratio against the chlorophyll a concentration (see Figure 12.5f) shows a higher standard error and a reduced correlation coefficient. Using these same bands and moving to the image retrieved spectra once again showed an increase in the standard error and reduction in correlation coefficient. A similar result was found in the case of the simple band ratio. It was found that a ratio of the 680 nm and 690 nm wavelengths gave the minimum standard error. When this approach was applied to the MERIS convolved spectra, it was necessary to use Band 8 (681.25 nm nominal center) and Band 9 (708.75 nm nominal center). Notwithstanding the increases in standard error and the correlation coefficient reduction, the expected relationship has been robust enough to cope with the change in data in the last two cases. Equations 12.6 and 12.7 are for the line of best fit using the MERIS image for the Yang approach ($R_{8,7,9}$) and the two-band approach ($R_{7,8}$):

$$Chl = 20.4\,R_{8,7,9} + 1.3 \tag{12.6}$$

and

$$Chl = -6.0\,R_{7,8} + 12.5. \tag{12.7}$$

Figure 12.6 shows the results of applying the three approaches to the August 2009 image. In all cases, there is no discernable correlation between the ratio and the in situ–measured chlorophyll a concentrations. The lack of correlation meant that it was not appropriate to use the 2008 model to retrieve chlorophyll a concentration estimates.

12.3.2 Semianalytical approach

The MIM was applied to the atmospherically corrected October 2008 and August 2009 MERIS images using the MER_BU_RAN2 weighting scheme described in Campbell et al. (2011b). The resulting plots that compare the in situ–measured chlorophyll a concentrations with the image-retrieved concentrations are shown in Figure 12.3.

The application of the MIM algorithm for the 2008 image and 2009 image had means of the absolute value of difference between the

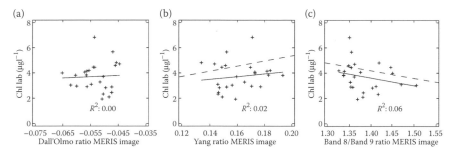

Figure 12.6 The ratio versus in situ chlorophyll *a* measurements for the three semiempirical algorithms for the 2009 image. In panels (b) and (c), the dashed line shows the line of best fit calculated from the 2008 image. The line is omitted in panel (a) as the model derived from the 2008 image was considered unreliable.

laboratory-measured concentrations and chlorophyll *a* retrieval, and those retrieved from the image of 1.66 µgl⁻¹ and of 0.78 µgl⁻¹, respectively. The 2008 value is not directly comparable to results of the semiempirical approaches. The results of the MIM inversion are far less directly dependent on the in situ chlorophyll *a* concentration measurements. In the case of the MIM, there are steps in the image-processing chain such as the atmospheric correction, air-water interface correction, and the weighting scheme used that are independent of the in situ–measured water quality parameters. In addition, the relationship that relates reflectance to the final parameter has been developed from a physical model of how the light interacts with the color-producing agents in the water. A better comparison may be to look at the standard error of a line of best fit between the in situ and image-retrieved chlorophyll *a* concentrations. For the 2008 image, this standard error is 1.41 µgl⁻¹.

The 2008 image inversion returned a physically impossible negative value for Station 4. Investigation of the whole image revealed that approximately 5% of the 1,727 pixels have returned physically impossible negative concentrations of chlorophyll *a*. Inspection of the original image showed that the negative concentrations coincided with those areas of the image in which the water appeared the darkest, suggesting the possibility of over-correction for the atmospheric effects or merely of low chlorophyll *a* and tripton concentration.

The semianalytical approach selects the most likely SIOP set. The selected SIOP set images are shown in Figure 12.7.

12.3.3 Discussion

The dynamic nature of the aquatic environment makes it difficult to establish the retrieval accuracy for water quality constituent concentrations

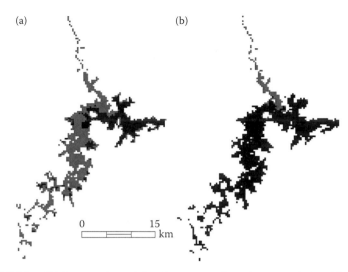

Figure 12.7 The SIOP set selected by the semianalytical approach for (a) the 2008 image and (b) the 2009 image. The upper-basin SIOP set is colored gray and the lower basin SIOP set is colored black.

because unquantifiable errors are associated with spatial and temporal patterns in the water as well as limitations on representative sampling. In addition, it is not possible to be definitive about the chlorophyll *a* concentration retrieval accuracy when there is notable uncertainty in the ground truth values. An attempt to quantify this uncertainty was made in the 2009 field campaign, which showed the mean difference between the two replicate groups of 0.3 μgl^{-1} (8%) and a standard deviation of 0.35 μgl^{-1} (9%), and the maximum difference was 1.36 μgl^{-1} (33%).

The measured laboratory concentrations come from a 10 L sample of water taken from the surface. It is not unreasonable to suspect that there may be some stratification of the water constituents. It has been shown for Lake Constance (Europe) that the assumption of a constant concentration of phytoplankton and suspended matter with depth meant the irradiance reflectance was underestimated by 12%–15% for the range of 2–5 μgl^{-1} of chlorophyll and 2–5 mgl^{-1} for suspended material (Albert and Mobley, 2003). This work did not indicate whether this difference was spectrally constant, but Kutser et al. (2008) did show that the simulated reflectance spectra varied in shape with changes in the vertical distribution of cyanobacteria. Griffiths and Faithful (1996) measured profiles of turbidity at three positions in the dam (corresponding to the stations BFD1 and BFD2 [river sites] and BFD7 [lake site]). They found that during periods of high river inflow, the greater turbidity values occurred in the lower part of the water column for the river sites, whereas the opposite was true for the lake site. After a long period of no flow, the stratification of turbidity disappeared in the river sites, but the turbidity slowly increased with depth at the lake site. The general applicability of these results to the optically active part of the water column can be questioned as that paper measured the turbidity at 2 meter (m) depth intervals and the maximum Secchi disk depth measured in October 2008 was 1.3 m.

12.3.3.1 In situ reflectance versus image reflectance

The direct comparison of chlorophyll a concentration retrieval algorithms that have been applied to in situ measurements and remote image pixel is difficult because of complications arising from the image correction, the difference in the instantaneous field of view of the sensors, adjacency effects, and the inherent noise in the image.

As 90% of the total radiance from a scene over a water body entering a sensor comes from the atmospheric path radiance (Vidot and Santer, 2005), the accuracy of the atmospheric correction has a significant effect on the final accuracies of estimated water quality parameter concentrations. The purpose of the atmospheric correction is to convert top of atmosphere (TOA) radiances into water-leaving radiance, then into above-water reflectance (R_{app}), and finally into the below–surface water reflectance $R(0^-)$. There are many ways to approach the correction procedure, but they all rely on a priori knowledge of the atmospheric properties, the water-leaving radiance, or some pseudo-invariant feature in the image. The approach used in this paper assumes that the reflectance of the DDV in three MERIS bands is known and that the selected aerosol model has the appropriate Ångström exponent, which describes the spectral dependence of AOT, as well as suitable absorption and scattering properties. Uncertainty in the DDV reflectance manifests in a consistent over- or underestimation of the water reflectance spectrum. This should have a limited effect in the case of the semiempirical algorithms, but it will affect the semianalytical algorithm. The spectral dependence of the path radiance conforms to a power law so an inappropriate aerosol model results in a spectrally variable error. This will affect both the semiempirical and semianalytical approaches. Another difficulty to consider is that of representative scale. The MERIS satellite spectra and subsequent inversion are the integration of a 290 m × 260 m area of water that is being compared with an in situ submeter radiometric measurement and 10 L sample of water. Kutser (2004) found variations in chlorophyll a concentrations within a MERIS-sized pixel of two orders of magnitude, and he attributed the errors in chlorophyll a concentration estimation in past studies to the patchiness of cyanobacterial blooms. Because the reflectance spectra are dependent on the concentration of the water quality parameters, the result can be extended to the within-pixel variation of the reflectance spectra. In addition, there is the potential that the satellite-measured spectra have been "contaminated" by other objects, such as the standing timber above and below the water level, within the pixel's point spread function.

Adjacency effects occur when atmospheric multiple scattering makes photons reflected from the area around the target pixel appear to be originating from the target pixel. This is particularly pronounced when the target pixel is much darker than the surrounding area and the aerosol loading in the atmosphere is high. Because of the size and shape of the target water body and inland waters in general, a substantial proportion of pixels can be contaminated by the adjacency effect. In the case of Burdekin Falls Dam, however, the gentle surrounding topography and low aerosol loading that typifies the site was likely to keep any adjacency effect to a minimum. Even if it is assumed that the adjacency effect has been fully accounted for, the MERIS image pixel still represents an average spectrum for an area of 290 m × 260 m. The environmental noise-equivalent reflectance difference $(NE\Delta R(0^-)_E)$ is a measure of the inherent noise in an image and is calculated as the standard deviation of the subsurface reflectance in each band over a homogeneous area of optically deep water (Brando and Dekker, 2003). Using a MERIS full-resolution image acquired at another Australian inland water body on July 2, 2007, corrected using c-WOMBAT-c (Brando and Dekker, 2003) $(NE\Delta R(0^-)_E)$ was estimated to be a constant 0.1% in all bands.

Previous work with semiempirical approaches has found that the largest relative errors in chlorophyll a concentration retrieval occur below 20 µgl⁻¹ (Dall'Olmo and Gitelson, 2005; Gitelson et al., 2008; Le et al., 2009). In this case study, the range of water quality parameter concentrations measured in the two field campaigns was limited to low values. For example, the measured in situ chlorophyll a values ranged from 1.8–7.7 µgl⁻¹, but 80% of the values were within the range of 2.7–5.5 µgl⁻¹. Therefore, it might be said that the results describe the application of the approach at a suboptimal level. One of the most immediate concerns for water quality managers, however, is the ability to monitor for harmful algal blooms. In the Australian context, it has been asserted that a practical remote-sensing system must be able to measure accurately to below 10 µgl⁻¹ in turbid waters to be useful in prebloom conditions (Jupp, Kirk, and Harris, 1994). Moreover, the application of any empirical model should be restricted to the range of measurements over which the model was developed. That means to parameterize an operational algorithm, reflectance spectra need to be sourced over sufficient ranges of the three independently varying color-producing agents in the water.

12.3.3.2 Algorithm assumptions

Dall'Olmo and Gitelson (2005) made six explicit assumptions related to their three-band algorithm. The assumptions are related to the IOPs of the color-producing agents in the water. The parameterization of the semi-analytical method relied on a Hydrolight® simulation made with varying

Sun elevation angles and concentrations of the water quality parameters (Chlorophyll a 0–20 μgl^{-1}, tripton 0–20 mgl^{-1}, and CDOM 0–1.6 m^{-1}). Theses simulations for the Sun elevation closest to the 2008 image were used to examine the validity of the assumptions in relation to the Burdekin Falls Dam IOP sets.

1. The ratio of the proportionality factor (f) to the radiance-to-irradiance conversion factor (Q) is spectrally invariant in the range of 650–750 nm. The maximum variation of the f/Q ratio between λ_1 and λ_3 was 8.5% with a mean of 3.8% and a standard deviation of 2.2%.
2. The sum of the absorption due to CDOM and tripton is approximately equal at λ_1 and λ_2. The absorption due to CDOM and tripton at λ_1 was on average 25% larger than the absorption due to CDOM and tripton at λ_2 for the upper-basin SIOP set and 30% greater for the lower-basin SIOP set.
3. The absorption due to phytoplankton is much greater at λ_1 than λ_2. In both cases, the absorption due to phytoplankton is actually 20% larger at λ_1 than λ_2.
4. The absorption due to water at λ_3 is much larger than the absorption due to the other color-producing agents. In both cases, the absorption due to the other color-producing agents is only 6% of absorption due to water at λ_3.
5. The backscattering due to the color-producing agents is constant across all three wavelengths. The backscattering in this part of the spectrum is almost exclusively due to tripton, and the variation between λ_1 and λ_3 is approximately 9%.
6. The absorption due to all the color-producing agents at λ_3 is much greater than the backscattering due to those same agents. The ratio of backscattering to absorption at λ_3 is linearly related to the tripton concentration. At the highest tripton concentration, the backscattering is as much as 60% of the absorption for the upper-basin SIOP set and 90% of the absorption for lower-basin SIOP set.

On the basis of this example, it can be said that the first, fourth, and fifth assumptions are appropriate for Burdekin Falls Dam, and the second assumption is not grossly inappropriate. The second assumption is dependent of the chlorophyll a–specific absorption of phytoplankton in the dam (see Figure 12.8). The absorption maxima observed (ca. 630 nm) indicate that the phytoplankton assemblages in Burdekin Falls Dam are dominated by cyanobacteria (Richardson, 1996). The expected absorption peak due to chlorophyll a is distorted by other accessory pigments present in the phytoplankton. Moving λ_2 to Band 9 (708.75 nm nominal center), to replicate the MERIS band combination used by Gitelson et al. (2009), made this assumption appropriate, but it made no discernable difference in the result.

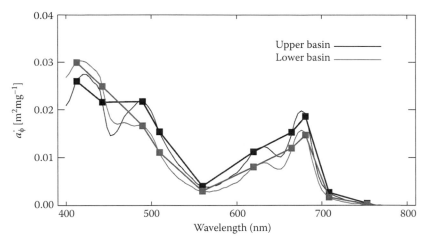

Figure 12.8 The chlorophyll *a*–specific absorption spectra of phytoplankton convolved onto the MERIS spectral bands for the upper basin (black) and lower basin (gray) SIOP sets.

That the backscattering at λ_3 is much less that the absorption is the assumption that is least applicable to the water in Burdekin Falls Dam. Like the waters investigated by Le et al. (2009) and Yang et al. (2010), the backscattering of tripton is not negligible in the NIR in Burdekin Falls Dam. Their modification of the original algorithm delivered an enhanced correlation when using the MERIS bands for the 2008 image (see Figure 12.5), but it had a negligible effect on the 2009 image (see Figure 12.6).

The two band ratio is similar to the ratio used by Gons (1999). It relies on the large change in the chlorophyll *a*–specific absorption of the phytoplankton between two closely placed bands. The short spectral distance between the bands means that the absorption and scattering due to the other water quality parameters is close to constant. Gons found that the result was improved when backscattering was not presumed to be constant but rather was derived from a wavelength in the NIR. To achieve this, he made assumptions in relation to the proportionality factor and the radiance-to-irradiance conversion factor.

For the latter two semiempirical approaches, the algorithms perform effectively on the 2008 image. There is, however, no discernable correlation between the ratios and the chlorophyll *a* concentration for the 2009 image. Care should be exercised when considering the significance of this result. Unlike the 2008 image, it was not possible to have an independent verification of the accuracy of the atmospheric correction. As mentioned, the correction is tuned by adjusting the AOT value until the reflectance of the DDV around the water body matches as close as possible to an a priori value. The errors associated with the atmospheric correction distort the magnitude and shape of the corrected spectrum. An incorrect estimate of the AOT will result in a corrected spectrum that is over- or underestimated by a uniform amount. Because the semiempirical methods rely on ratios, rather than absolute values, the effect is likely to be negligible. As the spectral dependence of the path radiance conforms to a power law, however, a poor choice of the aerosol types or their mixing ratio will result in a corrected spectrum that is "tilted." The reflectance of the DDV is the most stable at the blue end of the spectrum, so the comparison is made at that end of the spectrum. A poor choice of aerosol model will not necessarily be detected. An error of this type should have a greater effect in the area of the spectrum that the semiempirical methods utilize. The effect will be felt more if the bands being used are widely spaced, however, so it should have a limited effect on the latter two approaches.

One of the implicit assumptions of all of the approaches is that the SIOP of the water quality constituents does not change between the parameterization and the implementation. The semianalytical approach allows multiple SIOP sets to be used in the retrieval. Figure 12.7 shows the SIOP set used by the algorithm based on an optical closure criterion. It shows that in the parameterization image, the measurement stations are dominated by the upper-basin SIOP set. In contrast, the 2009 image, taken at a different time of the year, shows the water to be dominated by the lower-basin SIOP set. It would be expected that this would affect the accuracy of the retrieval to some degree. This may be why in Figure 12.6b and 12.6c using the previous line of best fit would have resulted in an overestimation of chlorophyll a. Nevertheless, it is difficult to see a reason why the different SIOP should have such an effect on the coefficient of correlation between the ratios and the in situ chlorophyll a measurements. The second assumption required that the sum of the absorption due to CDOM and tripton is approximately equal at λ_1 and λ_2. The SIOP measurements show that this assumption was less appropriate for the lower-basin SIOP set. The predominance of the lower-basin SIOP set in the water body at the time of the 2009 image may be a concern; but, if anything, the effect of tripton should be smaller in the 2009 image because the in situ–measured tripton values and the values retrieved by the semianalytical approach show lower concentrations of tripton with respect to 2008. The main difference between the two SIOP sets is that the tripton in the lower basin exhibited both a greater tripton-specific backscattering value as well as a greater spectral slope (Campbell, Phinn, and Daniel, 2011). The semiempirical algorithms assume that the backscattering is constant or at least spectrally flat. The SIOP sets were sampled in Burdekin Falls Dam over 3 days, before the start of the annual wet season. The fieldwork in 2008 found that there was sufficient intraimpoundment variation in the specific absorption and specific scattering of phytoplankton and tripton to require a well-distributed network of measurement stations to fully characterize the SIOPs of the water body. There is no reason to assume that the SIOPs of the optical water quality parameters vary only spatially, and the tripton scattering does not have a seasonal variation as well. The best explanation for poor performance of the Dall'Olmo et al. (2003) algorithm was the high tripton backscattering value in the NIR. If the endemic tripton backscattering had a greater spectral slope in 2009 than in 2008, the failure of the constant backscattering assumption may explain why the latter two semiempirical approaches do not work as well for the 2009 image.

12.4 CONCLUSION AND FUTURE WORK

The measurement of water quality parameters for inland waters from remotely sensed images for monitoring cannot be achieved by purely empirical or analytical methods. The semiempirical approach is mathematically simple and requires unsophisticated processing. In contrast, the semianalytical approach relies on modeling the interaction of the light field with the optical properties of the water. The fact that the anisotropy factor is related to the SIOP values of the water quality parameters means that a very large computational overhead is imposed every time the demonstrated MIM approach is used to target a new environment. Although this is inconvenient, it does mean that a smaller set of measurements can be leveraged by simulations to parameterize the algorithm for all the possible water quality parameter combinations. In the case of a semiempirical approach, the parameterization measurement set needs to be broad enough to encompass not only the full range of possible chlorophyll *a* values but also the full range of values for the other color-producing agents in the water. It is unlikely that this breadth of data will be able to be obtained without multiple measurement campaigns and coincident satellite images. If interest in the water body extends beyond chlorophyll *a* concentration—as a proxy for algal distribution—to sediment and dissolved organic matter, semianalytical approaches have the advantage of simultaneously delivering concentrations of three water quality parameters.

Whether the adopted approach is semiempirical or semianalytical, limitations are imposed when the approach is used for monitoring rather than a single application. Using the semiempirical approach, it is possible to extend a limited number of in situ measurements to a synoptic view of the water body without necessarily having to correct the image for the atmospheric effects. This is possible if the assumption is made that the effects are constant over the spatial scale of the water body. To parameterize the model with more than one image or to apply it over time, however, it is necessary to correct the images before applying the algorithm. The atmospheric correction of images of case-II waters has been, and continues to be, a difficult problem because of the complexity of the reflectance and water quality parameter relationship. Any long-term or archival monitoring project needs to come to terms with how the images can be corrected in the absence of in situ data. The image-based approaches all rely on the assumption that there is some characteristic of the water reflectance spectrum that is known or invariant. A limit on developing generally applicable methods is the variation in catchment soil, land cover and land use conditions, and impoundment substrate mineralogy that result in significantly different SIOP sets for inland waters (Campbell, Phinn, and Daniel, 2011). Another implicit assumption on long-term monitoring of a water body is the necessity to assume that SIOPs are constant over the monitoring period and that the parameterization of the model has taken into account the full variation of these values. The validity of this assumption will be site specific, but for water bodies that have significant development and land use changes in their catchment, it will be the most tentative.

Notwithstanding all the site-specific limitations, the fact that the inversion problem for inland water remote sensing is ill posed limits its reliability (Defoin-Platel and Chami, 2007). The solution for the inverse problem is ambiguous because multiple combinations of water quality parameter concentrations can lead to the same or very similar reflectance spectra. When random measurement noise is superimposed on this already ill-posed problem, then the retrieval uncertainties are exacerbated.

This chapter has given an overview of methods used to retrieve the chlorophyll *a* concentration in inland waters. It applied a number of semiempirical algorithms and a semianalytical approach to a tropical inland water impoundment, Burdekin Falls Dam. It found that all the semiempirical algorithms could be successfully applied to in situ radiometric observations, but the Dall'Olmo et al. (2003) and Gitelson et al. (2008) approaches failed when applied to simulated MERIS bands and a MERIS image. The other two semiempirical approaches were successfully applied to the 2008 MERIS image, but they failed when applied to another MERIS image. The semianalytical approach resulted in a less accurate retrieval of chlorophyll *a* from the 2008 image but managed to successfully invert the 2009 image.

ACKNOWLEDGMENTS

I would like to thank the *Commonwealth Scientific and Industrial Research Organisation* Water for a Healthy Country Flagship for supporting the fieldwork and the laboratory work. In particular, I would like to thank Professor Arnold Dekker and Dr. Vittorio Brando. I would like to thank Professor Stuart Phinn of Queensland University for reviewing the manuscript. I would like to thank the European Space Agency for providing the MERIS FR images under the AO595 agreement. I would like to express our gratitude to the water-body operators Sunwater for their assistance in the field operations.

REFERENCES

Albert, A., and C. D. Mobley. 2003. An analytical model for subsurface irradiance and remote sensing reflectance in deep and shallow case-2 waters. *Optics Express* 11(22): 2873–2890.

Arar, E. J., and G. B. Collins. 1997. *Method 445.0, In Vitro Determination of Chlorophyll a and Pheophytin a in Marine and Freshwater Algae by Fluorescence.* Revision 1.2 edition. Cincinnati, OH: National Exposure Research Laboratory, Office of Research and Development, U.S. Environmental Protection Agency.

Australian State of the Environment Committee. 2001. *Australia, State of the Environment 2001 Inland Waters Theme Report.* Melbourne: CSIRO Publishing.

Baruah, P. J., M. Tamura, K. Oki, and H. Nishimura. 2001. Neural network modeling of lake surface chlorophyll and suspended sediment from Landsat TM imagery. Paper read at 22nd Asian Conference of Remote Sensing, November 5–9, Singapore.

Brando, V. E., and A. G. Dekker. 2003. Satellite hyperspectral remote sensing for estimating estuarine and coastal water quality. *IEEE Transactions on Geoscience and Remote Sensing* 41(6): 1378–1387.

Bukata, R. P. 2005. *Satellite Monitoring of Inland and Coastal Water Quality: Retrospection, Introspection, Future Direction.* Boca Raton, FL: Taylor & Francis.

Campbell, G., and S. R. Phinn. 2010. An assessment of the accuracy and precision of water quality parameters retrieved with the Matrix Inversion Method. *Limnology and Oceanography Methods* 8: 16–29.

Campbell, G., S. R. Phinn, and P. Daniel. 2011a. The specific inherent optical properties of three sub-tropical and tropical water reservoirs in Queensland, Australia. *Hydrobiologia* 658(1): 233–252.

Campbell, G., S. R. Phinn, A. G. Dekker, and V. E. Brando. 2011b. Remote sensing of water quality in an Australian tropical freshwater impoundment using matrix inversion and MERIS images. *Remote Sensing of Environment* 115(9): 2402–2414.

Chorus, I., I. R. Falconer, H. J. Salas, and J. Bartram. 2003. Health risks caused by freshwater cyanobacteria in recreational waters. *Journal of Toxicology and Environmental Health Part B: Critical Reviews* 3(4): 323–347.

Cole, J. J., Y. T. Prairie, N. F. Caraco, W. H. McDowell, L. J. Tranvik, R. G. Striegl, C. M. Duarte, et al. 2007. Plumbing the global carbon cycle: integrating inland waters into the terrestrial carbon budget. *Ecosystems* 10(1): 171–184.

Dall'Olmo, G., and A. A. Gitelson. 2005. Effect of bio-optical parameter variability on the remote estimation of chlorophyll-a concentration in turbid productive waters: experimental results. *Applied Optics* 44(3): 412–422.

Dall'Olmo, G., A. A. Gitelson, and D. C. Rundquist. 2003. Towards a unified approach for remote estimation of chlorophyll-a in both terrestrial vegetation and turbid productive waters. *Geophysical Research Letters* 30(18): 1938.

Defoin-Platel, M., and M. Chami. 2007. How ambiguous is the inverse problem of ocean color in coastal waters? *Journal of Geophysical Research-Oceans* 112(C3): C03004.

Dekker, A. G. 1993. Detection of optical water quality parameters for eutrophic waters by high resolution remote sensing. PhD diss., Vrije Universiteit, Amsterdam, The Netherlands.

Doerffer, R., and H. Schiller. 2007. The MERIS Case 2 water algorithm. *International Journal of Remote Sensing* 28(3): 517–535.

Fraser, R. N. 1998. Hyperspectral remote sensing of turbidity and chlorophyll a among Nebraska Sand Hills lakes. *International Journal of Remote Sensing* 19(8): 1579–1589.

Garnett, C., G. Shaw, D. Moore, P. Florian, and M. Moore. 2003. *Impact of Climate Change on Toxic Cyanbacterial (Blue Green Algal) Blooms and Algal Toxin Production in Queensland*. Queensland: National Research Centre for Environmental Toxicology.

George, D. G., and T. J. Malthus. 2001. Using a compact airborne spectrographic imager to monitor phytoplankton biomass in a series of lakes in north Wales. *The Science of the Total Environment* 268(1/3): 215–226.

Giardino, C., G. Candiani, and E. Zilioli. 2005. Detecting chlorophyll-a in Lake Garda using TOA MERIS radiances. *Photogrammetric Engineering and Remote Sensing* 71(9): 1045–1051.

Gitelson, A. A., G. Dall'Olmo, W. Moses, D. C. Rundquist, T. Barrow, T. R. Fisher, D. Gurlin, and J. Holz. 2008. A simple semi-analytical model for remote estimation of chlorophyll-a in turbid waters: validation. *Remote Sensing of Environment* 112(9): 3582–3593.

Gitelson, A. A, D. Gurlin, W. J. Moses, and T. Barrow. 2009. A bio-optical algorithm for the remote estimation of the chlorophyll- a concentration in case 2 waters. *Environmental Research Letters* 4(4): 045003.

Gons, H. J. 1999. Optical teledetection of chlorophyll a in turbid inland water. *Environmental Science & Technology* 33(7): 1127–1133.

Gordon, H. R., O. B. Brown, and M. M. Jacobs. 1975. Computed relationships between the inherent and apparent optical properties of a flat homogeneous ocean. *Applied Optics* 14(2): 417–427.

Gower, J. F. R., R. Doerffer, and G. A. Borstad. 1999. Interpretation of the 685nm peak in water-leaving radiance spectra in terms of fluorescence, absorption and scattering, and its observation by MERIS. *International Journal of Remote Sensing* 20(9): 1771–1786.

Griffiths, D. J., and J. W. Faithful. 1996. Effects of the sediment load of a tropical North-Australian river on water column characteristics in the receiving impoundment. *Archive für Hydrobiolie (Supplement 113), Large Rivers* 10(1/4): 147–157.

Hoge, F. E., and P. E. Lyon. 1996. Satellite retrieval of inherent optical properties by linear matrix inversion of oceanic radiance models: an analysis of model and radiance measurement errors. *Journal of Geophysical Research* 101(C7): 16631–16648.

Hoogenboom, H. J., A. G. Dekker, and J. F. de Haan. 1998. Retrieval of chlorophyll and suspended matter from imaging spectrometry data by matrix inversion. *Canadian Journal of Remote Sensing* 24(2): 144–152.

Hotzel, G., and R. Croome. 1999. *A Phytoplankton Methods Manual for Australian Freshwaters*. Canberra: Land and Water Resources Research and Development Corporation.

Jupp, D. L. B., J. T. O. Kirk, and G. P. Harris. 1994. Detection, identification and mapping of cyanobacteria-using remote sensing to measure the optical quality of turbid inland waters. *Australian Journal of Marine and Freshwater Research* 45(5): 801–828.

Kaufman, Y. J., and D. Tanre. 1992. Atmospherically resistant vegetation index (ARVI) for EOS-MODIS. *IEEE Transactions on Geoscience and Remote Sensing* 30(2): 261–270.

Keller, P. A. 2001. Comparison of two inversion techniques of a semi-analytical model for the determination of lake water constituents using imaging spectrometry data. *The Science of the Total Environment* 268(1/3): 189–196.

Koponen, S., J. Pulliainen, H. Servomaa, Y. Zhang, M. Hallikainen, K. Kallio, J. Vepsalainen, T. Pyhalahti, and T. Hannonen. 2001. Analysis on the feasibility of multi-source remote sensing observations for chl-a monitoring in Finnish lakes. *The Science of the Total Environment* 268(1/3): 95–106.

Kosten, S., F. Roland, D. Marques, E. H. Van Nes, N. Mazzeo, L. D. L. Sternberg, M. Scheffer, and J. J. Cole. 2010. Climate-dependent CO_2 emissions from lakes. *Global Biogeochemical Cycles* 24: GB2007.

Kutser, T. 2004. Quantitative detection of chlorophyll in cyanobacterial blooms by satellite remote sensing. *Limnology and Oceanography* 49(6): 2179–2189.

Kutser, T., L. Metsamaa, and A. G. Dekker. 2008. Influence of the vertical distribution of cyanobacteria in the water column on the remote sensing signal. *Estuarine Coastal and Shelf Science* 78(4): 649–654.

Le, C. F., Y. M. Li, Y. Zha, D. Y. Sun, C. C. Huang, and H. Lu. 2009. A four-band semi-analytical model for estimating chlorophyll a in highly turbid lakes: the case of Taihu Lake, China. *Remote Sensing of Environment* 113(6): 1175–1182.

Lee, Z. P., K. L. Carder, and R. A. Arnone. 2002. Deriving inherent optical properties from water color: a multiband quasi-analytical algorithm for optically deep waters. *Applied Optics* 41(27): 5755–5772.

Leroy, M., V. Bruniquel-Pinel, O. Hautecoeur, F. M. Bréon, and F. Baret. 1998. Corrections atmosphériques des données MERIS/ENVISAT: Caractérisations de la BRDF de surfaces "sombres." European Space Agency final report. Noordwijk, the Netherlands: ESA Communications.

Maritorena, S., D. A. Siegel, and A. R. Peterson. 2002. Optimization of a semianalytical ocean color model for global-scale applications. *Applied Optics* 41(15): 2705–2714.

Matarrese, R., M. T. Chiaradia, V. De Pasquale, and G. Pasquariello. 2004. Chlorophyll-a concentration measure in coastal waters using MERIS and MODIS data. Paper read at IGARSS 2004. *IEEE International Geoscience and Remote Sensing Symposium Proceedings*. Vols. 1–7of Science for Society: Exploring and Managing a Changing Planet, 3639–3641. New York: IEEE.

Matthews, M. W., S. Bernard, and K. Winter. 2010. Remote sensing of cyanobacteria-dominant algal blooms and water quality parameters in Zeekoevlei, a small hypertrophic lake, using MERIS. *Remote Sensing of Environment* 114(9): 2070–2087.

Mittenzwey, K. H., A. A. Gitelson, and K. Y. Kondratiev. 1992. Determination of Chlorophyll-a of Inland Waters on the Basis of Spectral Reflectance. *Limnology and Oceanography* 37(1): 147–149.

Mobley, C. D., L. Sundman, C. O. Davis, J. H. Bowles, T. V. Downes, R. A. Leathers, M. J. Montes, et al. 2005. Interpretation of hyperspectral remote-sensing imagery by spectrum matching and look-up tables. *Applied Optics* 44(17): 3576–3592.

Morel, A. 1974. Optical properties of pure water and pure seawater. In *Optical Aspects of Oceanography*, edited by N. G. Jerlov and E. Steeman Nielsen. London: Academic Press Inc.

Morel, A., and L. Prieur. 1977. Analysis of variations in ocean color. *Limnology and Oceanography* 22(4): 709–722.

O'Reilly, J. E., S. Maritorena, B. G. Mitchell, D. A. Siegel, K. L. Carder, S. A. Garver, M. Kahru, et al. 1998. Ocean color chlorophyll algorithms for SeaWiFS. *Journal of Geophysical Research* 103(C11): 24937–24953.

Pope, R. M., and E. S. Fry. 1997. Absorption spectrum (380 -700 nm) of pure water: integrating cavity measurements. *Applied Optics* 36(33): 8710–8723.

Richardson, L. L. 1996. Remote sensing of algal bloom dynamics. *BioScience* 46(7): 492–501.

Rijkeboer, M., A. G. Dekker, and H. J. Gons. 1997. Subsurface irradiance reflectance spectra of inland waters differing in morphometry and hydrology. *Aquatic Ecology* 31(3): 313–323.

Santini, F., L. Alberotanza, R. M. Cavalli, and S. Pignatti. 2010. A two-step optimization procedure for assessing water constituent concentrations by hyperspectral remote sensing techniques: an application to the highly turbid Venice lagoon waters. *Remote Sensing of Environment* 114(4): 887–898.

Schaale, M., J. Fischer, and C. Olbert. 1998. Quantitative estimation of substances contained in inland water from multispectral airborne measurements by neural networks. Paper read at ASPRS-RTI Annual Conference, 30 March–3 April 1998, Tampa, FL.

Schalles, J. F., A. A. Gitelson, Y. Z. Yacobi, and A. E. Kroenke. 1998. Estimation of chlorophyll a from time series measurements of high spectral resolution reflectance in an eutrophic lake. *Journal of Phycology* 34(2): 383–390.

Smith, R. C., and K. S. Baker. 1981. Optical properties of the clearest natural waters (200–800 nm). *Applied Optics* 20(2): 177–184.

Su, F. C., C. R. Ho, Q. Zheng, N. J. Kuo, and C. T. Chen. 2006. Satellite chlorophyll retrievals with a bipartite artificial neural network model. *International Journal of Remote Sensing* 27(8): 1563–1579.

Thuillier, G., M. Hersé, D. Labs, T. Foujols, W. Peetermans, D. Gillotay, P. C. Simon, et al. 2003. The solar spectral irradiance from 200 to 2400 nm as measured by the SOLSPEC spectrometer from the Atlas and Eureca missions. *Solar Physics* 214(1): 1–22.

Vidot, J., and R. Santer. 2005. Atmospheric correction for inland waters: application to SeaWiFS. *International Journal of Remote Sensing* 26(17): 3663–3682.

Vos, R. J., J. H. M. Hakvoort, R. R. W. Jordans, and B. W. Ibelings. 2003. Multiplatform optical monitoring of eutrophication in temporally and spatially variable lakes. *The Science of the Total Environment* 312(1/3): 221–243.

Yacobi, Y. Z., A. Gitelson, and M. Mayo. 1995. Remote sensing of chlorophyll in Lake Kinneret using high-spectral-resolution radiometer and Landsat TM: spectral features of reflectance and algorithm development. *Journal of Plankton Research* 17(11): 2155–2173.

Yang, W., B. Matsushita, J. Chen, T. Fukushima, and R. Ma. 2010. An Enhanced Three-Band Index for Estimating Chlorophyll-a in Turbid Case-II Waters: case Studies of Lake Kasumigaura, Japan, and Lake Dianchi, China. *IEEE Geoscience and Remote Sensing Letters* 7(4): 655–659.

Chapter 13

Advanced geospatial techniques for mapping and monitoring invasive species

Le Wang and Amy E. Frazier

CONTENTS

In this chapter, we examine the recent trends in remote sensing for mapping and monitoring species invasions. With the rapid pace of globalization, plant and animal species are being introduced into new territories at an alarming rate. Many of these species not only have devastating impacts on the ecosystems in which they are introduced but also generate huge economic losses through decreased ecosystem services and costly eradication measures. Tackling invasions can be challenging because of their dynamic nature and large spatial extents. Aerospace imagery has proven to be a key tool for mapping, monitoring, and eradicating these nuisance species. This chapter introduces the problem of invasive species and the need for early detection and widespread monitoring through remote sensing. We then explore a case study of saltcedar (*Tamarix* spp.) to review the most advanced methods for discerning invasive species from both high- and medium-resolution data sets, discuss the development of a novel classification technique that is particularly well suited for discriminating the species-level variation needed to discern invasions, and conclude with a discussion on how to interpret these types of advanced remote-sensing classifications to derive useful and pertinent ecological information that can be applied to target eradication efforts.

13.1 INTRODUCTION

Amid a rapidly globalizing world, the movement of people through countries and across continents is often accompanied, both purposefully and unintentionally, by the coincidental movement of plants and animals. As exotic species are released into new territories, they can quickly establish and become permanent fixtures of the landscape. Sometimes these alien species are hailed for their economic or aesthetic benefits (e.g., much of the food we eat comes from introduced species), but often times they have dire consequences for the environment into which they are released (Sax et al., 2005). For example, there are more than 2,000 species of alien plants in the continental United States (Vitousek et al., 1997) such as European cheatgrass (*Bromustectorum*), which has disrupted the fire regime in the intermontane West, causing irreversible ecological damages. Similar examples are common, yet despite firsthand knowledge of the negative consequences of releasing alien species into new territories, globalization has prevailed, and introductions continue to occur at an alarming rate (Vitousek et al., 1997). Not only do invasive species threaten environmental systems and ecosystem health by disrupting natural landscape patterns and interfering with ecosystem processes (Dukes and Mooney, 2004), but they also cause significant biological and economic losses and indirectly affect global climate change (Vitousek et al., 1997).

Early detection is one of the most crucial steps for effective management of invasive species. Detecting the presence of a population in the initial stages of an invasion can reduce control costs and increase the likelihood of successful eradication (Rejmánek and Pitcairn, 2002). The areal extent of an invasion can cover multiple states or regions, even during the initial stages, and tools are needed to track species progression across wide areas. Remote-sensing technology is well suited for monitoring these considerable spatial extents and has proven successful at detecting a wide array of invasive species in a variety of environments, including forests, rangeland, and grasslands (Lass et al., 2005). The earliest uses of remote sensing to study invasive species were based on aerial photographs and relied on manual interpretation of visual differences between species. Aerial photography was limited by the small extent that could be flown by plane, high costs, and the significant manpower needed to interpret and characterize the photographs. With the advent of multispectral satellite platforms, it became possible to record the intensity of reflected light in multiple wavelengths, including those outside the visual range. Furthermore, as technology advanced to the point at which multispectral images could be processed digitally, the power to detect invasive species from satellite remote sensing rapidly surpassed manual photograph interpretation (Lass et al., 2005).

Since the advent of spaceborne sensors, geospatial technologies have become pivotal in the study of plant invasions. In this chapter, we discuss how advanced remote-sensing imagery and classification techniques are becoming even more vital for mapping and managing species invasions as the rate of introductions increases worldwide. Specifically, we explore the use of aerospace imagery to investigate the invasion of saltcedar (*Tamarix* spp.), a weedy shrub that was introduced to the United States from Europe and Asia in the mid-nineteenth century. Saltcedar is one of two dominant invasive vegetation species in the southwestern United States (Reynolds and Cooper, 2010) and provides an excellent example of the role remote-sensing data and geospatial techniques have played in managing ecosystem operations and understanding the invasion process (Hunt et al., 2003; Lass et al., 2005). We begin by providing a brief introduction to the ecology of saltcedar and its impacts on environmental conditions. We then examine how various types of high-spatial and hyperspectral imagery can be integrated to differentiate saltcedar from other riparian vegetation species in the Rio Grande basin. Because these types of detailed data sets are not always readily available, in the following section, we discuss the application of a relatively new classification technique for detecting invasions using medium spatial and spectral resolution data sets. In the final section, we introduce a new method for interpreting ecological data acquired from aerospace imagery that utilizes geospatial tools, such as landscape metrics, to examine the spatial patterns of invasions. We conclude by highlighting the importance of developing accurate and easy-to-implement interpretation techniques to analyze data derived from aerospace imagery in an ecological context to support multidisciplinary efforts for solving invasion problems.

13.2 MAPPING AND MONITORING THE REGIONAL INVASION OF SALTCEDAR

Saltcedar (*Tamarix* spp.) is the common name given to eight species of *Tamarix* that were introduced to the United States to provide windbreaks and erosion control along riverbanks (Baum, 1967). Saltcedar quickly escaped cultivation and has become one of the worst invasive vegetation species in the United States (Morisette et al., 2006) and a great threat to western riparian ecosystems (Dudley and DeLoach, 2004). The plants outcompete native vegetation species for water and other resources but offer fewer foraging, habitat, and food source benefits to wildlife (Everitt and

DeLoach, 1990). Saltcedar's current spatial extent is estimated to exceed 600,000 hectares (ha; 1.5 million acres) in the southwestern United States, but its range has been expanding southward into northwestern Mexico and northward into Montana and Canada (Pearce and Smith, 2003), even reaching elevations as high as 2,135 meters (m) in the Rocky Mountains (Everitt, 1980).

Described as an "opportunistic colonizer" (Everitt, 1998), saltcedar is able to quickly establish itself in new areas. It rapidly inhabits stream banks by producing an abundance of small, light seeds (up to 0.5 million per plant) that are easily disseminated by wind and water (DiTomaso, 1998). Seed germination typically occurs in bare, moist, exposed areas commonly found at high floodwater marks where the ground is saturated for an extended period (Everitt, 1980). As floodwaters recede, germination penetrates into the waterway, slowly narrowing the stream channel and choking water flow (DiTomaso, 1998). This channel narrowing has many distressing hydrological implications. It typically causes depletion of water flow, a lowering of the water table, an increase in sediment production, and an increase in the occurrence of flooding, which subsequently promotes germination of another generation of plants (DiTomaso, 1998; Smith et al., 1998). In addition to altering stream morphology, saltcedar plants consume large amounts of water, which can contribute to decreased water availability and low stream flow. Estimates vary based on location, maturity, density, and depth to groundwater, but water consumption by individual plants has been measured as high as 760 liters per day (L/day) (DiTomaso, 1998).

Saltcedar plants can quickly form dense stands, precluding understory vegetation and outcompeting native plants for sunlight, moisture, and nutrients (Everitt et al., 1996; Everitt et al. 2007). It is widely acknowledged that saltcedar is replacing native vegetation species along major rivers in the U.S. southwest at an alarming rate (Dudley and DeLoach, 2004; Glenn and Nagler, 2005; Nagler et al., 2009). It is depleting already-scarce water supplies, endangering fish, and diminishing recreational opportunities for human enjoyment (Tamarisk Coalition, 2003). With annual economic losses estimated at $133–$285 million (Zavaleta, 2000), tackling the salt-cedar invasion issue is a high priority for riparian ecosystem management (Hultine et al., 2010).

Our study focuses on a stretch of the Rio Grande in southwestern Texas known as the Forgotten River Reach (FRR) where the progressive replace-ment of the native floodplain vegetation by saltcedar is of particular concern (U.S. Army Corps of Engineers, 2008). Along this section of the Rio Grande (see Figure 13.1), saltcedar has overtaken riverbanks, and the native cotton-wood (*Populus* spp.) plants that once dominated the area are completely absent (Everitt, 1998). The vegetation on both sides of the river is currently composed mostly of a mixture of saltcedar and willow (*Salix* spp.), with mesquite (*Prosopis* spp.) occupying the uplands. The spatial distribution of

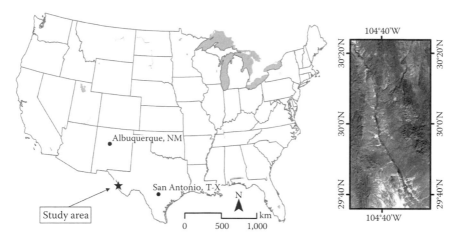

Figure 13.1 Location of the Forgotten River Reach (FRR) study site along the Rio Grande in southwest Texas with the cropped Landsat image showing the vegetated riparian buffer that makes up the area of investigation.

saltcedar along the FRR varies, partly because of differences in the local hydrologic system (Silván-Cárdenas and Wang, 2010), and we will explore how those distributions are being monitored using aerospace imagery.

13.2.1 Detecting invasions using high-resolution satellite imagery

Early detection of saltcedar is one of the most crucial components for mapping and monitoring its spread. Exposing a population during the initial stages of establishment, before plants have had the opportunity to expand and cause irreparable ecosystem damage, can significantly improve the likelihood of successful eradication. During this nascent stage, plant density is typically low because saltcedar is a "slow starter" (Everitt, 1980, 80), and plants prefer to establish in open, uncrowded areas. These early plants do not typically form dense stands but rather grow as part of the heterogeneous mixture of other vegetation species. Therefore, they may contribute only a limited number of plants to the overall mix. These beginning stages of the invasion are difficult to detect from medium and coarse resolution platforms, such as Landsat Thematic Mapper ™ and the Moderate Resolution Imaging Spectroradiometer (MODIS), because of the disparity between pixel and object sizes. Traditional pixel-based classification methods are not sufficient for detecting the early presence of saltcedar because it is unlikely that saltcedar will constitute the majority of any image pixel. Therefore, the beginning stages of an invasion can be completely overlooked when using medium and coarse resolution imagery.

In addition to spatial resolution, the spectral resolution of a sensor also plays an important role in detecting invasions. Multispectral platforms, such as Landsat TM, collect data in a limited number of discrete bands, making it difficult to distinguish fine variations in the spectral signatures of plant species. Unlike general land cover studies that differentiate broad classes (e.g., forest, wetlands, grasslands, etc.), vegetation must be discerned at the species level to map an invasion. This level of detection usually is not possible with multispectral platforms, especially if cooccurring or competing species have similar phenological traits or spectral signatures. In contrast, hyperspectral sensors are capable of capturing a broader spectrum of reflectance and thus can add to the power of species discrimination. To date, many invasive species have been successfully detected with hyperspectral sensors, including iceplant (*Carpobrotusedulis*), jubata grass (*Cortaderiajubata*) (Underwood et al., 2003), leafy spurge (O'Neill, 2000; William and Hunt, 2002; Glenn et al., 2005), Brazilian pepper (Lass and Prather, 2004), spotted knapweed, and baby's breath (Lass et al., 2005). As far as saltcedar is concerned, however, a systematic evaluation of the respective potential of emerging high spatial and hyperspectral resolution data sets will aid in ecosystem management.

High-resolution remote-sensing data became available from commercial satellites, such as IKONOS, QuickBird, and the Airborne Imaging Spectroradiometer for Applications (AISA) beginning in the late 1990s. These platforms collect data at high spatial resolutions (1–4 m) and can distinguish individual plant crowns. They therefore have the potential to detect the initial stages of an invasion when plant cover is low. Both IKONOS and QuickBird are multispectral sensors that have been used to detect invasive species, including giant reed and spiny aster in southern Texas (Everitt et al., 2004) and melaleuca trees (*Melaleuca quinquinervia*) in south Florida (Fuller, 2005). AISA is a hyperspectral sensor that has been used frequently for separating various invasive species growing in common areas (Narumalani et al., 2009; McGwire et al., 2011).

We carried out a comprehensive test to examine the ability of integrating various types of high spatial and hyperspectral imagery for differentiating saltcedar from other riparian vegetation species in the Rio Grande basin (Wang et al., in press). A spaceborne QuickBird image and an airborne AISA image were acquired over a portion of the FRR in southwest Texas where saltcedar has become a threat to the native ecosystem. The images were evaluated using various classification schemes to determine the most appropriate methods for discerning saltcedar. We first address the issue of band selection for hyperspectral imagery and then provide a brief description for each of the five classification methods tested. We conclude this section by presenting the results for the imagery and classification methods that performed best for detecting saltcedar.

13.2.1.1 Band selection

When working with multiple bands of imagery, particularly with hyper-spectral data, it is important to select the most appropriate combination of bands for successful classification. The QuickBird image contains four multispectral bands, and all of the bands were included in each classification. The AISA image was acquired with 61 bands in the visual near infrared (VNIR) region, so several band combinations were tested. The first approach used all 61 bands. The second approach selected four bands: Band 11 (481.39 nanometer [nm]), Band 20 (563.17 nm), Band 30 (656.99 nm), and Band 46 (823.25 nm) with wavelength centers coinciding with those of the QuickBird imagery (485 nm, 560 nm, 660 nm, and 830 nm, respectively). The third band selection strategy was based on a linear discriminant analysis (LDA), and the objective was to determine the optimal bands that maximized the linear separability of training data, computed as linear combinations of the original bands. This method produced seven bands. The fourth selection strategy used the minimum noise fraction (MNF) transform to select a subset of bands. Like principal component analysis (PCA), MNF transform is used to determine the intrinsic dimensionality of the data. Unlike PCA, however, the axes of the MNF transform are aligned along the axis of MNF (maximum signal-to-noise ratio), rather than along the principal components (directions of maximum variance). The first 10 bands were selected based on MNF.

13.2.1.2 Image classification

Five pixel-based methods were tested to discriminate saltcedar: maximum likelihood classifier (MLC), neural network (NN) classifier, support vector machine (SVM), spectral angle mapper (SAM), and maximum matching feature (MMF). The MLC method maximizes the posterior probability that a pixel is covered with a land cover class, given the observed spectral information of the pixel (Richards and Jia, 1999). The method assumes that all classes have a Gaussian distribution. This assumption simplifies the classification problem so only the mean vector and covariance matrix are required for each class. The final output from MLC is the class with the highest probability. This method is generally acceptable for multispectral data sets for which the number of bands is below 10. With a larger number of bands, this method becomes impractical as it requires a considerable number of training samples.

The SAM method is an alternative for compiling a large number of bands (Kruse et al., 2003). In this case, the spectral information of each pixel is considered as a vector with a dimension equal to the number of bands. The geometric concept of an angle between vectors is then extended to a higher dimensional space. The SAM method uses the "spectral" angle as a similarity measure to classify every pixel. The advantages of SAM are that it can operate with a large number of bands and is insensitive to the scaling of spectral values caused by changes in illumination. The major limitations are that the angle has no physical interpretation and all classes are treated alike (e.g., having the same variability).

The third method, NN, is generally suggested when the discriminating surfaces in the feature space are complex or the Gaussian assumption is violated (Benediktsson et al., 1990; Kavzoglu and Mather, 2003). This can occur when two or more categories with different spectral signatures are aggregated into a single class. An NN is composed of basic processing elements called neurons, usually arranged in layers. In isolation, neuron units perform linear combinations of the input and compare the results by differencing them with a threshold (also known as an activation level). The residual from the comparison is evaluated through a transfer function (also known as anactivation function) that has the shape of a sharp step. The result is a binary value that can be interpreted as either belonging to (high output) or not belonging to (low output) a class. Weight and threshold parameters are adjusted by a process known as training. The NN used here was a feed-forward network with two hidden layers and was trained in MATLAB (The Mathworks, Inc.).

SVM is a classification method derived from statistical learning theory. It was originally designed for binary classification, but it can function as a multiclass classifier by combining several binary classifiers (Wu et al., 2004). Each binary SVM classifier separates two classes with a decision surface that maximizes the margin between the classes. The surface is often called the optimal hyperplane, and the data points closest to the hyperplane are called support vectors. The support vectors are the critical elements of the training set, as they alone define the optimal hyperplane. If the two classes are not linearly separable, the method uses a kernel function to map the data to a higher dimension where the data are linearly separable. The hyperplane from the higher dimensional space results in a complex nonlinear boundary when projected to a lower dimensional space. As with neural networks, processing large data sets through SVM requires a large amount of computation time.

The MMF selects the class that has the maximum abundance through the matched filtering (MF) technique (Boardman et al., 1995), which finds the abundances of user-defined endmembers (pure spectral signatures of a class) using partial unmixing (a detailed description of unmixing is provided in the next section). This technique maximizes the response of the known endmembers and suppresses the response of the composite unknown background, thus matching the known signature. It provides a rapid means of detecting specific materials based on matches to library or image endmember spectra and does not require knowledge of all endmembers within an image scene.

Table 13.1 Classification accuracies for quickbird imagery

Method	Overall accuracy	Kappa statistic	Producer's accuracy	User's accuracy
MLC	70.9	0.69	77.7	71.8
NN	66.9	0.65	69.3	69.3
SVM	70.2	0.68	70.3	79.5
MLC PAN	70.6	0.68	72.5	68.3
MLC QB$_{SYN}$	82.0	0.81	89.6	93.9

MLC = maximum likelihood; NN = neural network; SVM = support vector machine; MLC PAN = MLC including the panchromatic band; MLC QB$_{SYN}$ = MLC for a synthetic QuickBird image derived from airborne imaging spectroradiometer for applications. Accuracies are for detecting saltcedar only.

13.2.1.3 Results

Classification results for the QuickBird image are summarized in Table 13.1. Three classifiers (MLC, NN, and SVM) performed similarly, yet MLC achieved the highest overall accuracy. Given its simplicity, the MLC method was selected for testing the incorporation of the panchromatic band. The results indicate that inclusion of the panchromatic band does not improve classification accuracy. Instead, it tends to enlarge the spectral confusion, resulting in lower user and producer accuracies. The MLC method was also applied to a synthetic QuickBird image derived from the higher spatial resolution AISA image. The results indicate that atmospheric effects play a significant role in the classification accuracy of the acquired QuickBird. Specifically, the overall accuracy from the synthetic image increased more than 10% with respect to the satellite acquired QuickBird.

Classification results for the AISA image are summarized in Table 13.2. Because AISA contains more detailed spectral information, higher overall and individual classification accuracies were expected. Results indicate, however, that the processing of hyperspectral data must be done carefully. The methods that utilized all of the hyperspectral bands (SAM and

Table 13.2 Classification accuracies for AISA

Method	Overall accuracy	Kappa statistic	Producer's accuracy	User's accuracy
SAM AISA$_{61}$	68.6	0.66	84.8	94.7
MMF AISA$_{61}$	45.0	0.42	45.8	78.6
MLC AISA$_{4}$	84.5	0.83	89.4	91.4
NN AISA$_{4}$	63.9	0.62	95.1	87.1
SVM AISA$_{7(LDA)}$	85.8	0.85	91.9	94.8
MLC AISA$_{10(MNF)}$	87.6	0.87	93.7	94.6
SVM AISA$_{10(MNF)}$	87.7	0.87	93.9	96.5

SAM = spectral angle mapper; MMF = maximum matching feature; MLC = maximum likelihood classification; NN = neural network; SVM = support vector machine; AISA = airborne imaging spectroradiometer for applications. Subscript numbers indicate the number of AISA bands used and whether they were selected by the linear discriminant analysis or minimum noise function.

MMF) performed poorly when compared with the results from QuickBird. Band selection generally increased the performance even when only the four narrow bands that aligned with the QuickBird band-pass filters were selected. The best band-selection method was MNF. The SVM method produced the highest accuracy when the MNF band selection was adopted. Yet, the MLC performed very close to SVM. A reason for this may be due to the fact that all of the classes defined in this study do not exhibit complex spectral variability. In fact, the spectral variability of saltcedar was largely avoided by differentiating three subclasses—saltcedar green-brown, saltcedar orange-brown, and giant saltcedar (evergreen)—which present distinct spectral characteristics. In summary, this study investigated some of the best capabilities of contemporary aerospace imagery for assisting with the reconnaissance of saltcedar and found that AISA hyperspectral imagery outperformed QuickBird imagery in differentiating saltcedar from other riparian vegetation species.

13.2.2 Detecting invasions from medium spatial and spectral resolution data

Although the high resolution data sets discussed above are optimal for pixel-based classifications, acquiring these types of images over wide geographic areas is expensive and therefore prohibitive for large-scale investigations. Most studies must rely on widely available medium and coarse spatial resolution images from satellites, such as Landsat TM (30 m spatial resolution), MODIS (250 m, 500 m, or 1,000 m spatial resolution, depending on the data set), and the Advanced Very High Resolution Radiometer (AVHRR; 1,000 m spatial resolution). When using these types of images, target objects (i.e., individual plants) are usually smaller than the pixel size, resulting in pixels that contain a mixture of land cover components.

These mixed pixels present one of the most difficult challenges for deriving land cover from remote sensing to map invasions. Pixel-based classifications do not provide enough detailed information on the abundance and location of saltcedar, so advanced classification techniques are needed. Spectral unmixing has emerged as a leading method for classifying mixed pixels. The rationale behind spectral unmixing, also referred to as subpixel classification or spectral mixture analysis, is that the spectral reflectance of a pixel results from the combination of multiple component spectra (endmembers) present in a sensor's instantaneous field of view. Linear spectral unmixing (LSU) methods assume that light reacts with only a single component (Roberts et al., 1993), and therefore the relative spectral contribution of each endmember is theoretically proportional to its areal extent. LSU does not account for multiple scattering from vegetation surfaces. This scattering is a real possibility for saltcedar because of its dominance of the canopy and occlusion of other native vegetation species, which can lead to an increase of leaf transmittance. Because LSU methods are the simplest way to estimate fractional cover components with reasonable accuracy, however, they are preferred over nonlinear mixture models. Subpixel classification techniques, such as LSU, are superior to traditional pixel-level classification techniques because they can estimate the proportions of multiple species or land covers within a single pixel. This is particularly important when trying to assess the local presence of an invasive species across a large area where it may not necessarily constitute a majority of the pixel.

In this section, we discuss an innovative method for subpixel classification, the tessellated linear spectral unmixing (TLSU) technique (Silván-Cárdenas and Wang, 2010) that is particularly well suited for detecting saltcedar invasion. The major advantage of TLSU over other spectral unmixing methods for discriminating competing vegetation species is the incorporation of detailed spectral variability into the classification scheme. We also discuss the improved accuracy of this technique for detecting low levels of land cover during the nascent stages of invasion. In the following section, we will interpret these classification results in an ecological context to understand the spatial patterns of saltcedar invasion in the FRR.

13.2.2.1 Incorporating species variability

Subpixel classification techniques are highly influenced by the definition of endmembers, or spectrally pure features (Silván-Cárdenas and Wang, 2010). Therefore, determining appropriate endmembers is an important step toward capturing inter- and intraspecies variability. The subpixel classification technique discussed here relies on a two-level endmember selection approach, which allows a greater amount of species variability to be incorporated into the unmixing scheme. This increased variability ultimately contributes to more accurate classifications. First, 17 Level II endmembers were collected for individual species and specific land cover types in the study area (see Table 13.3). Next, the Level II endmembers were aggregated into three Level I classes: (a) saltcedar, (b) native woody riparian vegetation, and (c) other vegetated and nonvegetated land cover types. The generalization of Level II endmembers into Level I classes is pertinent because the maximum number of endmembers that can be considered by an unmixing scheme is constrained by the intrinsic dimension of the data (i.e., the number of spectral bands). Multispectral images have a limited number of bands and therefore typically can consider only a limited number of endmembers, but the two-level scheme allows fine spectral variations between saltcedar and the cooccurring native vegetation to be incorporated into the mixture model without requiring hyperspectral image data.

TLSU is developed around the two-level endmember selection strategy, and a full description of the method can be found in Silván-Cárdenas and Wang (2010). In brief, TLSU builds a Delaunay tessellation of the entire set of endmembers by considering each spectrum as a point in the n-dimensional

Table 13.3 Level I classes and corresponding level II endmembers for the two-set endmember classification system

Level I classes	Level II endmembers
Saltcedar (*Tamarix*)	Green saltcedar
	Senescent saltcedar
	Dry saltcedar
Native woody riparian vegetation	Willow (leaf on)
	Willow (leaf off)
	Mesquite (leaf on)
	Mesquite (leaf off)
	Poverty weed
Other nonwoody riparian vegetation land cover types	Creosote bush
	Herbaceous (dry)
	Herbaceous (green)
	Wetland
	Water
	Soil
	Gravel
	Road
	Roof

Euclidian space. The mixing space is partitioned by tessellations into simplices (e.g., a triangle in two-dimensional space, a tetrahedron in three-dimensional space, etc.). Mixed pixels are points in the *n*-dimensional space, and the vertices of the simplex closest to the point represent the best candidates for endmembers. In this manner, a pixel can be unmixed using any combination of Level II endmembers.

TLSU was rigorously tested and outperformed all other linear mixture models tested for detecting saltcedar (Silván-Cárdenas and Wang, 2010). The reason for its exceptional performance lies in the increased number of species (i.e., variability) that can be incorporated into the classification. Most studies seek to detect only broad categories (e.g., photosynthetic vegetation, nonphotosynthetic vegetation, and bare soil), but these general groupings are not sufficient to discern the species-level information needed to study plant invasions. With the TLSU approach, the number of species that can be incorporated is limitless, as long as the number of Level I classes conforms to the band constraints. This allows the model to be transferable to a wide range of study sites and research problems.

13.2.2.2 Improved accuracy for detecting incipient invasions

The complexity of dealing with multiple land covers within each pixel is complicated further by the uncertainty of unmixing accuracy. Several techniques have been proposed to determine the accuracy of subpixel classifications (Gopal and Woodcock, 1994; Binaghi et al., 1999; Green and Congalton, 2004; Pontius and Cheuk, 2006; Silván-Cárdenas and Wang, 2008), but none have yet been adopted as a standard accuracy reporting measure. If not assessed properly, classification errors can be propagated into spatial analyses performed on the data (discussed in the next section) and will lead to inaccurate results. Because detecting the initial stages of an invasion is highly critical for successful management and eradication, we were specifically interested in testing whether TLSU accuracy varied with land cover proportion. We used the root mean square error (RMSE) operator to assess unmixing performance at specific fractional abundance proportions. RMSE expresses the magnitude of the average error generated by the classification and is commonly used to validate spectral unmixing results (Mishra et al., 2009; Foody et al., 2010; Pacheco and McNairn, 2010).

The SAM classification of the AISA image served as reference data. A single Level II class was assigned by SAM to each AISA pixel (1 m) and the pixels were aggregated to 30 m (hereafter referred to as AISA$_{30m}$) to match the spatial resolution of the TLSU results. Next, Level II endmembers were aggregated to Level I classes, and exact reference land cover proportions for the three Level I classes were calculated for each AISA$_{30m}$ pixel. One thousand pixels were randomly selected from AISA$_{30m}$ and segregated into 10 distinct ranges based on their land cover proportion. The ranges are hereafter referred to as proportional ranges (PR). The pixel proportions were then compared with the results from TLSU, and RMSE values were calculated for each PR using the following equation:

$$\text{RMSE}_{\text{PR}} = \sqrt{\frac{\sum_{i=1}^{n} (Ri - Ai)^2}{n - 1}}, \tag{13.1}$$

where R_i is the proportion of saltcedar in the reference pixel (AISA$_{30m}$), A_i is the proportion of saltcedar in the TLSU pixel, and n is the number of reference pixels in each range. Lower RMSE values indicate better accuracy. Results (see Table 13.4) show that TLSU performed best when detecting low–fractional cover proportions (0.0–0.4). Highest RMSE values (i.e., lowest accuracies) occurred in PR 0.8–0.9 and PR 0.9–1.0.

These findings are encouraging. There is evidence to suggest that detection and control of invasive species is most critical during the early stages of the invasion process (Hamada et al., 2007; Walsh et al., 2008). During these initial stages, fractional cover is sparse, which makes detection and monitoring difficult for traditional pixel-based classification methods. The innovative TLSU technique incorporates a greater amount of species variability than typical unmixing methods and achieves higher accuracies when detecting sparse fractional cover. This improvement makes TLSU a useful weapon in the battle against saltcedar invasion.

Table 13.4 Accuracy assessment of the fractional cover proportional ranges from the TLSU using root mean square error

Proportional range (PR)	AISA$_{30m}$ reference pixels (n)	RMSE
0.0–0.1	238	0.280
0.1–0.2	90	0.206
0.2–0.3	110	0.192
0.3–0.4	85	0.214
0.4–0.5	94	0.236
0.5–0.6	85	0.320
0.6–0.7	86	0.359
0.7–0.8	55	0.450
0.8–0.9	66	0.580
0.9–1.0	91	0.579

AISA = airborne imaging spectroradiometer for applications; RMSE = root mean square error.

13.2.3 Ecological interpretation of geospatial data

Differentiating saltcedar from the cooccurring native vegetation and achieving accurate classification has been the main focus of remote-sensing efforts to study invasion, but without appropriate tools for interpretation, these classifications are of little value to ecologists studying the issue. To evaluate classification maps and answer questions regarding species expansion, landscape ecologists typically employ a set of tools called landscape metrics. Landscape metrics are a group of algorithms developed to quantify the spatial characteristics of categorical map patterns to analyze the patches, classes, and mosaic structure of landscapes (McGarigal and Marks, 2005). For images classified using pixel-based techniques, the resulting categorical map can be statistically analyzed through landscape metrics for a variety of spatial patterns, including area, density, and shape characteristics. These spatial metrics are an important step toward understanding ecological processes, and interpretation of them can provide insight into the process of invasion (Walsh et al., 2008).

Soft classifications, which are the best option for mapping saltcedar invasion, do not fit well with the traditional patch-pattern-mosaic fundamentals of landscape ecology (Forman and Godron, 1986) because they do not have discrete boundaries. Without discrete boundaries, classifications cannot be translated into categorical maps and therefore cannot be assessed using conventional landscape metric analysis and tools. We thus find ourselves at an impasse: Classification science has evolved beyond the capabilities for interpretation, leaving a gap between the remote-sensing data and the methods to analyze them for invasion studies.

To bridge this gap, we developed a new approach, the threshold continuum approach (Frazier and Wang, 2011) for reclassifying subpixel data that delineates clear boundaries for soft classifications; these discretized classes can then be used to calculate landscape metrics. Our approach relies on setting multiple thresholds along the continuum of possible proportions (0–1.0) of saltcedar land cover within a pixel. At each threshold, every pixel is assessed to determine whether the proportion of saltcedar satisfies or exceeds the threshold value. Pixels that meet the criteria are included in the class, and pixels that do not are omitted. This process is repeated for each threshold, partitioning the soft classification into multiple categorical maps that are analyzed through landscape metrics. Thresholds can be set at any increment, with smaller steps retaining more of the spatial and spectral heterogeneity. Landscape metrics are calculated for each map using FRAGSTATS (McGarigal et al., 2002), a publicly available and widely used landscape metric software program. Figure 13.2 shows a simple example of how a spectrally unmixed image is subjected to reclassification using the threshold continuum approach at four discrete thresholds.

The major benefit of the threshold continuum method is that it generates multiple metric values for each soft classification (a separate value for each threshold). With traditional pixel-based classifications, only a single metric value can be calculated. With the advent of the threshold continuum approach, however, the landscape can now be analyzed and interpreted as a function of fractional cover. This added dimension is crucial for invasion studies because the progress of an invasion is often measured by the density of the species.

To illustrate the threshold continuum concept for landscape metrics, let us assume that in the year 2000, a pixel contained 60% saltcedar coverage. By 2008, saltcedar had expanded and covered 100% of the pixel. A hard classification scheme would not capture this change because saltcedar accounts for the majority of the pixel in both years of imagery. Both pixels would be classified as saltcedar, and any ecological analysis performed on the classified data would treat both pixels as equivalent. Ecologically, however, there

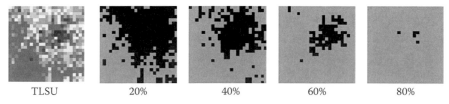

TLSU   20%   40%   60%   80%

Figure 13.2 **(See color insert.)** Example showing the results of a threshold continuum reclassification. The tessellated spectral unmixing (TLSU) results are shown on the far left (red indicates high saltcedar proportions, green indicates low proportions). The four gray-scale images show threshold reclassifications at 20%, 40%, 60%, and 80% saltcedar land cover. For each threshold, all pixels equal to or greater than the threshold are in black. All other pixels are shown in gray.

is a notable difference between 60% and 100% land cover. In the former case, 40% of the pixel (360 m² for a Landsat TM pixel) is available to be occupied by native vegetation species, but at 100% cover, native vegetation is completely excluded by saltcedar. Capturing these differences is the motivation behind subpixel classifications, and modeling these differences through landscape metrics is the motivation for this portion of our research to foster better ecological interpretation of remotely sensed data.

We applied the threshold continuum approach to categorize TLSU soft classifications of the FRR and completed metric analyses to illustrate how the additional landscape information derived from fractional land cover values can be interpreted. We examined a straightforward and commonly used metric, Number of Patches (NP), for a series of 19 images from 1982 to 2009. We set thresholds at every 0.05 increment along the continuum from 0 to 1.0 and reclassified a new categorical map at each of the 20 thresholds for each year of imagery. The results (see Figure 13.3) show fractional cover (x-axis) plotted against NP (y-axis) across the 19 years of data (z-axis). The response curve for NP versus fractional cover threshold (x- and y-axes) illustrates how the metric value changes as a function of land cover. When these response curves are combined for multiple years (z-axis), they can be interpolated as a surface to understand the dynamic nature of the invasion over time. Examination of the changes in surface slope

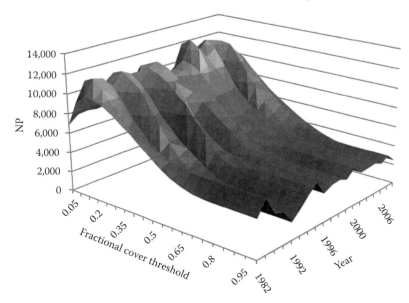

Figure 13.3 Example showing how ecological interpretation of sub-pixel data can introduce an additional dimension for data analysis. The Number of Patches (NP), a commonly used metric, is plotted against fractional cover proportion on the x-axis and the year of imagery on the z-axis to illustrate the changing landscape patterns over both space and time.

reveals that the largest values for *NP* occur at low thresholds, indicating a fragmented landscape at these lower land cover values. After the peak, which occurs approximately around 0.2, *NP* decreases rapidly as threshold increases before reaching an asymptote approximately around 0.65 to 0.70 (65%–70% cover). From interpretation of the surface, it is clear that above 70% cover, *NP* remains low and fairly stable, indicating that the landscape at these densities is more cohesive (less fragmented), and increases in inner-pixel density are not greatly affecting the overall number of patches. The ridge-like peaks in the surface result from annual fluctuations that occur because of local changes in the hydrologic system based on precipitation pulses. This type of detailed ecological analysis using a continuous discretization of soft classified data is not possible with pixel-based classifications and highlights the advancements being made to utilize advanced remote-sensing capabilities to study ecological invasions.

In summary, widely available remote-sensing data sets, such as Landsat TM, must be spectrally unmixed to extract valuable information regarding the abundance and distribution of saltcedar. These subpixel classifications, although highly accurate, do not adhere to the traditional categorical map format upon which most ecological analytical software programs for characterizing landscapes are built. We have shown in this section how a simple transformation of soft-classified remote-sensing data using a series of threshold steps can preserve much of the spatial and spectral heterogeneity captured by spectral unmixing and make these types of classifications interpretable for the wide range of disciplines involved in invasion research. The methods described here for discretization and interpretation are easily transferable to other types of invasive species or other ecological problems.

13.3 CONCLUSION

The case study presented in this chapter demonstrates how recent advances in aerospace imagery, remote-sensing classification techniques, and ecological interpretation of geospatial data can improve invasive species detection, particularly during the initially stages of an invasion when mapping and monitoring is most critical. Specifically, this study investigated many of the best capabilities of contemporary remote sensing for assisting with the reconnaissance of saltcedar, one of the most threatening invasive species in the southwestern United States. First, a comprehensive study was conducted using multiresolution, multisource remote-sensing imagery encompassing QuickBird, AISA, and Landsat TM. The results indicate that AISA hyperspectral imagery outperformed QuickBird imagery in differentiating saltcedar from other riparian vegetation species, mainly because of the removal of many atmospheric effects with airborne imagery. In terms of a cost-benefit analysis, the costs of acquiring data from three sensors

(QuickBird, IKONOS, and AISA) are relatively comparable, yet classification accuracy was much improved with AISA. The drawback to using AISA is that it is an airborne platform and therefore cannot be dispatched to survey any area of the globe. When far-reaching global data were needed, we found QuickBird outperformed IKONOS.

For situations when it is not feasible to acquire any type of high-resolution imagery, medium-resolution data sets, such as Landsat TM, are an attractive option. We highlighted the benefits of the TLSU technique for spectrally unmixing these types of images, including its increased ability to incorporate fine spectral variations between native and invasive species and its improved accuracy in detecting low–fractional cover proportions. Our results show that TLSU is particularly well suited for detecting the beginning stages of an invasion when land cover is still relatively low. These findings are encouraging for tackling the saltcedar invasion problem because early detection is the most crucial, but also the most difficult, aspect of monitoring invasions.

Finally, we subjected the classification results to a practical ecological assessment using landscape metrics, a commonly used tool for assessing changing land cover patterns. We illustrated how soft classifications can be reclassified into multiple discrete maps using the threshold continuum approach and how those maps can be analyzed using landscape metrics to generate responses of spatial patterns to changes in land cover. This type of ecological assessment is not possible using pixel-based classifications. In conclusion, with the rapid pace of globalization, the introduction of exotic plant and animal species into new host communities will likely continue to occur worldwide. Early detection is one of the most crucial steps toward effective management of invasive species, and continuing the development of aerospace techniques for detecting, mapping, and monitoring invasions is essential to protect natural ecosystems.

REFERENCES

Baum, B. R. 1967. Introduced and naturalized tamarisks in the United States and Canada. *Bayleya* 15: 19–25.

Benediktsson, J. A., P. H. Swain, and O. K. Ersoy. 1990. Neural network approaches versus statistical-methods in classification of multisource remote-sensing data. *IEEE Transactions on Geoscience and Remote Sensing* 4: 540–552.

Binaghi, E., P. A. Brivio, P. Chezzi, and A. Rampini. 1999. A fuzzy set-based assessment of soft classification. *Pattern Recognition Letters* 20: 935–948.

Boardman, J. W., F. A. Kruse, and R. O. Green. 1995. Mapping target signatures via partial unmixing of AVIRIS data. In *Summaries, Fifth JPL Airborne Earth Science Workshop*. JPL publication 95–1 no. 1, 23–26. Pasadena, CA: Jet Propulsion Laboratory.

DiTomaso, J. M. 1998. Impact, biology, and ecology of saltcedar (*Tamarix* spp.) in the Southern United States. *Weed Technology* 12: 326–336.

Dudley, T. L., and C. J. DeLoach. 2004. Saltcedar (*Tamarix* spp.), endangered species, and biological weed control: can they mix? *Weed Technology* 18: 1542–1551.

Dukes, J. S., and H. A. Mooney. 2004. Disruption of ecosystem processes in western North America by invasive species. *Revista Chilena De Historia Natural* 77: 411–437.

Everitt, B. 1980. Ecology of saltcedar: a plea for research. *Environmental Geology* 3: 77–84.

Everitt, B. 1998. Chronology of the spread of Tamarisk in the Central Rio Grande. *Wetlands* 18(4): 658–668.

Everitt, J. H., and C. J. Deloach. 1990. Remote-sensing of Chinese tamarisk (*Tamarix-chinensis*) and associated vegetation. *Weed Science* 38: 273–278.

Everitt, J. H., D. E. Escobar, M. A. Alaniz, M. R. Davis, and J. V. Richerson. 1996. Using spatial information technologies to map Chinese tamarisk (*Tamarix chinensis*) infestations. *Weed Science* 44: 194–201.

Everitt, J. H., C. Yang, and M. R. Davis. 2004. Remote mapping of two invasive weeds in the Rio Grande system of Texas. In *Proceedings of the 40th Annual American Water Resources Conference.* Middleburg, VA: American Water Resources Association.

Everitt, J. H., C. Yang, R. S. Fletcher, C. J. DeLoach, and M. R. Davis. 2007. Using remote sensing to assess biological control of Saltcedar. *Southwester Entomologist* 32(2): 93–103.

Foody, G., R. Lucas, P. Curran, and M. Honzak. 2010. Non-linear mixture modeling without end-members using an artificial neural network. *International Journal of Remote Sensing* 18(4): 937–953.

Forman, R. T., and M. Godron. 1986. *Landscape Ecology.* New York: Wiley & Sons.

Frazier, A. E., and L. Wang. 2011. Characterizing spatial patterns of invasive species using sub-pixel classifications. *Remote Sensing of Environment* 115: 1997–2007.

Fuller, D. O. 2005. Remote detection of invasive malameuca trees (*Malaleucaquin-quenervia*) in South Florida with multispectral IKONOS imagery. *International Journal of Remote Sensing* 26(5): 1057–1063.

Glenn, E. P., and P. L. Nagler. 2005. Comparative ecophysiology of *Tamarix ramosissima* and native trees in western U.S. riparian zones. *Journal of Arid Environments* 61: 419–446.

Glenn, N. F., J. T. Mundt, K. T. Weber, T. S. Prather, L. W. Lass, and J. Pettingill. 2005. Hyperspectral data processing for repeat detection of small infestations of leafy spurge. *Remote Sensing of Environment* 95: 399–412.

Gopal, S., and C. E. Woodcock. 1994. Theory and methods for accuracy assessment of thematic maps using fuzzy sets. *Photogrammetric and Engineering of Remote Sensing* 60(2): 181–188.

Green, K., and G. Congalton. 2004. An error matrix approach to fuzzy accuracy assessment: the NIMA geocover project. In *Remote Sensing and GIS Accuracy Assessment*, edited by R. S. Lunetta and J. G. Lyon, 163–172. Boca Raton, FL: CRC Press.

Hamada, Y., D. A. Stow, L. L. Coulter, J. C. Jafolla, and L. W. Hendricks. 2007. Detecting tamarisk species (Tamarix spp.) in riparian habitats of Southern California using high spatial resolution hyperspectral imagery. *Remote Sensing of Environment* 109: 237–248.

Hultine, K. R., P. L. Nagler, K. Morino, S. E. Bush, K. G. Burtch, P. E. Dennison, E. P. Glenn, and J. R. Ehleringer. 2010. Sap flux-scaled transpiration by tamarisk (Tamarix spp.) before, during and after episodic defoliation by the saltcedar leaf beetle (Diorhabdacarinulata). *Agricultural and Forest Meteorology* 150: 1467–1475.

Hunt, E. R., J. H. Everitt, J. C. Ritchie, M. S. Moran, D. T. Booth, G. L. Anderson, P. E. Clark, and M.S. Seyfried. 2003. Application and research using remote sensing for range management. *Photogrammetric Engineering and Remote Sensing* 69(6): 675–693.

Kavzoglu, T., and P. M. Mather. 2003. The use of backpropagation artificial neural networks in land cover classification. *International Journal of Remote Sensing* 24 (23): 4907–4938.

Kruse, F. A., J. W. Boardman, and J. F. Huntington. 2003. Comparison of airborne hyperspectral data and EO-1 Hyperion for mineral mapping. *IEEE Transactions on Geoscience and Remote Sensing* 41(6): 1388–1400.

Lass, L. W., and T. S. Prather. 2004. Detecting the locations of Brazilian pepper trees in the Everglades with hyperspectral sensor. *Weed Technology* 18: 437–442.

Lass, L. W., T. S. Prather, N. F. Glenn, K. T. Weber, J. T. Mundt, and J. Pettingill. 2005. A review of remote sensing of invasive weeds and example of the early detection of spotted knapweed (Centaureamaculosa) and babysbreath (Gypsophila paniculata) with a hyperspectral sensor. *Weed Science* 53(2): 242–251.

McGarigal, K., S. Cushman, M. Neel, and E. Ene. 2002. FRAGSTATS: spatial pattern analysis program for categorical maps. Computer software program produced by the authors at the University of Massachusetts, Amherst. http://www.umass.edu/landeco/research/fragstats/fragstats.html.

McGarigal, K., and B. J. Marks. 1995. FRAGSTATS: spatial pattern analysis program for quantifying landscape structure. U.S. Forest Service General Technical Report PNW 0(351), I–122. Portland, OR: U.S. Department of Agriculture.

McGwire, K. C., T. B. Minor, and B. W. Schultz. 2011. Progressive discrimination: an automatic method for mapping individual targets in hyperspectral imagery. *IEEE Transactions on Geoscience and Remote Sensing* 49(7): 2674–2685.

Mishra, V. D., H. S. Negi, A. K. Rawat, A. Chaturvedi, and R. P. Singh. 2009. Retrieval of sub-pixel snow cover information in the Himalayan region using medium and coarse resolution remote sensing data. *International Journal of Remote Sensing* 30(18): 4707–4731.

Morisette, J. T., C. S. Jarnevich, A. Ullah, W. J. Cai, J. A. Pedelty, J. E. Gentle, T. J. Stohlgren, and J. L. Schnase. 2006. A tamarisk habitat suitability map for the continental United States. *Frontiers in Ecology and the Environment* 4: 11–17.

Nagler, P. L., K. Morino, K. Didan, J. Erker, J. Osterberg, K. R. Hultine, and E. P. Glenn. 2009. Wide-area estimates of saltcedar (*Tamarix* spp.) evapotranspiration on the lower Colorado River measured by heat balance and remote sensing methods. *Ecohydrology* 2: 18–33.

Narumalani, S., D. R. Mishra, R. Wilson, P. Reece, and A. Kohler. 2009. Detecting and mapping four invasive species along the floodplain of North Platte River, Nebraska. *Weed Technology* 23(1): 99–107.

O'Neill, M., S. L. Ustin, S. Hager, and R. Root. 2000. Mapping the distribution of leafy spurge at Theodore Roosevelt National Park using AVIRIS. In: *Proceedings of Ninth JPL Airborne Visible Infrared Imaging Spectrometer (AVIRIS) Workshop*. Pasedena, CA: Jet Propulsion Laboratory.

Pacheco, A., and H. McNairn. 2010. Evaluating multispectral remote sensing and spectral unmixing analysis for crop residue mapping. *Remote Sensing of Environment* 114: 2219–2228.

Pearce, C. M., and D. G. Smith. 2003. Saltcedar: distribution, abundance, and dispersal mechanisms, northern Montana, USA. *Wetlands* 23(2): 215–228.

Pontius, R. G., and M. L. Cheuk. 2006. A generalized cross-tabulation matrix to compare soft-classified maps at multiple resolutions. *International Journal of Geographic Information Science* 20(1): 1–30.

Rejmánek, M., and M. J. Pitcairn.2002. When is eradication of exotic pest plats a realistic goal? In *Turning the Tide: The Eradication of Invasive Species*, edited by C. R. Veitch, and M. N. Clour. Gland, Switzerland: IUCN SSC Invasive Species Specialist Group.

Reynolds, L. V., and D. J. Cooper. 2010. Environmental tolerance of an invasive riparian tree and its potential for continued spread in the southwestern US. *Journal of Vegetable Science* 21: 733–743.

Richards, J. A., and X. Jia. 1999. *Remote Sensing Digital Image Analysis*, 3rd ed. Berlin: Springer-Verlag.

Roberts, D. A., M. O. Smith, and J. B. Adams. 1993. Green vegetation, non-photosynthetic vegetation and soils in AVIRIS data. *Remote Sensing of Environment* 44: 255–269.

Sax, D. F., J. J. Stachowicz, and S. D. Gaines. 2005. *Species Invasions: Insights into Ecology, Evolution, and Biogeography*. Sunderland, MA: Sinauer Associates Incorporated.

Silván-Cárdenas, J. L., and L. Wang. 2008. Sub-pixel confusion-uncertainty matrix for assessing soft classifications. *Remote Sensing of Environment* 112: 1081–1095.

Silván-Cárdenas, J. L., and L. Wang. 2010. Retrieval of subpixel Tamarix canopy cover from Landsat data along the Forgotten River using linear and nonlinear spectral mixture models. *Remote Sensing of Environment* 114: 1777–1790.

Smith, S. D., D. A. David, A. Sala, J. R. Cleverly, and D. E. Busch. 1998. Water relations of riparian plants from warm desert regions. *Wetlands* 18(4): 687–696.

Tamarisk Coalition. 2003. Impact of tamarisk infestation on the water resources of Colorado. Colorado Department of Natural Resources Colorado Water Conservation Board. http://cwcb.state.co.us/Resource_Studies/Tamarisk_Study_2003.pdf.

Underwood, E., S. Ustin, and D. DiPietro. 2003. Mapping non-native plants using hyperspectralimagery. *Remote Sensing of Environment* 86: 150–161.

U.S. Army Corps of Engineers. 2008. *Forgotten Reach of the Rio Grande, Fort Quitman to Presidio Texas*. Section 729, January 2008. Albuquerque, NM: U.S. Army Corps of Engineers.

Vitousek, P. M., C. M. Dantonio, L. L. Loope, M. Rejmanek, and R. Westbrooks. 1997. Introduced species: a significant component of human-caused global change. *New Zealand Journal of Ecology* 21: 1–16.

Walsh, S. J., A. L. McCleary, C. F. Mena, Y. Shao, J. P. Tuttle, A. Gonzalez, and R. Atkinson. 2008. QuickBird and hyperion data analysis of an invasive plant species in the. Galapagos Islands of Ecuador: implications for control and land use management. *Remote Sensing of Environment* 112: 1927–1941.

Wang, L., J. Silvan, J. Yang, and A. Frazier. In press. Invasive Saltcedar spread mapping using multi-resolution remote sensing data. *Professional Geographer.*

William, P. A., and E. R. Hunt. 2002. Estimation of leafy spurge cover from hyper-spectralimagery using mixture tuned matched filtering. *Remote Sensing of Environment* 82: 446–456.

Wu, T.-F., C.-J.Lin, and R. C. Weng. 2004. Probability estimates for multi-class classification by pairwise coupling. *Journal of Machine Learning Research* 5: 975–1005.

Zavaleta, E. 2000.The economic value of controlling and invasive shrub. *Ambio: A Journal of the Human Environment* 29: 462–467.

Chapter 14

Surface deformation mapping with persistent scatterer radar interferometry

Guoxiang Liu, Lei Zhang, Xiaoli Ding,
Qiang Chen, Xiaojun Luo, and Guolin Cai

CONTENTS

Surface deformation resulting from anthropic activities or crustal motion is a major concern for land use planning and engineering risk assessment. Spaceborne interferometric synthetic aperture radar (InSAR) has evolved in the past decades as a unique tool for quantitative measurements of regional deformation with large coverage and a high accuracy of centimeters to millimeters. To mitigate the technical limitations of InSAR, persistent scatterer InSAR (PSI) techniques have been proposed in recent years to extract deformation signals from a set of interferograms and to estimate the atmospheric phase screen and digital elevation model (DEM) errors. In this chapter, we first review the basic concepts and principles related to PSI. We then present the mathematical models and the data reduction procedures of the PSI methodologies that have been developed by our research group for regional deformation detection. The experimental results derived by the two approaches using the European Remote Sensing (ERS)-1/2 SAR imagery are also reported for mapping spatiotemporal deformation over the two study areas: Phoenix, Arizona, and the Los Angeles basin, California, in the United States.

14.1 INTRODUCTION

Interferometric synthetic aperture radar (InSAR) has proven very useful in extracting surface displacements for a large area with an accuracy of centimeters to millimeters (Gabriel et al., 1989). Several imaging SAR sensors onboard satellites such as European Remote Sensing (ERS)-1/2, ENVISAT, Advanced Land Observing Satellite (ALOS), RADARSAT-1/2, TerraSAR-X, and COSMO-SkyMed have provided huge amounts of data to investigate surface deformation. The displacements in the radar line of sight (LOS) are principally estimated by phase differencing between two SAR images collected at different times over an area of interest. In recent years, InSAR has been widely used to investigate various deformation events, such as earthquakes (e.g., Massonnet et al., 1993; Zebker et al., 1994; Liu et al., 2004; Zhang et al., 2008; Liu et al., 2010; Feng et al., 2010), volcanoes (e.g., Lu, 1998; Lu et al., 2003), glaciers (e.g., Mattar et al., 1998; Strozzi et al., 2008), and subsiding activities (e.g., Buckley, 2000; Buckley et al., 2003; Ding et al., 2004; Liu, 2006).

The quality of InSAR measurements, however, is highly dependent on the features of the imaged surfaces and on the meteorological conditions. Two types of major factors are impeding InSAR applications: spatial and temporal decorrelation (Zebker and Villasensor, 1992) and atmospheric artifacts (Zebker and Rosen, 1996; Ding et al., 2008; Li et al., 2010). In general, the longer the time interval between two SAR acquisitions, the lower the signal-to-noise ratio (SNR) of the interferometric phase because of random surface change over time. This may lead to failure in detecting deformation, particularly in slowly deforming and heavily vegetated areas (Ding et al., 2004; Liu, 2006). The signal-to-noise ratio (SNR) also drops with the longer orbital separation between the two SAR acquisitions because of the diverse radar look angles. Moreover, the atmospheric inhomogeneity in time and space can cause varying phase delays that cannot be cancelled out by phase differencing, thus contaminating deformation measurements.

Several advanced methodologies have been proposed in recent years by various research groups to mitigate the technical limitations of InSAR. Although these proposed methods differ in the mathematical models or the data reduction procedures, they share the same ideas on the data use and the objects tracked for deformation analysis. In these methods, the time-series of SAR images collected over the same area are used to analyze the temporal behavior of deformation only for hard objects (e.g., rocks, roofs, bridges) with temporal stability of radar reflectivity. Such objects are often called temporally coherent targets or persistent scatterers (PS). Since Usai (1997) first found that deformation can be derived from the points that retain high radar coherence over time (see Figure 14.1), many efforts have been made to develop robust algorithms for mapping spatiotemporal defor-mation in urban and nonurban areas. Such an advanced InSAR technique is often termed PS InSAR (PSI).

Technical progress regarding PSI can be witnessed from two aspects: (a) PS identification (e.g., Ferretti et al., 2001; Werner et al., 2003; Hooper, 2004; Adam and Bamler, 2005; Shanker and Zebker, 2007; Zhang et al., 2011a), and (b) data modeling and parameter estimation (e.g., Usai, 2000; Ferretti et al., 2000; Berardino et al., 2002; Mora et al., 2003; Lanari et al., 2004a; Kampes and Hanssen, 2004; Hooper and Zebker, 2007; Hooper, 2008; Liu et al., 2008; Adam and Parizzi, 2009; Liu et al., 2009; Liu et al., 2011; Chen et al., 2010; Shanker and Zebker, 2010; Lauknes et al., 2011; Zhang et al., 2011b). Indeed, by exploring the coherent points in SAR images (see Figure 14.1) and analyzing their phases as a function of time and

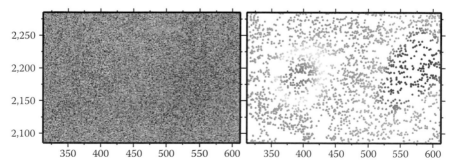

Figure 14.1 **(See color insert.)** An example showing that the coherent points identified from (a) the highly noisy phases can be used to estimate (b) the deformation pattern.

space, one can precisely estimate the spatiotemporal patterns of deformation signals at expense of spatial resolution. The unique capability of PSI methods has given rise to new tools for tracking the dynamic process of crustal movement or land subsidence because of anthropic activities.

This chapter is concerned with the mathematical models and the data reduction procedures of the PSI technique for regional deformation mapping. In Section 14.2, we give a brief review of the basic principles behind the classical PSI technique for which both PS detection and parameter estimation are addressed. We then describe in Sections 14.3 and 14.4 the mathematical models and the data-reduction procedures of the two new alternative PSI methods, respectively, which have been developed by our group to partially overcome the limitations of classical PSI. The two case studies using the time-series of ERS-1/2 SAR images have been carried out to verify the performance of the proposed methods for tracking the spatiotemporal deformation pattern over Phoenix, Arizona, and the Los Angeles basin, California, in the United States, respectively.

14.2 BASIC PRINCIPLES OF PSI

14.2.1 PS identification

A PS is a natural or manmade target with interpretable phase characteristics in time. It means that a PS can maintain the stability of radar reflectivity over months to years, thus resulting in a high SNR of phase observations at a PS and benefiting the deformation extraction. Such a PS may also be termed a permanent scatterer or a coherent point in the literature. The detection of PS is usually the first step and plays an important role in PSI analysis because the quality and density of PSs strongly affect the subsequent estimation of parameters. Several criteria proposed in recent years for screening out PSs will be discussed as follows.

14.2.1.1 Amplitude dispersion index

The amplitude dispersion index (ADI) was first introduced by Ferretti et al. (2001) for their patented permanent scatterer InSAR technique, which employs a single master image for generating interferograms without baseline thresholding. As the interferograms are highly affected by spatial decorrelation, it is impossible to use coherence criteria to select pixels with good phase quality. Theoretical analysis showed that the phase dispersion (σ_ϕ) can be estimated using the time-series of amplitude values of a pixel, which is expressed by (Ferretti et al., 2001) as follows:

$$\sigma_\phi \cong \frac{\sigma_A}{m_A} = D_A, \qquad (14.1)$$

where m_A and σ_A are the mean and the standard deviation of the time-series of amplitude values, respectively. The ratio σ_A/m_A is termed ADI. Simulation studies indicated that the ADI is a good approximation for phase dispersion of a pixel with a high SNR. In the case of the availability of sufficient SAR images (typically >25) after radiometric calibration, a pixel with ADI below a given threshold (e.g., 0.25) is identified as a PS candidate.

14.2.1.2 Signal-to-clutter ratio

The signal-to-clutter ratio (SCR) approach was first suggested by Adam (2004) to select coherent points for PSI analysis. Assuming that the phase observation of a PS includes the deterministic signal that is disturbed by random circular clutter with Gaussian distribution, the SCR can be estimated by calculating the ratio of the power of a PS candidate to that of its adjacent pixels. The relation between SCR and the phase standard deviation (σ_ϕ) can be defined follows (Adam, 2004):

$$\sigma_\phi \cong \frac{1}{\sqrt{2 \cdot SCR}}, \; SCR = \frac{s^2}{c^2}, \tag{14.2}$$

where s is the amplitude of the dominant scatterer and c the clutter of the adjacent pixels.

Equation 14.2 can be used to determine a reasonable threshold of SCR for PS selection. For example, if the phase standard deviation for a PS candidate should not be greater than 0.5 radians, one can determine that the threshold of the SCR should be 2. The SCR estimation is generally performed with a single SAR image. A pixel with high SCR (greater than the threshold) through the entire time span of all SAR acquisitions is selected as a PS candidate.

14.2.1.3 Phase stability

To identify coherent points in nonurban areas in which scatterers usually have low SNR in phase measurements, Hooper et al. (2004) proposed a new selection method based on the concept of phase stability. The analysis of phase stability is performed under the assumption that deformation is spatially correlated. Given a set of interferograms after removal of topographic contribution, a measure of phase stability for a pixel can be defined as follows

$$\gamma_x = \frac{1}{N} \left| \sum_{i=1}^{N} \exp\left\{ j \left(\Phi_{int,x,i} - \overline{\Phi}_{int,x,i} - \Delta\hat{\Phi}_{\varepsilon,x,i} \right) \right\} \right|, \tag{14.3}$$

where N is the number of interferograms, $\Phi_{int,x,i}$ is the differential phase of the pixel x from the ith interferogram, $\overline{\Phi}_{int,x,i}$ is the mean phase of the pixels within a circular patch (with radius L) centered on the pixel x, and $\Delta\hat{\Phi}_{\varepsilon,x,i}$ is the phase component due to errors in the digital elevation model (DEM) used in differential processing.

The threshold γ_x is determined in a probabilistic fashion with the assumption that a non-PS pixel has a coherence value of smaller than 0.3 (Shanker and Zebker, 2007). To decrease the computational burden of calculating $\bar{\Phi}_{int,x,i}$ for all pixels, the preliminary PS candidates (PPSC) are first selected by ADI thresholding. The true PSs are then screened out from the PPSC set with iterative procedures based on the phase stability analysis.

14.2.2 Modeling and parameter estimation

Although several typical PSI approaches (e.g., Ferretti et al., 2000; Berardino et al., 2002; Mora et al., 2003; Hooper and Zebker, 2007) have been proposed in recent years, we will describe the basic ideas of PSI modeling and parameter estimation for deformation extraction by following the work by Ferretti et al. (2000) and Colesanti et al. (2003a).

14.2.2.1 Phase modeling

With $N + 1$ SAR images available for deformation analysis, one can obtain N full-resolution differential interferograms formed by sharing the same sole master image. For a PS candidate (PSC) at the pixel x in the ith interferogram with temporal baseline t_i, the differential phase can be written as follows:

$$\Phi(x, t_i) = \phi_{top}(x, t_i) + \phi_{def}(x, t_i) + \phi_{atm}(x, t_i) + \phi_{noise}(x, t_i), i = 1\dots N, \quad (14.4)$$

where $\phi_{top}(x, t_i)$, $\phi_{def}(x, t_i)$, $\phi_{atm}(x, t_i)$, and $\phi_{noise}(x, t_i)$ stand for the phase components due to topographic error, surface motion in radar LOS, atmospheric delay in radar LOS, and decorrelation noise, respectively. The topographic phase can be expressed as follows:

$$\phi_{top}(x, t_i) = \partial(x, t_i) \cdot \Delta h_x, \quad (14.5)$$

where $\partial(x, t_i) = 4\pi \cdot B_{x,i}^{\perp}/(\lambda \cdot R_x \cdot \sin\theta_x)$, that is, the conversion factor accounting for the perpendicular baseline $B_{x,i}^{\perp}$ of the ith interferogram and the radar parameters, including the radar wavelength λ, the radar look angle θ_x and the sensor-to-PS range R_x. Δh_x in Equation 14.5 is the error in elevation of the DEM used. The deformation phase can be expressed as follows:

$$\phi_{def}(x, t_i) = \beta(t_i) \cdot v_{xi} + \phi_{nldef}(x, t_i), \quad (14.6)$$

where $\beta(t_i) = 4\pi \cdot t_i / \lambda$, v_x is the deformation rate in radar LOS at the pixel x, and $\phi_{nldef}(x, t_i)$ is the phase component due to nonlinear motion. Therefore, the differential phase can be rewritten as follows:

$$\Phi(x, t_i) = \partial(x, t_i) \cdot \Delta h_x + \beta(t_i) \cdot v_x + \omega(x, t_i), \qquad (14.7)$$

where $\omega(x, t_i)$ is the sum of three phase components related to atmospheric delay, noise, and nonlinear motion. The difference of the differential phases of two adjacent PSCs (at the pixels x and y, and their connection is often referred to as an arc) can be expressed as follows:

$$\Delta\Phi(x, y, t_i) = \partial(x, t_i) \cdot \Delta h_{x,y} + \beta(t_i) \cdot \Delta v_{x,y} + \Delta\omega(x, y, t_i), \qquad (14.8)$$

where $\Delta h_{x,y}$, $\Delta v_{x,y}$, and $\Delta\omega(x, y, t_i)$ are the differential elevation error, the differential deformation rate, and the differential part of the remaining components (as indicated by Equation 14.7), respectively.

14.2.2.2 Estimation of linear parameters

The estimation of the two linear parameters (i.e., $\Delta h_{x,y}$ and $\Delta v_{x,y}$) of an arc can be performed by searching through a given solution space. Under the condition of $|\Delta\omega_i| < \pi$, the coherence $\hat{\gamma}_{x,y}$ ($\in [0, 1]$) of an arc can be measured from the entire interferometric data set as follows:

$$\hat{\gamma}_{x,y} = \left| \frac{1}{N} \sum_{i=1}^{N} e^{j\Delta\omega(x,y,t_i)} \right|, \qquad (14.9)$$

where $j = \sqrt{-1}$, and $\Delta\omega(x, y, t_i)$ is derived by the following

$$\Delta\omega(x, y, t_i) = \Delta\Phi(x, y, t_i) - \partial(x, t_i) \cdot \Delta h_{x,y} - \beta(t_i) \cdot \Delta v_{x,y}. \qquad (14.10)$$

As indicated by Ferretti et al. (2000), the best solution for $\Delta h_{x,y}$ and $\Delta v_{x,y}$ can be derived by maximizing $\hat{\gamma}_{x,y}$. In practice, the process is conducted by sampling the two-dimensional (2D) solution space with a given resolution and bounds and evaluating the coherence for every possible solution (Kampes, 2006). The final result of the two linear parameters should correspond to the maximal coherence in the solution space. The higher the final coherence value of an arc, the more accurate the two linear parameters derived in this way.

The quality check for all arcs in the study area must be made after the solution of all arcs based on Equations 14.9 and 14.10. Ferretti et al. (2000) indicated that any arc with a maximized coherence value smaller than 0.75 should be viewed as an unreliable arc, thus eliminating it from the list of arcs for subsequent analysis. The absolute parameters (elevation errors and deformation rates) at the valid PSCs can then be derived by integrating the differential elevation errors and deformation rates of the arcs with the use of a reference point.

14.2.2.3 Estimation of atmospheric phase screen

After removing the phase components contributed by DEM error and linear motion of every arc, the residual phases at the valid PSCs can be unwrapped by the weighted least squares approach. The resultant absolute residual phase $\omega(x, t_i)$ of each PSC (at pixel x) is the sum of several components due to atmospheric delay, nonlinear motion, and random noise. The atmospheric phase screen (APS) behaves randomly in time and correlatively in space. On the other hand, the nonlinear motion generally has a shorter spatial correlation distance and low temporal frequency. Therefore, the APS of each SAR acquisition may be isolated from the other components by low-pass filtering in the spatial domain and high-pass filtering in the temporal domain (Ferretti et al., 2000).

The APS at the time of the master acquisition can be derived by averaging the residual phases of all the interferometric pairs. As the master-related atmospheric phase $\bar{\omega}(x)$ will not pass the high-pass filter, it should be subtracted from the residual phase $\omega(x, t_i)$, that is,

$$\omega'(x, t_i) = \omega(x, t_i) - \bar{\omega}(x). \tag{14.11}$$

Then the temporal high-pass filtering is performed to remove the temporally correlated motion from the residual phase. Finally, the spatial low-pass filter is applied to the temporally filtered residuals to remove the random noise part (Kampes, 2006). The order of filtering steps can be interchanged. The estimated atmospheric phase $(\hat{\phi}_{atmo}(x, t_i))$ at the pixel x in the ith interferogram can be symbolically expressed as follows

$$\hat{\phi}_{atmo}(x, t_i) = [[\omega'(x, t_i)]_{HP\_time}]_{LP\_space} + [\bar{\omega}(x)]_{LP\_space}. \tag{14.12}$$

For the filtering implementation, Ferretti et al. (2000) utilized a triangular window with a length of 300 days for the temporal filter and a window of $2 \times 2 \ km^2$ for the spatial filter.

14.2.2.4 Repeated estimation and limitation

After estimating the atmospheric phases at the discrete PSCs, the full resolution of APS for each interferometric pair can be computed by Kriging interpolation (Olea, 1999). With the availability of all the APSs, each differential interferogram can be calibrated by removing the relevant APS. The procedures as discussed in Sections 14.2.2.2 and 14.2.2.3 can be repeated once again for better solution of all the parameters of interest. In this way, more true PSs can be obtained, and the deformation parameters can be refined.

The aforementioned algorithms for PS selection and parameter estimation have been successfully applied to a number of case studies of deformation tracking. There is still room, however, for improvement regarding the implementation of the algorithms. For example, the PS identification cannot be performed satisfactorily with a small set of SAR images, thus increasing the cost for SAR data use. Although the SCR-based method for PS detection can work with only one image, it may result in failure for screening out PSs in urban areas in which the PSs are densely distributed. Moreover, as the crucial steps in the PSI analysis, both the phase unwrapping and the signal decomposition are always the challenging tasks, and this more or less decreases the reliability of the PS solution. For these reasons, the two methodologies developed by our group will be presented in Sections 14.3 and 14.4, respectively, for the purpose of improving deformation extraction. Unlike the classical PSI technique using a single master image for interferometric combination, our two approaches employ the interferograms generated with multiple master images for PS solution.

14.3 FCN-BASED PSI

This section describes a PS-networking method in which all PS points within a given distance threshold are connected to form a freely connected network (FCN) for deformation analysis. Only interferometric combinations that have both small spatial and temporal baselines are used in the PS solution. The linear deformation rates and DEM errors are estimated based on the least squares (LS) principle. The time-series of phase measurements is reconstructed by the singular value decomposition (SVD) as done in the small baseline subset (SBAS) technique (Berardino et al., 2002) and then decomposed into nonlinear deformation and atmospheric signals by using empirical mode decomposition (EMD) (Huang et al., 1998). The obvious feature of this approach is that the PS solution is based on the use of FCN and EMD, which improves the extraction and separation of the signals of interest. Although the FCN-based PSI method has been explained by Liu et al. (2009), we briefly describe here the key procedures, which include PS detection, PS networking, data modeling and parameter estimation, and the separation of nonlinear deformation and atmospheric artifacts.

14.3.1 PS networking and differential phase extraction

Before detecting the PS candidates with M SAR images, the SAR image acquired in the middle of all SAR acquisition dates is selected as the reference image space, and the remaining M-1 SAR images are coregistered and resampled into the reference image space (Liu et al., 2009). The temporal effects in image matching can be minimized in this way. As the PSs will be used to form an observation network for further analysis, they need to be screened out from the decorrelated pixels or areas. Basically, we follow the strategy proposed by Ferretti et al. (2001) to identify the PS candidates on a pixel-by-pixel basis. Using the time-series of M SAR amplitude samples at each pixel, both the amplitude mean and standard deviation are first calculated, and the ADI is then derived. According to the theoretical analysis by Ferretti et al. (2001), any pixel with ADI less than 0.25 can be identified as a PS candidate.

After selecting all of the PS candidates, the neighboring PSs are connected to form a strong network. Unlike a triangular irregular network (TIN) as applied by Mora et al. (2003), we freely link the adjacent PSs using a given Euclidian distance threshold. Any two PSs are linked only if their distance is less than a given threshold (e.g., 1 kilometer [km]), thus forming an FCN. For example, Figure 14.2 shows the comparison between a TIN and an FCN formed with the same PSs. As indicated by Liu et al. (2009), such an FCN with more arcs can provide a better framework for the subsequent data modeling and parameter estimating by LS solution.

For each interferometric pair, the initial interferogram is derived by a pixel-wise conjugate multiplication between the master and slave SLC image. Both the precise orbital data and the external DEM are then used to remove the flat-Earth trend and topographic effects from each initial interferogram, thus resulting in N differential interferograms. No filtering is performed during these procedures to avoid deterioration of phase data

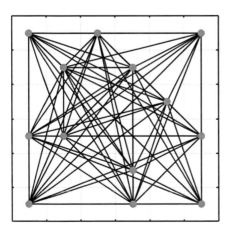 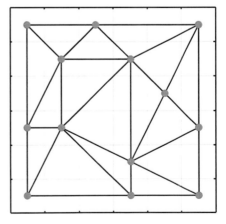

Figure 14.2 Examples of PS networking: (a) freely connected network and (b) triangular irregular network. (From Liu, G. X., S. M. Buckley, X. L. Ding, Q. Chen, and X. J. Luo, *IEEE Transactions on Geoscience and Remote Sensing* 47: 3209–3219, 2009. With permission.)

at PSs. The N differential interferometric phases at each PS candidate can be extracted according to its pixel coordinates.

14.3.2 FCN-based modeling and parameter estimation

The data modeling for deformation analysis is based on the idea of neighborhood differencing along each arc of the FCN (see Figure 14.2a). Being similar to the differential global positioning system (GPS), the differential modeling in the PS network is important for largely canceling out the spatially correlated biases or errors resulting from atmospheric phase delay and uncertainty in orbital data. For each interferometric pair, the differential interferometric phase between two PSs of each arc can be modeled according to Equation 14.8. Taking an arc (connecting PS x and PS y) in Figure 14.2a as an example, N observation equations can be written with the N interferometric pairs as follows:

$$\begin{cases} \Delta\Phi(x, y, t_1) = \partial(x, t_1) \cdot \Delta h_{x,y} + \beta(t_1) \cdot \Delta v_{x,y} + \Delta\omega(x, y, t_1) \\ \Delta\Phi(x, y, t_2) = \partial(x, t_2) \cdot \Delta h_{x,y} + \beta(t_2) \cdot \Delta v_{x,y} + \Delta\omega(x, y, t_2) \\ \qquad\qquad \cdots \\ \Delta\Phi(x, y, t_N) = \partial(x, t_N) \cdot \Delta h_{x,y} + \beta(t_N) \cdot \Delta v_{x,y} + \Delta\omega(x, y, t_N). \end{cases} \quad (14.13)$$

Although the phase measurement $\Delta\Phi(x, y, t_i)$ is ambiguous to within the integer multiples of 2π, both the differential elevation error $\Delta h_{x,y}$ and the differential deformation rate $\Delta v_{x,y}$ can be solved from the N equations given in Equation 14.13 by following Equations 14.9 and 14.10. The two linear parameters $(\Delta h_{x,y}, \Delta v_{x,y})$ can be derived by maximizing the following objective function:

$$\hat{\gamma}_{x,y} = \left| \frac{1}{N} \sum_{i=1}^{N} [\cos \Delta\omega(x, y, t_i) + j \cdot \sin \Delta\omega(x, y, t_i)] \right| \Rightarrow \text{maximum}, \quad (14.14)$$

where $\Delta\omega(x, y, t_i)$ is written as follows:

$$\Delta\omega(x, y, t_i) = \Delta\Phi(x, y, t_i) - \partial(x, t_i) \cdot \Delta h_{x,y} - \beta(t_i) \cdot \Delta v_{x,y}. \quad (14.15)$$

In our implementation, both $\Delta h_{x,y}$ and $\Delta v_{x,y}$ are determined by searching a predefined solution space (e.g., $\Delta v_{x,y} \in [-6, 6]$ millimeters per year [mm/yr], $\Delta h_{x,y} \in [-20, 20]$ meters [m]) to maximize the coherence value. Using the steps characterized by Equations 14.13–14.15, we can estimate the deformation-rate and elevation-error increment of each arc in the FCN.

Once the linear parameters of all arcs are determined, the FCN-based PS network can be treated by the weighted LS adjustment in which the coherence value of each arc is used for weighting (Liu et al., 2009). Before further analysis, the quality of the PS network must first be assessed using the coherence value of each arc as the measure. If the coherence value is less than 0.45 (Liu et al., 2009), the relevant arc is discarded from the LS solution, thus possibly resulting in some isolated PSs that are also removed from the list of PS candidates. The geometric inconsistency in the PS network can be eliminated by using the LS solution to derive the most probable values of deformation rates and elevation errors in an absolute sense at all the true PS points. A detailed description of the LS-based adjustment can be found in Liu et al. (2009). A reference point (e.g., a point located in a stable area) needs to be chosen and used in the LS solution. The LOS deformation rates of all other PS points are estimated relative to this reference point.

14.3.3 Separation of signatures

Further analysis concentrates on isolating nonlinear deformation from atmospheric artifacts. For each interferometric pair, the residual phase at each of the true PSs can be derived after removing the deformation-rate and elevation-error components (Ferretti et al., 2000; Liu et al., 2009). The residual phases are then unwrapped by the LS-based method (Ghiglia and Pritt, 1998), and thus the absolute residual phases are obtained at all the true PSs for each interferometric pair. The residual phases are due to nonlinear subsidence and atmospheric delay. Although a perfect separation of the two terms is a challenging task, we attempt to achieve this by introducing an approach that is referred to as EMD (Huang et al., 1998).

To decouple signatures at each PS, we first reconstruct the time-series of unwrapped residual phases corresponding to all the SAR acquisition dates using the SVD method (Berardino et al., 2002). After the SVD-based reconstruction, the M residual phases at a PS corresponding to the dates of SAR acquisitions $(d_1, d_2, ..., d_M)$ can be expressed as follows:

$$\psi(d) = [\psi(d_1), \psi(d_2), \cdots, \psi(d_M)]. \tag{14.16}$$

The signal separation is then conducted by the EMD method using the time-series of unwrapped residual phases (Liu et al., 2009). The derived two time-series of phases for nonlinear motion and atmospheric delay at each PS are expressed as follows:

$$\psi^{nldef}(d) = [\psi^{nldef}(d_1), \psi^{nldef}(d_2), \cdots, \psi^{nldef}(d_M)], \tag{14.17}$$

$$\psi^{atm}(d) = [\psi^{atm}(d_1), \psi^{atm}(d_2), \cdots, \psi^{atm}(d_M)]. \tag{14.18}$$

The overall deformation at a PS x can be computed by summing the linear and nonlinear components, that is,

$$def(x, d_i) = (d_i - d_1) \cdot \hat{v}_x + \frac{\lambda}{4\pi} \cdot \psi^{\text{nldef}}(x, d_i). \tag{14.19}$$

14.3.4 Case study for subsidence detection by FCN-based PSI

14.3.4.1 Study area and data set used

The western part of Phoenix (in Arizona, United States) consists of Deer Valley and the west Salt River Valley (Buckley, 2000), including several towns (i.e., Glendale, Peoria, and Sun City). The excessive pumping of groundwater in the areas has resulted in a large extent of land subsidence. The early interferometric study with a limited set of ERS C-band SAR images reported that several subsiding bowls were exhibited across the towns (Buckley, 2000). Although the PSI analysis over this study area was reported in Liu et al. (2009), we summarize the primary results of the case study for subsidence detection.

We attempted to detect the spatiotemporal subsidence distribution in Phoenix by the method presented in Sections 14.3.1–14.3.3. For this study, we used 39 ERS-1/2 SAR images acquired from 1992 to 2000, which were combined to form 86 interferometric pairs by thresholding spatial and temporal baselines shorter than 120 m and 4 years, respectively. Figure 14.3 shows the spatiotemporal baselines of all the 86 interferometric pairs used for the PS solution. We only concentrated on a patch of 27 km by 15 km within the SAR frame. Figure 14.4 shows the SAR amplitude image of the area, averaged from all the images. Most of the area is urban suburbs with the major streets appearing as dark lines and buildings as bright spots, and a small part of the area is farmland located in the lower-left corner. The detected PS candidates (to be discussed later) that are basically within the urban area were superimposed onto the amplitude image.

14.3.4.2 Experimental results and analysis

The 86 differential interferograms were generated with the "two-pass" method (e.g., Buckley, 2000; Liu, 2006) by using the Gamma DIFF software. Precise orbit state vectors computed by the Technical University of Delft and the Shuttle Radar Topography Mission (SRTM) DEM data with a height accuracy of about 10 m were used to remove the flat-Earth trend and the topographic components. Both the SAR images and the differential interferograms were processed with a multilooking factor of 5 in azimuth and 1 in range, resulting in pixel dimension of about 20×20 m.

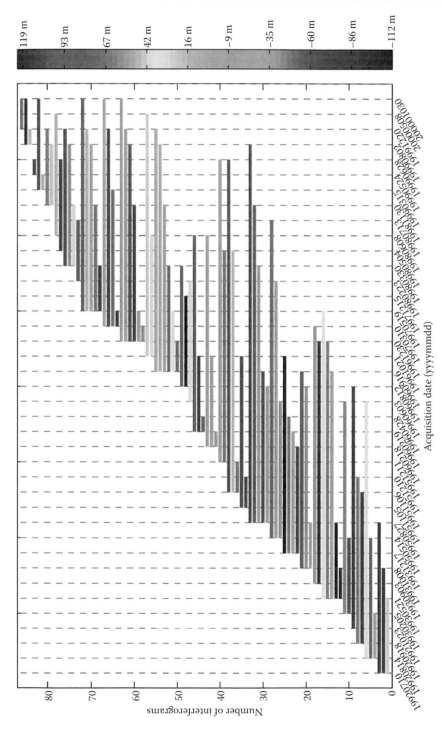

Figure 14.3 **(See color insert.)** Spatiotemporal baselines of the 86 ERS-1/2 interferometric pairs used for PS solution.

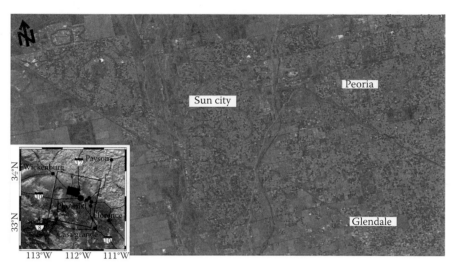

Figure 14.4 **(See color insert.)** SAR amplitude image of the study area averaged from all the images. The PS candidates are superimposed as red points. The inset shows the geographic setting of the study area (tilted dark box) in Phoenix. The tilted rectangle identifies the SAR frame. (From Liu, G. X., S. M. Buckley, X. L. Ding, Q. Chen, and X. J. Luo, *IEEE Transactions on Geoscience and Remote Sensing* 47: 3209–3219, 2009. With permission.)

As shown in Figure 14.4, the 14,618 temporal-coherent PS candidates were selected by applying ADI thresholding. As expected, the density of the PS points appears high in the urban areas, 48 square kilometers (km²) on average, whereas that in the farmland is very low. A strong network of PS points is formed by freely connecting the PS to all those that are within 1 km in distance, resulting in 1,463,306 connected PS pairs in total. The increments of both linear deformation rates and elevation errors along each of the pairs were then estimated by maximizing the coherence with Equations 14.13–14.15. A coherence threshold of 0.45 was used to reject pairs with low coherence and nonconnected PS candidates. As a result, 14,493 PS pixels and 1,433,233 connected PS pairs remain as the input for the subsequent LS network adjustment computation. The linear deformation rates and DEM errors at 14,428 PS points were eventually estimated from 1,225,265 "clean" PS pairs.

Figure 14.5 shows the linear subsidence rate (in mm/yr) map derived after Kriging interpolation and by projecting the LOS displacements onto the vertical direction. It is clear that a subsiding bowl with a diameter of about 5 km appears in Glendale that has a peak subsidence rate of 54 mm/yr, while a wider subsiding bowl with a diameter of about 15 km spans Glendale, Peoria, and Sun City and has a peak subsidence rate of 30 mm/yr. It can be inferred from the results that the accumulative linear subsidence over the time period of the SAR acquisitions (about 8 years) is up to

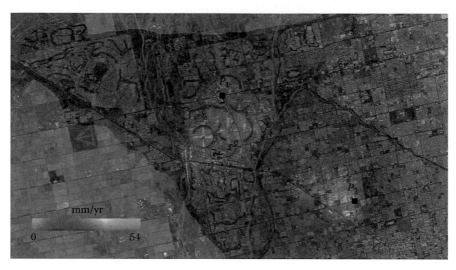

Figure 14.5 **(See color insert.)** Distribution of linear subsidence rates in millimeters per year. The subsidence in the farmland is not available because of the lack of PS points in the area. P1 and P2 are two PS points to be analyzed subsequently. (From Liu, G. X., S. M. Buckley, X. L. Ding, Q. Chen, and X. J. Luo, *IEEE Transactions on Geoscience and Remote Sensing* 47: 3209–3219, 2009. With permission.)

43 centimeters (cm) and 24 cm, respectively, at the two peak points. The eastern part of the study area appears basically stable. In the farmland to the northwest of Sun City (see Figure 14.4), the linear subsidence rates derived with the Kriging interpolator may be less reliable because of the sparsity of PS points in the area.

The fidelity of the estimated subsidence rates has been checked by visually comparing the observed differential interferograms with those simulated using the subsidence-rate map. For example, Figure 14.6 shows such a comparison for the differential interferograms with a time interval of about 4 years. It is evident that they are in good agreement. Some minor inconsistencies in some areas can be ascribed to atmospheric artifacts, topographic errors, and nonlinear motion. It also can be seen that the smaller but deeper subsiding bowl in Glendale can be completely revealed by the PS networking method. Its complete shape and extent, however, do not show in any observed individual differential interferograms because of temporal decorrelation. All these not only verify that the estimation approach is powerful and reliable but also suggest that the linear subsidence in the study area dominates the nonlinear component.

The nonlinear subsidence was separated from the atmospheric artifacts by both the SVD and EMD methods. Figure 14.7 shows the temporal evolution of atmospheric delay in the LOS direction, and nonlinear and total subsidence at two PS points (P1 and P2) near the centers of two subsiding

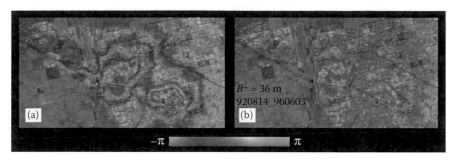

Figure 14.6 **(See color insert.)** Comparison between (a) simulated and (b) observed differential interferograms with a time interval of about 4 years.

bowls (see Figure 14.5). The atmospheric variation is evidently random in time. The atmospheric artifacts at P2 range from –2.0 cm to 2.1 cm, which are slightly higher than those at P1. Point P2 presents a dynamic range of –2.5 cm to 2.2 cm nonlinear subsidence, whereas point P1 has a narrower range of nonlinear subsidence (–2.0 cm to 1.4 cm). Additionally, it can be seen that point P1 located near the deeper subsiding bowl exhibits more seasonal undulation than point P2 located near the shallow subsiding bowl. From the two profiles of total subsidence, we stress once again that the linear trend of subsidence dominates the nonlinear component in this study area.

We have carried out the solution using a TIN formed by linking PS pixels with Delaunay triangles and deleting PS pairs larger than 1 km apart. It was

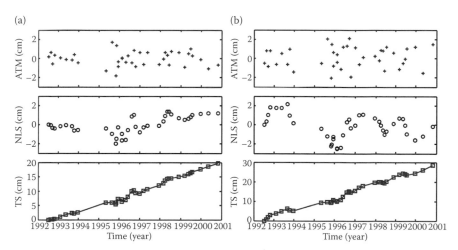

Figure 14.7 (a)–(b) Temporal variation of atmospheric (ATM) delays in the LOS direction, nonlinear subsidence (NLS), and total subsidence (TS) at two PS points as marked in Figure 14.4. (a) is for P1 and (b) for P2. (From Liu, G. X., S. M. Buckley, X. L. Ding, Q. Chen, and X. J. Luo, *IEEE Transactions on Geoscience and Remote Sensing* 47: 3209–3219, 2009. With permission.)

found that more crucial PS candidates are removed from the TIN when carrying out coherence thresholding, as the weak TIN results in more isolated PS points. Therefore, at the end, only 12,550 valid PS points remain in the TIN compared with 14,428 in the FCN. As the computer time of a PS solution is primarily governed by the numbers of Δh and Δv to be estimated, the computer time of the FCN-based PS solution can be potentially much higher than that of the TIN-based PS solution because FCN usually has many more PS connections than a TIN. For the study area, the computer time of the FCN-based PS solutions is about 30 times greater than that of the TIN-based PS solutions. It can be concluded that the FCN is much stronger and more advantageous in terms of both accuracy and reliability, although the FCN-based PSI analysis requires more computer time.

14.4 TEMPORARILY COHERENT POINT InSAR

The temporarily coherent point (TCP) InSAR is alternative implementation of PSI, which was proposed by Zhang et al. (2011a, 2011b). It can be applied to estimate the deformation rates without the need of phase unwrapping. The TCPInSAR approach proposed is characterized by several innovations in TCP identification, TCP networking, and LS-based parameter solution.

Because TCPInSAR focuses only on the interferometric pairs with reasonably short spatial and temporal baselines, the first step of the methodology deals with the selection of such pairs. After calculating the spatiotemporal baselines of all possible combinations from the SAR images available for the study area, we select the interferometric pairs for TCP solution using the baseline criteria. The criteria are generally determined according to the interferometric quality and the maximum magnitude of deformation rates in the study area, which can be inferred by the fringe density in interferograms. The SAR images related to the selected pairs are the basic observations for extracting deformation rates by TCPInSAR.

Both the modeling and data-reduction procedures of this approach will be described shortly. TCP identification and coregistration and TCP networking are addressed in Section 14.4.1. The mathematical framework underlying TCPInSAR is explained in Section 14.4.2 for the purpose of estimating deformation parameters. A case study for subsidence detection by TCPInSAR is reported in Section 14.4.3.

14.4.1 TCP identification, coregistration, and networking

TCPs are defined as the surface targets that can maintain the satisfactory radar coherence in one or several interferograms with different time intervals. This means that some surface objects may remain stable in radar reflectivity for some time spans but not for others. We identify such TCPs by statistical calculation of the offsets in range and azimuth directions between two SAR images of each interferometric pair. Such offsets can be estimated through image matching on a pixel-by-pixel basis. Bamler (2000) indicated that the standard deviation of the offsets estimated for a point-like scatterer with strong radar reflection is less sensitive to the window size and the oversampling factor used for image matching than that for a distributed scatterer with weak radar reflection. Therefore, it is possible to distinguish the point-like scatterers from the distributed scatterers by offset estimation.

The method of TCP identification based on offset estimation can be found in Zhang et al. (2011a). This section briefly describes the key procedures of this method. The TCP candidates are first selected based on spectral diversity (Werner et al., 2003). The strongly backscattering points are expected to retain similar energies when processed by different looks with fractional azimuth and range bandwidth. The points with low spectral diversity are identified as the TCP candidates. The TCP candidates are further evaluated by changing the size of the image patches (e.g., varying from 5×5 to 125×125) and the oversampling factor used for image matching to find the offsets at a subpixel level. A fixed oversampling factor can be used for simplicity. A set of offsets (OT_j) at a given TCP candidate (j) can be derived for the N set of different parameters used in image matching. Any TCP candidate with a standard deviation of less than 0.1 pixels in offsets is selected as a TCP, that is,

$$OT_j = [ot_{j1}\ ot_{j2}\ ... \ ot_{jN}]$$
$$std\,(OT_j) < 0.1.$$

(14.20)

A high-order polynomial is finally applied to fit the offsets of TCPs screened out by Equation 14.20. The final list of TCPs is refined by discarding the pixels whose offsets do not fit the polynomial well. The method proposed is useful in areas in which the conventional differential InSAR technique does not work properly because of decorrelation effects and in the case of insufficient SAR images being available for PS identification by ADI thresholding.

Although the coregistration step is critical in the PSI analysis, there is usually no special consideration for PS precise coregistration in the current PSI methods. All the slave SAR images are typically coregistered to a single master SAR image based on range and azimuth offsets of distributed windows over the images. It was reported that the majority (typically 90%) of pixels in an SAR image correspond to the distributed scatterers (i.e., non-PS pixels) (Kampes, 2006). The error in offsets estimated at a non-PS pixel for coregistration, in general, is larger than that in offsets estimated at a PS pixel. Therefore, caution should be taken in critically utilizing the information from offsets during coregistration. To determine the accurate polynomial for coregistration, we only use the set of offsets at the TCPs determined by Equation 14.20. A six-point truncated sinc interpolator kernel (Hanssen, 2001) is employed to resample the pixels corresponding to the TCPs in the slave image using the accurate polynomial.

Once the TCPs are identified, a network is generated to connect the adjacent TCPs (each connection is referred to as an arc) for differential modeling. As shown in Figure 14.8a, the Delaunay triangulation has been used widely for this purpose. It can be deduced from Section 14.3.4.3, however, that such a networking method cannot provide a satisfactory framework for the TCP solution. To balance between the computation burden and reliability for the TCP solution, we connect the TCPs using the local Delaunay triangulation method. A grid with a given spacing (e.g., 100 m) is first set onto the entire study area, and the local Delaunay triangulation is then performed around each grid node. All the TCPs within the circle of a given radius (e.g., 750 m) centered on the grid node are used for the local Delaunay triangulation. Figure 14.8b shows such a TCP network that can keep a sufficient number of arcs and an acceptable computation complexity.

14.4.2 TCP solution by LS estimator

As the interferometric pairs chosen for deformation analysis have short spatial and temporal baselines, it can be observed that the differential interferometric phase between two TCPs of an arc with a short length generally has no phase ambiguity. In this situation, Zhang et al. (2011b) indicated that the deformation rates at TCPs can be estimated by the LS approach. Suppose that N differential interferograms with temporal baselines of t_i ($i = 1, 2, ..., N$) have been generated from a set of M SAR

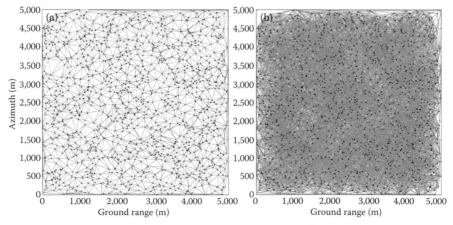

Figure 14.8 TCP networking by (a) global triangulation and (b) local triangulation.

images over the study area. For an arc formed by connecting PS x and PS y, N observation equations can be written with the N interferometric pairs as follows:

$$\begin{cases} \Delta\Phi(x, y, t_1) = \partial(x, t_1) \cdot \Delta h_{x,y} + \beta(t_1) \cdot \Delta v^1_{x,y} + \Delta\omega(x, y, t_1) \\ \Delta\Phi(x, y, t_2) = \partial(x, t_2) \cdot \Delta h_{x,y} + \beta(t_2) \cdot \Delta v^2_{x,y} + \Delta\omega(x, y, t_2) \\ \qquad\qquad \cdots \\ \Delta\Phi(x, y, t_N) = \partial(x, t_N) \cdot \Delta h_{x,y} + \beta(t_N) \cdot \Delta v^N_{x,y} + \Delta\omega(x, y, t_N). \end{cases} \tag{14.21}$$

Equation 14.21 is very similar to Equation 14.13. The differential deformation rates of the arc for all the temporal baselines are treated as nonconstant, however, that is,

$$\Delta V_{x,y} = [\Delta v^1_{x,y} \ \Delta v^2_{x,y} \ \cdots \ \Delta v^N_{x,y}]^T, \tag{14.22}$$

where $\Delta v^i_{x,y}$ stands for the differential rate related to the ith interferometric pair with a temporal baseline t_i. Equation 14.21 can be rewritten in the matrix form as follows:

$$\Delta\Phi = A \begin{bmatrix} \Delta h_{x,y} \\ \Delta V_{x,y} \end{bmatrix} + W, \tag{14.23}$$

where

$$\Delta\Phi = [\Delta\Phi(x, y, t_1) \ \Delta\Phi(x, y, t_2) \cdots \Delta\Phi(x, y, t_N)]^T, \tag{14.24}$$

$$W = [\Delta\omega(x, y, t_1) \ \Delta\omega(x, y, t_2) \cdots \Delta\omega(x, y, t_N)]^T, \tag{14.25}$$

$$A = \begin{bmatrix} \partial(x, t_1) & \partial(x, t_2) & \cdots & \partial(x, t_N) \\ \beta(t_1) & \beta(t_2) & \cdots & \beta(t_N) \end{bmatrix}^T. \tag{14.26}$$

The unknown parameters $\hat{X}$ can be solved by the LS approach, which can be expressed as follows:

$$\hat{X} = \begin{bmatrix} \Delta\hat{h}_{x,y} \\ \Delta\hat{V}_{x,y} \end{bmatrix} = (A^T P^{dd} A)^{-1} A^T P^{dd} \Delta\Phi, \tag{14.27}$$

where denotes the estimated quantities, and P^{dd} is the a priori weight matrix, which can be obtained by taking the inverse of the variance matrix of the double-difference phases. The matrix $A^T P^{dd} A$ may be singular.

A pseudo-inverse of the matrix obtained by SVD is used instead, that is, $(A^T P^{dd} A)^{-1} = (A^T P^{dd} A)^+$. The estimated quantities for $\Delta\Phi$ can be derived as follows:

$$\Delta\hat{\Phi} = A(A^T P^{dd} A)^{-1} A^T P^{dd} \Delta\Phi. \tag{14.28}$$

The residuals of the measurements can be derived as follows:

$$r = \Delta\Phi - A(A^T P^{dd} A)^{-1} A^T P^{dd} \Delta\Phi. \tag{14.29}$$

The variance matrices of the estimated quantities can be derived as follows:

$$\begin{aligned}
Q_{\hat{x}\hat{x}} &= (A^T P^{dd} A)^{-1} \\
Q_{\Delta\hat{\Phi}\Delta\hat{\Phi}} &= A(A^T P^{dd} A)^{-1} A^T \\
Q_{rr} &= Q^{dd} - A(A^T P^{dd} A)^{-1} A^T.
\end{aligned} \tag{14.30}$$

After the LS solution for each arc, the outlier detector proposed by Zhang et al. (2011b) is applied to check whether the phase measurements for the arc have phase ambiguities. Both the LS residuals and variance components for the arc are used for such a quality check. For subsequent analysis, the unacceptable arcs should be discarded. After deriving the parameters (i.e., elevation-error and deformation-rate increments) of all valid arcs, we can determine the parameters (i.e., elevation errors and deformation rates) in an absolute sense of all the valid TCPs by using a spatial integration method with a given reference point (Zhang et al., 2011b).

For the full time-series analysis, we usually estimate the linear deformation component as well as the DEM error by first updating the design matrix in Equation 14.27. After the linear deformation rates have been solved for the local Delaunay triangulation network, we then concentrate on retrieving the time-series of nonlinear deformations at each of the valid TCPs from the LS residuals. Before further analysis, space-time filtering is first applied to the LS residuals to suppress the possible atmospheric artifacts. The design matrix and observations can thus be updated for the repeated LS solution as shown in Equations 14.27–14.30. The time-series of nonlinear deformations at each TCP can be obtained in this way.

14.4.3 Case study for deformation detection by TCPInSAR

14.4.3.1 Study area and data set used

The Los Angeles (LA) basin located in southern California possesses complicated geological settings. It is bounded by the Newport-Inglewood Fault

(NIF) along the western margin (Hauksson, 1987; Hauksson et al., 1995). This main boundary fault is subparallel to the San Andreas Fault, and strikes roughly northwest. It has been identified as an active fault with the potential to trigger large earthquakes (Shen et al., 1996). Since 1920, at least five earthquakes with magnitude of 4.9 or larger have struck the regions, which were characterized by the focal mechanism of right-lateral strike slip. Although there were no surface ruptures generated by any historical earthquakes, the NIF and other blind thrust faults represent a major threat to the Los Angeles metropolitan area (Hauksson, 1987; Argus et al., 2005). Therefore, it is critical to investigate the surface deformation associated with the seismotectonic activity in the LA basin.

As a case study, we apply the TCPInSAR method proposed in Sections 14.4.1 and 14.4.2 to the LA basin for deformation tracking. The LA basin is suitable for testing a novel InSAR technique thanks to the moderate tectonic motion and the sparse vegetation coverage (i.e., resulting in high SNR for interferometric phases) in this region (Colesanti et al., 2003b; Lanari et al., 2004b). We generated 55 interferograms using 32 ERS-1/2 images (track 170, frame 2925) collected between 1995 and 2000 for this case study. Figure 14.9 shows the spatiotemporal baselines of all the 55 interferometric pairs used for the TCP solution.

14.4.3.2 Experimental results and analysis

Figure 14.10 shows the distribution of deformation rates in radar LOS estimated by the TCPInSAR method over the LA basin. The red triangle stands for the reference point for the TCP solution, and the red dots indicate the GPS sites with annotations of site names, which are used for validation. The GPS data set was taken from the Southern California Integrated GPS Network website (http://www.scign.org). The red squares correspond to the TCPs around oilfields. The inset shows the deformation rates around the oilfield indicated by the gray rectangle. The deformation rates along line AB are also shown. The TCP analysis takes the GPS site ELSC as the reference point.

The deformation rates as shown in Figure 14.10 are in good agreement with those derived using the SBAS method proposed by Casu et al. (2006) and Lanari et al. (2004b). Figure 14.10 shows that the deformation rates vary between –16 mm/yr and 12 mm/yr. Several factors are known to contribute to the surface deformation, such as oil and gas extraction, changes in groundwater storage, and the movement of active faults (Argus et al., 2005). The maximum deformation rate of 16 mm/yr occurred in the Wilmington oilfield, which is the largest oilfield in the LA basin. Figure 14.11 shows a comparison between the LS residual phases derived after the deformation-rate estimation and those derived after the nonlinear deformation-rate estimation for the 200th arc. The relatively small and uniform undulation

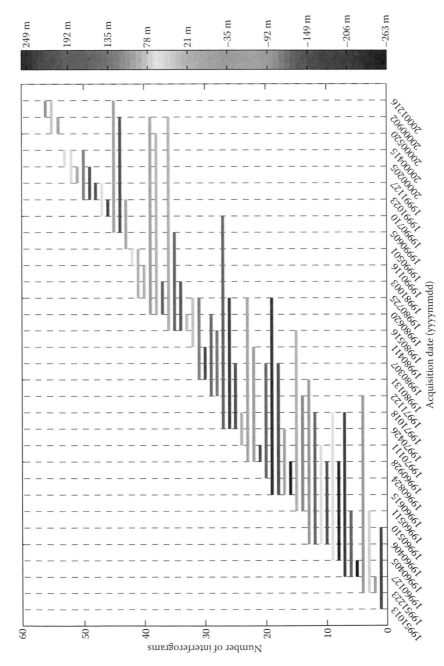

Figure 14.9 **(See color insert.)** Spatiotemporal baselines of the 55 interferometric pairs used in the study.

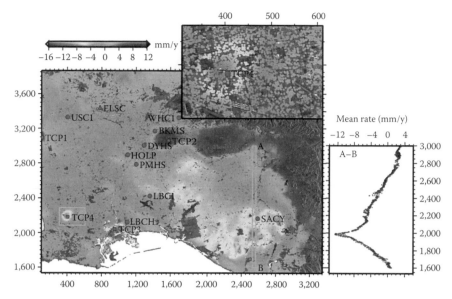

Figure 14.10 **(See color insert.)** The distribution of deformation rates estimated by the TCP-based InSAR method over the Los Angeles basin.

for the latter indicates that the nonlinear deformation components at the arc have been retrieved successfully.

For validation, we selected eight GPS sites from the SCIGN network, several of which were used by Casu et al. (2006) and Lanari et al. (2004b). The left and middle columns of Figure 14.12 clearly show that the deformation time-series derived by the GPS-based solution at the eight GPS sites are in very good agreement with those derived by the TCP-based solution; these time-series are also comparable with the results derived by the SBAS

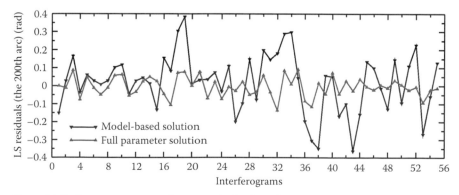

Figure 14.11 A comparison of phase residuals after the mean rate estimation and nonlinear rates deformation.

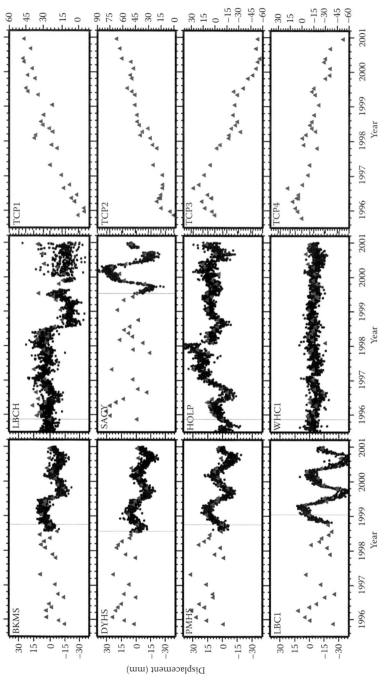

Figure 14.12 Comparison between the deformation time-series derived by the TCP-based solution and those derived by the GPS-based solution at the eight GPS sites (left and middle columns). The vertical line shows the reference time for both the GPS- and TCP-based results. The deformation time-series derived by the TCP-based solution at four TCPs (see Figure 14.11) is shown in the right column.

method (Casu et al., 2006; Lanari et al., 2004b). The deformation time-series at most of the GPS sites vary between −30 mm and 30 mm. It can be found from our TCP-based results that the relatively larger magnitude of deformation appears around the oilfields because of oil exploitation.

The right column of Figure 14.12 shows the deformation time-series derived by the TCP-based solution at the four TCPs located in the three oilfields, respectively. The upward ground motion resulting from an injection process occurred at TCP1, which is located in the Inglewood oilfield in the Baldwin Hills. The uplift trend also can be observed on TCP2, which is located in the Santa Fe Springs oilfield, and its fluctuations can be attributed to the variation of injection rates and the reduction of oil withdrawal. It is estimated that TCP3 and TCP4 in the Wilmington oilfield experienced an elevation loss of about 60 mm from October 1995 to December 2000. The subsidence rates at the two TCPs were occasionally mitigated during the time span of the entire SAR acquisitions, which might have been caused by increasing and realigning water injection (for more details, see http://www.conservation.ca.gov).

14.5 CONCLUSION

The deformation measurements derived by InSAR are of great interest to geophysicists because of their high spatial resolution, competent accuracy, and wide coverage. InSAR technology can greatly complement point-based geodetic methods, such as leveling and GPS surveying. InSAR-based deformation mapping can be carried out without much fieldwork, and this enables monitoring of regional deformation fields related to, for example, ice sheet motion, tectonic movement, volcanic dynamics, and land subsidence. InSAR is thus viewed as one of the most exciting members of the space-geodesy family. The full operational capability of InSAR in deformation mapping, however, has not yet been achieved because of the negative influences of spatiotemporal decorrelation and atmospheric artifacts. Advanced PSI techniques have been proposed in recent years to overcome the drawbacks in the conventional InSAR methodology.

The basic concepts and principles of the classical PSI are first reviewed in this chapter. We then present the mathematical models and the data-reduction procedures of the two new alternative PSI methods—FCN-based PSI and TCPInSAR—which have been developed by our group to partially overcome the limitations of the classical PSI. The experimental results of FCN-based PSI and TCPInSAR are also reported for deformation mapping over the two study areas, that is, Phoenix, Arizona, and the Los Angeles basin, California, in the United States, respectively, with the use of the time-series of ERS-1/2 SAR images. The obvious feature of FCN-based PSI is that the PS solution is based on the use of FCN and EMD, which benefits the extraction and separation of the signals of interest. FCN can provide a very good framework for the PS solution and thus improve both accuracy and reliability in spatiotemporal deformation measurements, although the FCN-based PSI analysis requires much more computer time. The proposed TCPInSAR features TCP identification based on offset statistics, local Delaunay triangulation, and LS-based parameter solution. TCPInSAR is more advantageous when few SAR images are available over an area of interest for deformation mapping.

ACKNOWLEDGMENTS

This work was jointly supported by the National Natural Science Foundation of China under Grant 41074005, the R&D Program of Railway Ministry under Grant 2008G031-5, the Program for New Century Excellent Talents in University under Grant NCET-08-0822, the National Basic Research Program of China (973 Program) under Grant 2012CB719901, and the Fundamental Research Funds for the Central Universities under Grants (SWJTU11CX139, SWJTU10ZT02, SWJTU11ZT13). The authors would like to thank Infoterra GmbH and U.S. Geological Survey for providing TerraSAR-X synthetic aperture radar images and Shuttle Radar Topography Mission digital elevation models, respectively.

REFERENCES

Adam, N. 2004. Development of a scientific permanent scatterer system: modifications for mixed ERS/ENVISAT time series. In *ENVISAT & ERS Symposium, Salzburg, Austria, September 6–10, 2004*, 1–9.

Adam, N., and R. Bamler. 2005. Parametric estimation and model selection based on amplitude-only data in PS-interferometry. In *Proceedings of Fringe 2005 Workshop, Frascati, Italy, November 28–December 2, 2005*.

Adam, N., and A. Parizzi. 2009. Practical persistent scatterer processing validation in the course of the Terrafirma project. *Journal of Applied Geophysics* 69(1): 59–65.

Argus, D. F., M. B. Heflin, G. Peltzer, F. Crampe, and F. H. Webb. 2005. Interseismic strain accumulation and anthropogenic motion in metropolitan Los Angeles. *Journal of Geophysical Research-Solid Earth* 110: B04401.

Bamler, R. 2000. Interferometric stereo radargrammetry: absolute height determination from ERS-ENVISAT interferograms. In *IEEE Proceedings of IGARSS*. Vol. 2, *Honolulu, HI, July 24–28, 2000*, 742–745.

Berardino, P., G. Fornaro, R. Lanari, and E. Sansosti. 2002. A new algorithm for surface deformation monitoring based on small baseline differential SAR interferograms. *IEEE Transactions on Geoscience and Remote Sensing* 40(11): 2375–2383.

Buckley, S. M. 2000. *Radar Interferometry Measurement of Land Subsidence.* Austin, TX: The University of Texas at Austin.

Buckley, S. M., P. A. Rosen, S. Hensley, and B. D. Tapley. 2003. Land subsidence in Houston, Texas, measured by radar interferometry and constrained by extensometers. *Journal of Geophysical Research* 108(B11): 2542–2554.

Casu, F., M. Manzo, and R. Lanari. 2006. A quantitative assessment of the SBAS algorithm performance for surface deformation retrieval from DInSAR data. *Remote Sensing of Environment* 102: 195–210.

Chen, Q., G. X. Liu, X. L. Ding, L. G. Yuan, J. C. Hu, P. Zhong, and M. Omura. 2010. Tight integration of persistent scatterer InSAR and GPS observations for detecting vertical ground motion in Hong Kong. *International Journal for Earth Observation and GeoInformation* 12(6): 477–486.

Colesanti, C., A. Ferretti, F. Novali, C. Prati, and F. Rocca. 2003a. SAR monitoring of progressive and seasonal ground deformation using the permanent scatterers technique. *IEEE Transactions on Geoscience and Remote Sensing* 41(7): 1685–1701.

Colesanti, C., A. Ferretti, C. Prati, and F. Rocca. 2003b. Monitoring landslides and tectonic motions with the Permanent Scatterers Technique. *Engineering Geology* 68: 3–14.

Ding, X. L., G. X. Liu, Z. W. Li, and Y. Q. Chen. 2004. Ground subsidence monitoring in Hong Kong with satellite SAR interferometry. *Photogrammetric Engineering & Remote Sensing* 70(10): 1151–1156.

Ding, X. L., Z. W. Li, J. J. Zhu, G. C. Feng, and J. P. Long. 2008. Atmospheric effects on InSAR measurements and their mitigation. *Sensors* 8(9): 5426–5448.

Feng, G. C., E. A. Hetland, X. L. Ding, Z. W. Li, and L. Zhang. 2010. Coseismic fault slip of the 2008 Mw 7.9 Wenchuan earthquake estimated from InSAR and GPS measurements. *Geophysical Research Letters* 37: L01302.

Ferretti, A., C. Prati, and F. Rocca. 2000. Nonlinear subsidence rate estimation using permanent scatterers in differential SAR interferometry, *IEEE Transactions on Geoscience and Remote Sensing* 38(5): 2202–2212.

Ferretti, A., C. Prati, and F. Rocca. 2001. Permanent scatterers in SAR interferometry. *IEEE Transactions on Geoscience and Remote Sensing* 39(1): 8–20.

Gabriel, A. K., R. M. Goldstein, and H. A. Zebker. 1989. Mapping small elevation changes over large areas: differential radar interferometry. *Journal of Geophysical Research* 94(B7): 9183–9191.

Ghiglia, D. C., and M. D. Pritt. 1998. *Two-Dimensional Phase Unwrapping: Theory, Algorithms, and Software.* New York: Wiley & Sons.

Hanssen, R. F. 2001. *Radar Interferometry: Data Interpretation and Error Analysis.* Dordrecht: Kluwer Academic Publishers.

Hauksson, E. 1987. Seismotectonics of the Newport-Inglewood fault zone in the Los Angeles basin, southern California. *Bulletin of the Seismological Society of America* 77: 539–561.

Hauksson, E., L. M. Jones, and K. Hutton. 1995. The 1994 northridge earthquake sequence in California: Seismological and tectonic aspects. *Journal of Geophysical Research-Solid Earth* 100: 12335–12355.

Hooper, A. 2008. A multi-temporal InSAR method incorporating both persistent scatterer and small baseline approaches. *Geophysical Research Letters* 35(16): 1–5.

Hooper, A., and H. A. Zebker. 2007. Phase unwrapping in three dimensions with application to InSAR time series. *Journal of the Optical Society of America A: Optics, Image Science, and Vision* 24(9): 2737–2747.

Hooper A., H. Zebker, P. Segall, and B. Kampes. 2004. A new method for measuring deformation on volcanoes and other natural terrains using InSAR persistent scatterers. *Geophysical Research Letters* 31(23): 1–5.

Huang, N. E., Z. Shen, S. R. Long, M. L. Wu, H. H. Shih, Q. Zheng, N. C. Yen, et al. 1998. The empirical mode decomposition and Hilbert spectrum for nonlinear and nonstationary time series analysis. *Proceedings of the Royal Society of London, Series A: Mathematical, Physical and Engineering Sciences, March 8, 1998* 454(1971): 903–995.

Kampes, B. M. 2006. *Radar Interferometry: Persistent Scatterer Technique.* Berlin: Springer-Verlag.

Kampes, B. M., and R. F. Hanssen. 2004. Ambiguity resolution for permanent scatterer interferometry. *IEEE Transactions on Geoscience and Remote Sensing* 42(11): 2446–2453.

Lanari, R., P. Lundgren, M. Manzo, and F. Casu. 2004b. Satellite radar interferometry time series analysis of surface deformation for Los Angeles, California. *Geophysical Research Letters* 31: L23613.

Lanari, R., O. Mora, M. Manunta, J. J. Mallorqui, P. Berardino, and E. Sansosti. 2004a. A small-baseline approach for investigating deformations on full-resolution differential SAR interferograms. *IEEE Transactions on Geoscience and Remote Sensing* 42(7): 1377–1386.

Lauknes, T. R., H. A. Zebker, and Y. Larsen. 2011. InSAR deformation time series using an L1-norm small-baseline approach. *IEEE Transactions on Geoscience and Remote Sensing* 49(1): 536–546.

Li, Z. W., X. L. Ding, C. Huang, G. Wadge, and D. W. Zheng. 2010. Modeling of atmospheric effects on InSAR measurements by incorporating terrain elevation information. *Journal of Atmospheric and Solar-Terrestrial Physics* 68(11): 1189–1194.

Liu, G. X. 2006. *Monitoring of Ground Deformations with Radar Interferometry.* Beijing: Surveying and Mapping Publishing House.

Liu, G. X., S. M. Buckley, X. L. Ding, Q. Chen, and X. J. Luo. 2009. Estimating spatiotemporal ground deformation with improved persistent-scatterer radar interferometry. *IEEE Transactions on Geoscience and Remote Sensing* 47(9): 3209–3219.

Liu, G. X., X. L. Ding, Z. L. Li, Z. W. Li, Y. Q. Chen, and S. B. Yu. 2004. Pre- and co-seismic ground deformations of the 1999 Chi-Chi, Taiwan earthquake, measured with SAR interferometry. *Computers & Geosciences* 30(4): 333–343.

Liu, G. X., H. G. Jia, R. Zhang, H. X. Zhang, H. L. Jia, B. Yu, M. Z. Sang. 2011. Exploration of subsidence estimation by PS-InSAR on time series of high-resolution TerraSAR-X images. *IEEE Journal of Selected Topics in Applied Earth Observations and Remote Sensing* 4(1): 159–170.

Liu, G. X., J. Li, Z. Xu, J. C. Wu, Q. Chen, H. X. Zhang, R. Zhang, H. G. Jia, et al. 2010. Surface deformation associated with the 2008 Ms8.0 Wenchuan earthquake from ALOS L-band SAR interferometry. *International Journal of Applied Earth Observation and Geoinformation* 12(6): 496–505.

Liu, G. X., X. J. Luo, Q. Chen, D. F. Huang, and X. L. Ding. 2008. Detecting land subsidence in Shanghai by PS-networking SAR interferometry. *Sensors* 8: 4725–4741.

Lu, Z. 1998. Synthetic aperture radar interferometry coherence analysis over Katmai volcano group, Alaska. *Journal of Geophysical Research* 103(B12): 29887–29894.

Lu, Z., T. Masterlark, D. Dzurisin, R. Rykhus, and C. Wicks, Jr. 2003. Magma supply dynamics at Westdahl volcano, Alaska, modeled from satellite radar interferometry. *Journal of Geophysical Research* 108(B7): doi:10.1029/2002JB002311.

Massonnet, D., M. Rossi, C. Carmona, F. Adragna, G. Peltzer, K. Feigl, and T. Rabaute. 1993. The displacement field of the Landers earthquake mapped by radar interferometry. *Nature* 364(6433): 138–142.

Mattar, K. E., P. W. Wachon, D. Geudtner, A. L. Gray, I. G. Cumming, and M. Brugman. 1998. Validation of alpine glacier velocity measurements using ERS tandem-mission SAR data. *IEEE Transactions on Geoscience and Remote Sensing* 36(3): 974–984.

Mora, O., J. J. Mallorqui, and A. Broquetas. 2003. Linear and nonlinear terrain deformation maps from a reduced set of interferometric SAR images. *IEEE Transactions on Geoscience and Remote Sensing* 41(10): 2243–2253.

Olea, R. A. 1999. *Geostatistics for Engineers and Earth Scientists*. Dordrecht: Kluwer Academic Publishers.

Shanker, A. P., and H. Zebker. 2010. Edgelist phase unwrapping algorithm for time series InSAR analysis. *Journal of the Optical Society of America A* 27(3): 605–612.

Shanker, P., and H. Zebker. 2007. Persistent scatterer selection using maximum likelihood estimation. *Geophysical Research Letters* 34(22): 2–5.

Shen, Z. K., D. D. Jackson, and B. X. Ge. 1996. Crustal deformation across and beyond the Los Angeles basin from geodetic measurements. *Journal of Geophysical Research-Solid Earth* 101: 27957–27980.

Strozzi, T., A. Kouraev, A. Wiesmann, U. Wegmüller, A. Sharov, and C. Werner. 2008. Estimation of Arctic glacier motion with satellite L-band SAR data. *Remote Sensing of Environment* 112(3): 636–645.

Usai, S. 2000. An analysis of the interferometric characteristics of anthropogenic features. *IEEE Transactions on Geoscience and Remote Sensing* 38(3): 1491–1497.

Usai, S., and R. F. Hanssen. 1997. Long time scale InSAR by means of high coherence features. In *Proceedings of 3rd ERS Symposium-Space at the Service of our Environment, Florence, Italy, March 17–21, 1997*, 225–228.

Werner, C., U. Wegmuller, T. Strozzi, and A. Wiesmann. 2003. Interferometric point target analysis for deformation mapping. In *Proceedings of International Geoscience and Remote Sensing Symposium (IGARSS'03), Toulouse, France, July 21–25, 2003*, 4362–4364.

Zebker, H. A., and P. A. Rosen. 1996. Atmospheric artifacts in interferometric SAR surface deformation and topographic maps. *Journal of Geophysical Research Solid Earth* 99(19): 617–634.

Zebker, H. A., P. A. Rosen, R. M. Goldstein, A. Gabriel, and C. L. Werner. 1994. On the derivation of coseismic displacement fields using differential radar interferometry: the Landers earthquake. *Journal of Geophysical Research* 99(B10): 19617–19634.

Zebker, H. A., and J. Villasensor. 1992. Decorrelation in interferometric radar echoes. *IEEE Transactions on Geoscience and Remote Sensing* 30(5): 950–959.

Zhang, L., X. L. Ding, and Z. Lu. 2011a. Ground settlement monitoring based on temporarily coherent points between two SAR acquisitions. *ISPRS Journal of Photogrammetry and Remote Sensing* 66(1): 146–152.

Zhang, L., X. L. Ding, and Z. Lu. 2011b. Modeling PSInSAR time series without phase unwrapping. *IEEE Transactions on Geoscience and Remote Sensing* 49(1): 547–556.

Zhang, L., J. C. Wu, L. L. Ge, X. L. Ding, and Y. L. Chen. 2008. Determining fault slip distribution of the Chi-Chi Taiwan earthquake with GPS and InSAR data using triangular dislocation elements. *Journal of Geodynamics* 45(4/5): 163–168.

Chapter 15

Mapping marine oil spills from space

Tae-Jung Kwon and Jonathan Li

CONTENTS

Ocean pollution caused by oil spills is currently considered a significant environmental hazard. Synthetic aperture radar (SAR) carried by Earth-observing satellites have proven to be a powerful tool for monitoring oil spills in marine waters. This chapter provides a comprehensive overview of SAR image analysis approaches to oil-spill monitoring with a focus on dark-spot detection. Dark-spot detection is a fundamental step in computerized marine oil-spill detection systems. Automated detection of dark-spots using SAR imagery is a challenging task, primarily because of the presence of multiplicative noise known as speckle. Speckle noise must be treated appropriately and adequately to make a good use of SAR imagery for dark-spot detection. The results of the method introduced in this chapter demonstrate that dark-spots (possible oil slicks) can be effectively detected in an SAR image by integrating the total variation optimization segmentation approach. Also, the proposed dark-spot detection method presents better segmentation results and takes less computing time.

15.1 INTRODUCTION

One of the most significant factors contributing to the degradation of marine water quality is accidental and deliberate oil spills (Topouzelis, 2008). Roughly 10% of oil spills come from oil leak at the bottom of the ocean (Kubat et al., 1998). It is evident that many oil spills are deliberate, motivated by the desire to reduce dumping and filtering costs (Brekke and Solberg, 2005). It is estimated that 75% of the oil spilled from ships is a result of routine operations, including the transfer of oil, whereas only 25% of spills are a result of accidents (Grau and Groves, 1997).

Oil spills can threaten coastal and marine ecosystems and cause permanent environmental damage. Tiny species such as plankton that lives in the upper layers of the ocean are particularly at risk because they are highly exposed to floating oil spills (Gin et al., 2001). In addition, because planktons are at the bottom of the food chain, they will be eaten by various other small sea animals and species of fish. These species that are higher in the food chain will then accumulate oil and the associated toxins in their bodies. These species will eventually be eaten by species higher in the food chain, such as sea birds, seals, and people. Consumption of polluted seafood and animals can consequently bring human health risks. Oil spills can have large negative economic consequences on both the personal and national levels as well. The livelihood of people residing near coastal areas, particularly those who are solely dependent on fishing and tourism can be heavily affected by oil spills (NOAA, 2007).

The catastrophe that occurred on April 20, 2010, in the Gulf of Mexico is an extreme example of the kind of negative impacts, ranging from environmental to economic, that an oil spill can cause. This catastrophic event was largely deemed as the worst environmental disaster in U.S. history, with many casualties, excessive ecological damage, and profound economic consequences. Reportedly, the losses attributed to this incident cost approximately US$22.6 billion, not including long-term environmental and economic losses. Nearly 7,000 animals, including birds, turtles, dolphins, and other mammals, were killed in this incident (Park et al., 2010). Furthermore, a large number of fishermen lost their source of income, and their livelihoods are now at risk. The tourism industry also has been heavily affected, with collateral damage affecting neighboring areas, where many tourists are now hesitant to travel even though a direct environmental impact is not yet evident. In light of the frequency of these accidental oil spills, and to minimize the environmental and economic impacts, a continuous, effective, and responsive strategy for monitoring the oil spills in the oceans is necessary (Etkin, 2005).

This chapter is organized as follows: The principle of synthetic aperture radar (SAR) imaging is described in Section 15.2. Section 15.3 summarizes currently available satellite SAR sensors. The automated detection of oil spills using SAR imagery, and their challenges are explained in Section 15.4. The underlying theory of the proposed method to detect dark spots is described in Section 15.5. Section 15.6 presents the results obtained using both synthetic and real SAR images, and quantitative assessment and comparison with existing methods. Last, conclusions are addressed in Section 15.7.

15.2 PRINCIPLE OF SAR IMAGING

SAR is an active side-looking microwave sensor that has the capability of recording objects at any time of the day under any weather condition. SAR emits electromagnetic radiation pulses that travel to the target area and are scattered off. The SAR antenna then records these backscattered pulses as amplitudes and phases, which then get preprocessed by the SAR processor, producing an image. A notable distinction between SAR and typical radar systems is the use of a synthetic antenna. The diagram in Figure 15.1 better illustrates the basic principles of SAR. The notations, L_{SA}, R, V, λ, and Θ_A in Figure 15.1 represent synthetic aperture length (phased array), range, velocity, wavelength of emitted pulses, and aperture azimuth beamwidth, respectively, and their relationships can be expressed as follows (McCandless and Jackson, 2004):

$$L_{SA} = \Theta_A R = \frac{\lambda R}{D_{AT}}. \tag{15.1}$$

The actual size of the along-track antenna is expressed in a single dot (denoted as D_{AT}) that sums to form the synthetic aperture length. The range is determined by precisely measuring the total time from the transmission to the receipt of the pulses by the antenna. Velocity is the speed of the platform (e.g., aircraft, satellite) carrying the SAR sensor. Finally, the along track, or azimuth, is the dimension that is perpendicular to the range.

SAR is made possible through utilization of an extremely large antenna or aperture, electronically. This means that SAR's physical antenna (represented as a small dot in Figure 15.1) continuously records backscattered signals while traveling in the azimuth direction from t_1 (initial recording point) to t_2 (last recording point). The SAR processor stores and processes all of these returned signals and reconstructs them as though they were recorded by the synthetic antenna whose length is equivalent to L_{SA}. In other words, as the platform moves along its trajectory, a synthetic aperture is generated by signal processing, which then elongates the actual SAR's small physical antenna. By utilizing this unique technique, a higher resolution with a large swath width that maximizes the visibility and clarity of the subject target is obtained.

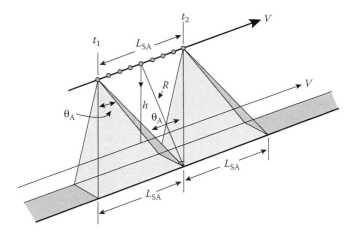

Figure 15.1 Basic principles of synthetic aperture radar. (Adapted from McCandless, S. W., & C. R. Jackson, *Synthetic Aperture Radar Marine User's Manual*, Department of Commerce, Washington, DC, 2004.)

15.3 SUMMARY OF RADAR-IMAGING SATELLITES

As emphasized, developing a proper system for monitoring marine environments to prevent the spreading of marine pollution caused by oil spills is of high importance. Hence, the key question now lies with choosing the optimal monitoring tool that is cost effective and efficient.

Satellite surveillance has been favorably chosen over other alternatives because of its superiority in covering a larger area at a relatively low cost. In particular, satellite SAR remote sensing has proven to be a powerful early warning tool in monitoring oil spills in coastal and marine waters. The use of satellite surveillance could reduce the potential damage and response time of cleanup operations (Solberg et al., 1999; del Frate et al., 2000; Marghany, 2001).

SAR is an active remote-sensing system that operates in the microwave regions of the electromagnetic spectrum (EMS) and measures the return energy that has been either reflected or scattered back from the target surface. These reflected or scattered pulses are then captured by the receiving antenna and are recorded to produce a two-dimensional (2D) image. In coastal water monitoring, SAR is preferred over optical sensing for the following reasons: (a) SAR is an active sensor that can provide its own source of illumination, and therefore it can operate at any time of the day; (b) SAR uses microwave electromagnetic radiation that can penetrate through rain, clouds, and other atmospheric substances providing good monitoring capabilities; and (c) SAR is capable of monitoring a wide range of area (including inaccessible areas) at a competitive cost compared with other ocean-monitoring tools, such as airborne ocean surveillance.

For more than three decades, satellite SAR data has been widely used to monitor marine environments. The various SAR sensors that are most suitable for ocean monitoring range from the Seasat, launched in 1978, to the most recent Canadian RADARSAT-2, launched in 2007. These SAR sensors are summarized in Table 15.1.

According to the literature, European Remote Sensing (ERS)-1 and -2, launched by the European Space Agency (ESA) in 1991 and 1995, respectively, were specifically targeting ocean environmental monitoring. Studies indicate that ERS-2 C-band SAR imagery has proven to be specifically suitable for monitoring oil spills (Brekke and Solberg, 2005). RADARSAT-1, launched by the Canadian Space Agency (CSA) in 1995, was the first SAR sensor capable of providing various types of information at multiple beam modes with varying resolution. Its ScanSAR Narrow (Wide) beam mode provides a spatial resolution of 50 m (100 m) with a swath width of 300 km (500 km), which is an ideal combination for monitoring much larger ocean areas. RADARSAT-1 was specifically designed to increase the monitoring capabilities in Canadian coastal and marine waters and has been used extensively in various studies (Monaldo et al., 2001; de Miranda et al., 2004). In 2002, ESA launched ENVISAT, carrying 10 sophisticated optical and radar sensors, among which the advanced synthetic aperture radar (ASAR) provides a spatial resolution of 30 m (150 m) with a swath width of 100 km (400 km) in its Precision Image (ScanSAR) mode. With its enhanced monitoring advantages, ASAR has been used specifically for oil-spill monitoring (Solberg et al., 2007; Brekke and Solberg, 2008). In 2007, the first X-band SAR satellites, German TerraSAR-X and Italian COSMO-SkyMed were flown into space and their images have been used for oil-spill monitoring (Trivero et al., 2007; Ciappa et al., 2009; Kim et al., 2010). In the same year, CSA launched RADARSAT-2, and its application in marine surveillance accounting for oil-spill monitoring has been studied (Bannerman et al., 2009).

Canada, as a leading country with possession of advanced RADARSAT-1/2, began a new mission development called the RADARSAT Constellation Mission (RCM) in 2005 and plans to launch three RCM satellites in 2014 and 2015. As shown in Figure 15.2, this program uses a three-satellite configuration that provides complete coverage of Canada's land and waters on a daily basis and provides daily access to 95% of the world (CSA, 2011). In addition, this program will enable enhanced monitoring capabilities, such as maritime surveillance, disaster management, and ecosystem monitoring.

15.4 AUTOMATED DETECTION OF OIL SPILLS IN SAR IMAGERY

The mechanism used to detect oil spills in SAR imagery is based on the fact that oil films floating on the ocean have a dampening effect on the short

Table 15.1 Summary of currently available satellite SAR sensors

Satellite	Agency	Operation	Frequency band	Polarization	Incidence angle (°)	Swath width (km)	Repeat cycle (days)	Best spatial resolution (m)
ERS-2	ESA	1995 to 2011	C	VV	20–59	100	35	30
RADARSAT-1	CSA	1995 to present	C	HH	20–49	50–500	24	8
ENVISAT (ASAR)	ESA	2002 to present	C	Dual-pol	15–45	100–400	35	30
ALOS (PALSAR)	JAXA	2006 to 2011	L	Quad-pol	8–60	70	46	10
TerraSAR-X	DLR	2007 to present	X	Dual-pol	20–55	10–100	11	1
COSMO-SkyMed	ASI	2007 to present	X	Quad-pol	20–60	10–200	16	1
RADARSAT-2	CSA/MDA	2007 to present	C	Quad-pol	10–60	25–500	24	1

ASI = Italian Space Agency; CSA = Canadian Space Agency; DLR = German Aerospace Centre; ESA = European Space Agency; JAXA = Japan Aerospace Exploration Agency; NASA = National Aeronautics and Space Administration, MDA = Missile Defense Agency.

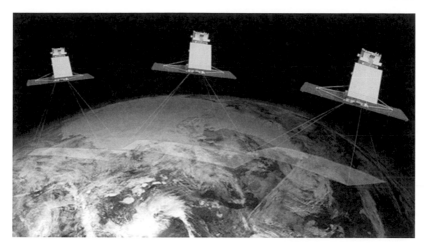

Figure 15.2 RADARSAT constellation mission. (CSA, RADARSAT Constellation; http://www.asc-csa.gc.ca/eng/satellites/radarsat/default.asp)

gravity-capillary waves. This dampening effect causes a reduction in the amount of backscattered signals recorded by the SAR antenna. Such a reduction then creates dark regions that highly contrast their surroundings in SAR intensity imagery (Topouzelis, 2008). The sea-surface roughness plays a significant role in detection because ocean phenomena, including oil spills, can never be detected without varying sea-surface roughness. The strength of the backscattered signal is a function of the viewing geometry of the SAR (i.e., incidence angle). Incidence angles represent the angular relationship between the ground target and the SAR beam (Topouzelis, 2008). SARs that are typically utilized for the purpose of monitoring coastal waters, particularly in searching for oil spills, operate at wide range of incidence angles between 20° and 70° so that the backscattering signals can be well observed. The backscatter coefficient decreases when the incidence angle increases (Brekke and Solberg, 2005). Another important factor that governs the detectability of oil spills in the ocean is the polarization of the incoming radar signal. The most appropriate configuration of SAR sensor for detecting oil spills is reported to be single-polarized VV SAR (i.e., the waves are vertically polarized upon both transmission and receipt) at incidence angles in the range of 20° to 45°, operating in C-band with 5.8 cm wavelength (Girard-Ardhuin et al., 2003). Moreover, because the damping of the Bragg resonant waves are better observed at shorter wavelength, X- and C-band-equipped SARs are preferred in terms of the ability of detecting oil spills than the SARs that are equipped with relatively longer wavelength bands, such as L- or P-band (Alpers and Espedal, 2004). In addition, utilization of cross-polarization such as HV (i.e., vertical transmission and horizontal receipt) or VH (i.e., horizontal transmission and vertical receipt) is not recommended because the backscattering signal from the water surface is reduced under those settings. Wahl et al. (1994) claimed that HH polarization would work poorly and oil-spill detection would not be possible for the largest incident angles, particularly when wind speed was low. No significant difference, however, has yet been reported in terms of its practical performance difference for operational oil-spill monitoring purposes when compared with VV polarization.

As mentioned, backscattering signals can be captured only when there is movement of waves. This indicates that the visibility of ocean phenomena is greatly affected by the speed of wind. It has been observed that wind speeds ranging from 3–10 m/s allow optimum visibility of oil spills (Brekke and Solberg, 2005). Having observed this backscattering phenomenon, challenges in detection are evident because of manmade or natural phenomena called look-alikes. These so-called look-alikes can cause a reduction in the backscattering signal (similar to that caused by oil spills) on SAR intensity imagery, thus making it difficult to discriminate between oil spills and look-alikes. Examples of look-alikes causing confusion in detection include low wind areas, organic films, rain cells, current shear zones, grease ice, and eddies (Topouzelis, 2008).

The amount of SAR images being produced is rapidly increasing. Manually analyzing these images to verify whether there are oil spills in the received images is time consuming (Brekke and Solberg, 2005). Thus, the development of an automated oil-spill detection method that provides fast and reliable results is in high demand (Nirchio et al., 2005; Karathanassi et al., 2006; Solberg et al., 2007; Karantzalos and Argialas, 2008). An algorithm for the detection of oil spills using SAR imagery is outlined in the three distinct steps shown in Figure 15.3.

In the illustrated oil-spill detection framework, an SAR image is input in the first step called the region selection or dark-spot detection step. In this step, only dark spots that contain actual oil spills and other natural phenomena such as look-alikes are extracted, whereas others are disregarded. Detected dark spots are then processed in the second step in which various features are extracted by analyzing the geometry and shape of the detected dark spots. In addition, the differences in the physical characteristics of the backscattered signals of the dark spot and neighboring areas are further investigated (Brekke and Solberg, 2005). With the aid of the extracted features, oil spills and look-alikes are classified in the final step to determine the presence of potential oil spills and decide whether to send the aircraft to verify the spill and take relevant action.

This chapter presents a state-of-the-art technique that tackles the preliminary task, which is to detect dark spots. Dark-spot detection has been regarded as the most critical and fundamental step because if no dark spots are detected, then the real oil spills cannot be found in later steps. An important concept to understand in dark-spot detection is that a low

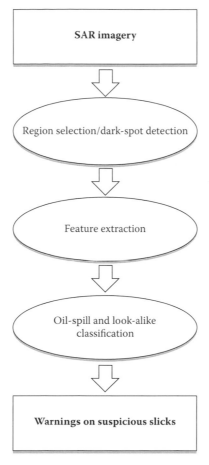

Figure 15.3 A common framework for detecting oil spills. (Adapted from Brekke, C., & A. H. S. Solberg, *Remote Sensing of Environment*, 95: 1–13, 2005.)

degree of accuracy of detected dark spots can negatively affect the feature extraction as well as the classification phase. These phases both require an accurate discrimination between oil spills and look-alikes. As such, it is important to develop a fast, reliable, and robust automated dark-spot detection algorithm that can accurately extract dark spots from the background so that the adverse effect in a later stage can be minimized and the likelihood of discriminating the actual oil spills can be maximized.

A common problem, yet the most fundamental and challenging task, when using SAR imagery is to eliminate speckle noise. Typically, SAR images contain vast amounts of multiplicative noise known as speckle. Speckle is caused mainly by constructive and destructive interference of reflective energy from a target surface (Richards and Jia, 2006). Transmitted and received SAR signals will not always be in-phase, even in the homogenous target surface. As a result, speckles may appear as slightly brighter or darker than the mean value of image intensity. The presence of speckle noise is problematic because it limits visibility of features and leads to misinterpretation when analyzing SAR imagery. Because SAR images always carry speckle noise, it remains a challenging task to develop a robust and efficient despeckling method to obtain desirable segmentation results.

In its fundamental form, automated segmentation approaches for SAR dark-spot detection can be realized by utilizing two distinctive methods: (a) global segmentation methods or (b) local segmentation methods.

Some of the global segmentation methods currently being used include k-means clustering (Shi et al., 2007), global histogram thresholding (Otsu, 1979), and Gaussian mixture model (GMM; Alf et al., 2008). These global methods extract the information from the entire image, and hence, they are more likely to ignore spatial relationships of neighboring pixels. Global segmentation methods do not perform well with highly speckled SAR images and tend to be less prone to under- or oversegmentation issues that consequently will lead to a better capability of separating dark spots from the background when speckles are treated appropriately. Many efforts have been put forth to develop a robust algorithm to effectively suppress speckle noise without changing the detailed feature characteristics within images (Lee, 1980; Frost et al., 1982; Kuan et al., 1987; Yu and Acton, 2002; Achim et al., 2003; Marques et al., 2004). These SAR denoising filters are commonly applied in the preprocessing step before the utilization of purely global segmentation. As a result, classes can be well separated without being significantly interrupted by the presence of speckle noise. It still remains as a challenging task, however, to develop a fast and robust denoising filter to obtain satisfactory results.

In contrast, local segmentation methods based on level sets (Chan and Vese, 2001), region growing (Lira and Frulla, 1998), Markov random fields (Pelizzari and Bioucas-Dias, 2007), neural networks (Topouzelis, 2008), and marked point processes (Li and Li, 2010) are more robust to the presence of speckle noise because they extract the spatial and tonal relationships between pixels locally. This is significant because SAR images always carry substantial amounts of speckle noise. These methods, however, are more prone to under- and oversegmentation problems. For example, the classical region-growing method does not always provide robust segmentation performance because this method is highly dependent on the initial selection of seed points. From a set of initial select seed points, the regions are grown to neighboring points as long as a predefined criterion (e.g., pixel intensity) is satisfied. This implies that having a bad selection of an initial seed point could possibly cause failure or region segmentation. Similarly, a conventional level-set segmentation method detects edges by solving a partial differential equation (PDE) but is not robust to the presence of boundary gaps and typically takes a large number of iterations to reach the final segmented result. Generally, local-based segmentation schemes are computationally expensive; for instance, a robust segmentation method using the marked point process takes approximately 30 minutes to complete the two-class segmentation task using a highly speckled SAR image whose size is 512 × 512 pixels (Li and Li, 2010).

Taking into account the abovementioned characteristics of both local- and global-based segmentation methods, the primary goal of this research is to combine the advantages of both global and local methods and develop a comprehensive segmentation method that minimizes the adverse effects created by speckle noise and better separates the dark spots from the surroundings.

15.5 PROPOSED METHODOLOGY

The proposed method is called total variation optimization segmentation (TVOS) and is motivated by the aforementioned challenging criteria that arise when dealing with SAR imagery to better handle the task of SAR image segmentation. The proposed TVOS algorithm is developed to allow for segmentation of an SAR image containing high speckle noise. To tackle this challenging task more efficiently, a two-phase algorithm is implemented. The proposed methodology is used to derive a generalized algorithm to deal with any highly speckled SAR images, regardless of their physical properties (i.e., polarization and spatial resolution).

15.5.1 Phase I: total variation optimization

The presence of multiplicative speckle noise and many other artifacts in SAR imagery greatly hinder the process of obtaining desirable segmentation results. Figure 15.4 illustrates an example of a typical SAR image, which will be used as a test image, and the corresponding histogram.

Although there is a clear visual distinction between dark spots and the background as shown in Figure 15.4a, its corresponding histogram in Figure 15.4b shows a unimodal pixel distribution, indicating that two classes are not properly recognized mainly because of the presence of speckle noise in the test SAR image. With this highly speckled SAR image, good segmentation results are almost impossible to achieve.

In its simplest form, speckle noise can be treated by taking either spatial or intensity differences into account. When only considering the spatial differences between the neighboring pixels and the center pixel in that neighborhood, it is assumed that pixel values in the images change slowly over time, and therefore it becomes appropriate to average them together. It is believed that taking the spatial characteristics would effectively eliminate noise because such characteristics are less correlated than the actual signal values, and hence this step eliminates the noise while preserving the actual

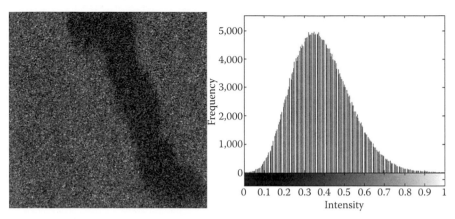

Figure 15.4 (a) A test SAR image containing a dark-spot with image size of 600 × 600 pixels and (b) its corresponding histogram.

signal. The assumption that pixel values vary slowly over time fails, however, when the edge of an object in the image is reached. The pixel values would be largely different around the edge, so making such an assumption is not valid. The question then arises of how to prevent the averaging across the edges while continuing to average in a smooth region where pixel values do slowly change over time. To overcome this challenge, the intensity difference is incorporated. The assumption made previously is ignored, and the averaging of pixel values that have similar intensity values is enforced. Therefore, pixels that have similar intensity values will be averaged. Previous research indicates that taking the gradient difference of pixels will also improve edge-preservation ability because pixels with large gradient differences are less likely to belong to the same region. As a result, the first phase of the proposed TVOS method was inspired by the observations made when combining three difference terms (spatial, intensity, and gradient differences) into account, so that noise can be reduced in a highly speckled image while effectively preserving edges by incorporating a nonlinear combination of neighboring pixel values.

Having acknowledged the importance of the aforementioned three difference terms, an optimization problem can be formulated to efficiently handle a SAR segmentation task. Let g be any observed SAR imagery that is defined on a continuous space, f be the piecewise smooth state of the observed g, and u be the residual state containing multiplicative noise. Their relationship can be expressed as follows:

$$g = f \times u. \tag{15.2}$$

One approach to solve this inverse problem can be realized by utilizing the Rudin-Osher-Fatemi Total Variation (ROFTV) model. Given the multiplicative relationship in Equation 15.2, the problem of image segmentation can be formulated into the minimization problem based on the existing ROFTV model (Rudin et al., 1992) shown by the following:

$$\hat{f} = \text{argmin}_f \left[\alpha \int_\Omega |\log f - \log g|^2 \, d\underline{x} + \beta \int_\Omega |\nabla f| \, d\underline{x} \right], \tag{15.3}$$

where Ω is an open set representing the image domain, and Δf represents the finite intensity difference between neighboring pixels. The first term of Equation 15.3 is the data fidelity term, whereas the second term is the total variation term that penalizes pixel intensity differences within regions to enforce piecewise smoothness in f. In the ROFTV model, the goal is to progressively evolve a noisy SAR imagery, g, whose imaging condition is in a nonpiecewise smooth state containing nonseparable classes into a piecewise smooth state, f. In the process of the evolution to achieve f, the total variation is minimized and the classes become well delineated.

As discussed earlier, incorporation of the intensity difference penalty term alone may not be sufficient, especially when an image is highly contaminated with speckle noise. As such, the existing ROFTV model has been modified by adding two additional total variation penalty terms: the spatial difference term, ∇x, and the gradient difference term, ∇k, to effectively and efficiently handle the SAR segmentation task. These additional total variation constraints work simultaneously to enforce a piecewise smooth state of the image, in which the edges or boundaries of subject targets are preserved by the intensity and gradient difference terms, while noise is treated by the spatial difference term. By observing the data histogram, it was found that the statistical distribution of pixels in the SAR imagery generally followed the Gaussian distribution. Therefore, the proposed total variation penalty terms were modeled with Gaussian functions. The first incorporated penalty term that enforces spatial difference between pixels can be expressed as follows (Tomasi and Manduchi, 1998):

$$x(\xi, \psi) = e^{-\frac{1}{2}\left(\frac{d(\xi, \psi)^2}{\sigma_x^2}\right)}, \tag{15.4}$$

where d is the Euclidean distance between the center pixel ψ, and the neighboring pixel, ξ. $x(\xi, \psi)$ measures the spatial closeness between pixels so that the homogeneity of surrounding pixels is enforced. σ_x is the standard deviation for pixel difference, and as its value gets larger, images will become smoother. This also implies that when a larger σ_x is used, pixels that are located farther away from the center pixel, ψ, are combined. This is mainly due to the fact that spatially nearby pixels are forced to merge together unless there is a pixel that has a large intensity difference when compared with the surrounding pixels.

The second penalty term that enforces the intensity difference between pixels is expressed as follows:

$$f(\xi, \psi) = e^{-\frac{1}{2}\left(\frac{\|s(\xi) - s(\psi)\|^2}{\sigma_f^2}\right)}, \tag{15.5}$$

where $f(\xi,\psi)$ measures the intensity differences at two interacting pixels ξ and ψ. When computing the intensity difference between pixels, a smaller intensity difference will cause pixels to be merged together, whereas pixels with large intensity differences will stay unchanged depending on the weight given by σ_f. Thus, pixels with the value of σ_f being closer to each other than σ_f are averaged, whereas pixels that are farther from each other than the value σ_f are disregarded.

The last penalty term that enforces the gradient difference between pixels is expressed as follows:

$$k(\xi, \psi) = e^{-\frac{1}{2}\left(\frac{\|\tau(\xi) - \tau(\psi)\|^2}{\sigma_k^2}\right)}, \tag{15.6}$$

where $\tau(\xi)$ and $\tau(\psi)$ are the gradients and their differences is denoted by $k(\xi,\psi)$. σ_k denotes the standard deviation of the gradient difference between pixels. Similar to the mechanism built for computing an intensity difference, the gradient difference is enforced between pixels because large gradient differences will have a small likelihood of belonging to the same class.

With these total variation constraints, the final formulation of the segmentation problem based on the modified ROFTV model can be rewritten as follows:

$$\hat{f} = \operatorname{argmin}_f \left[\alpha \int_\Omega |\log f - \log g|^2 \, d\underline{x} + \beta \int_\Omega \left(|\nabla f| + |\nabla x| + |\nabla k|\right) d\underline{x} \right]. \tag{15.7}$$

As can be seen in Equation 15.7, the additional total variation constraints, ∇x enforcing the spatial difference and ∇k enforcing the gradient difference, have been incorporated to extend the ROFTV model to better estimate the piecewise smooth state. This problem can be efficiently solved by incorporating an iterative weighted optimization strategy; the algorithm used in this chapter is called the diagonal normalized steepest descent (DNSD) to obtain the estimated f (Luenberger, 1987; Bertsekas, 1995).

From a theoretical perspective, running the total variation optimization phase iteratively would slowly evolve to a convergence in which the complete piecewise smooth state is realized. Achieving such a steady state can be expensive when considering the computational time (e.g., more than 1 hour), and thus it is not well suited for practical purposes. Through the course of numerous tests, it was discovered that even running the optimization strategy for a limited number of iterations produces a good approximation of the piecewise smooth state of the original noisy image in which classes are easily separable, as clearly illustrated in Figure 15.5.

As can be vividly seen in Figure 15.5, the two classes have become well delineated as represented by a bimodal shaped histogram that is generated using the test image. Motivated by this observation, a faster finite mixture-model-classification strategy is utilized in the second phase of the proposed TVOS algorithm to approximate the final segmented result by efficiently enforcing the complete piecewise smoothness of the observed SAR intensity imagery. The three standard deviation values (σ_x, σ_f, σ_k), which are associated with the spatial, intensity, and gradient differences, respectively, between pixels have to be carefully selected to obtain the best results (as shown in Figure 15.5). These values are selected based on the heuristic evaluations by cautiously analyzing the amount of speckles presented, size of the subject target (i.e., oil spills) to be detected, and spatial resolution of the images that are used. As briefly mentioned, when the large standard deviation value is selected, the target image will be smoother. This implies that for highly speckled images, some loss of detailed features is inevitable because the large value would be utilized to effectively eliminate the speckle noises. On the other hand, when dealing with relatively less speckled images, smaller standard deviation values can be used to accommodate the existing noises while sufficiently preserving the details of the subject target. It is common that SAR images are highly speckled and hence utilization

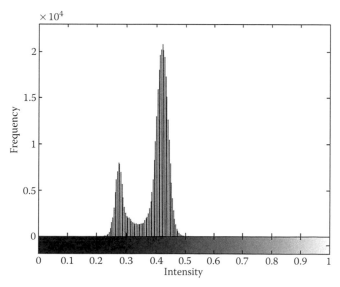

Figure 15.5 Histogram generated using the test image at the first iteration.

of higher spatial resolution image can be beneficial because larger values can be utilized to adequately suppress the noises without having to remove much of details. On the basis of the extensive amount of heuristic trial and error, it was found that the values of σ_x (6 to 12), σ_f (0.1 to 0.6), and σ_k (0.3 to 0.6) would produce the optimal results for highly speckled high-resolution SAR images whose image sizes ranged between 200×200 pixels and $1{,}500 \times 1{,}500$ pixels.

15.5.2 Phase II: finite mixture model classification

In the second phase of the proposed TVOS method, a finite mixture-model-classification strategy was applied to what was achieved in the first phase to obtain the final segmentation result. First, a GMM was utilized to estimate the unknown parameters to obtain a promising segmentation result using SAR intensity imagery.

Let n be the number of components within a GMM and let l be a class label where $l \in \{1, ..., n\}$. Furthermore, the set of unknown parameters to be estimated using a GMM is denoted as Θ,

$$\Theta = \{\mu_1, ..., \mu_n, \sigma_1, ..., \sigma_n, P\,(f=1), ... P\,(f=n)\}, \tag{15.8}$$

where μ, σ, and $P(f)$ denote the mean at the center of each Gaussian-distributed parabola, standard deviation, and prior probability of an observed subclass component within the mixture model, respectively. The goal is to precisely model the underlying distribution. With such a model, the probability of observing $\hat{f}$ is expressed as follows:

$$p\,(\hat{f}|\Theta) = \sum_{i}^{n} p\,(\hat{f}, l = i\,|\Theta). \tag{15.9}$$

The next step is to compute the maximum likelihood estimates of the unknown parameters that maximize the probability of obtaining the observed data, $\hat{f}$. Determination of unknown parameters, however, is often intractable to solve analytically.

To solve this problem in more efficient manner, an expectation maximization (EM) technique is utilized. EM is a popular method for finding the maximum likelihood estimates (MLE) of the unknown parameters and is used in various applications involving SAR imagery (Wang et al., 2005; Khan et al., 2007). EM consists of two steps that run in an iterative fashion until changes in the estimated parameters becomes marginal. The first step is called the expectation (E) step, where the log-likelihood function is used to predict the associated parameters in the mixture model as shown in the following equation (Moon, 1996):

$$E = \frac{p\,(\hat{f_j}|\Theta_i)\,P(i)}{\sum\limits_{u=1}^{n} p\,(\hat{f_j}|\Theta_u)\,P(u)}, \tag{15.10}$$

where each new data point $\hat{f}$ is generated by component i at the current estimates of Θ_i and $P(i)$. After the associated parameters in Θ are estimated, a subsequent step called maximization (M) is entered. In this step, parameters that maximize the expected log-likelihood function are estimated by updating the associated parameters that were initially determined in the previous E step, which is achieved by,

$$M = \sum\limits_{j=1}^{v} \log p\,(\hat{f}|\Theta). \tag{15.11}$$

The log-likelihood function is used because the logarithm is a function that increases monotonically, and hence when the logarithm of a function reaches its highest or maximum value, the function itself reaches the same maximum point (McLachlan and Krishnan, 1997). Furthermore, because calculating the maximum of a function typically involves the computation of a derivative, it is much easier to deal with a log-likelihood function. Thus, the combined expectation-maximization scheme to estimate a set of unknown parameters, Θ at an updated value at t can be realized by the following equation (Dempster et al., 1977):

$$\Theta_{t+1} = \mathrm{argmax}_\Theta \sum\limits_{j=1}^{v}\sum\limits_{i=1}^{n} p\,(l_j = i\,|\,\hat{f_j}, \Theta_t)\, \ln p\,(l_j = i\,|\,\hat{f_j}, \Theta), \tag{15.12}$$

where,

$$p\,(l_j = i\,|\,\hat{f_j}, \Theta_t) = \frac{p\,(\hat{f_j}|l_j = i, \Theta_t)\,P\,(l_j = i|\Theta_t)}{\sum\limits_{u=1}^{n} p\,(\hat{f_j}|l_j = u, \Theta_t)\,P\,(l_j = u|\Theta_t)}. \tag{15.13}$$

Finally, once unknown parameters are determined using EM, the maximum likelihood (ML) estimate of the final class l_f (target area or nontarget area) at pixel $\underline{x}$ can be obtained by calculating the following:

$$\hat{l}\,(\underline{x}) = \mathrm{argmax}_l\, p\,(\hat{f}\,(\underline{x})|l). \tag{15.14}$$

The final segmentation result achieved in the finite mixture-model-classification phase followed by the total variation optimization phase is illustrated in Figure 15.6.

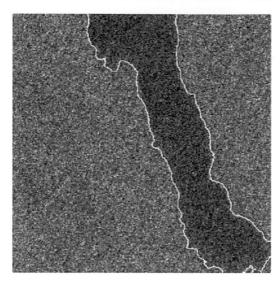

Figure 15.6 Final segmentation result of the test image achieved in a finite mixture model classification phase.

Dark spots have been outlined in white contour lines to increase the visibility of results where only dark spots have been detected. Such a result was obtained using the first iterated product from the first total variation optimization phase.

As shown in Figure 15.6, the segmentation result clearly shows the subject target being detected in the test image. Another notable achievement is a short processing time of 19 s in which the segmentation result was obtained. Thus, it can be concluded that the proposed comprehensive segmentation approach is fast, robust, and effective in handling the complicated SAR segmentation task.

15.6 RESULTS AND DISCUSSION

To show the effectiveness and robustness of the proposed TVOS method, two different experimental tests were conducted using (a) an artificially created synthetic image under noise and (b) a real COSMO-SkyMed SAR imagery containing verified oil spills. Synthetic testing using the artificially created image is regarded important, especially for quantitative analysis, as the results can be compared and validated with its true reference data, whereas such validation data generally are not available for real SAR data. In addition, three conventional methods including the purely global method based on GMM, the purely local method based on level set (Lu et al., 2009), and the combined global and local method based on Lee +

K-means (Marques et al., 2004) were used to compare the segmentation performance with the proposed method.

15.6.1 Experiment using synthetic imagery under noise

A synthetic test image was produced that was similar to the real SAR oil-spill imagery by creating two different classes: dark-spots and the seawaters. To investigate the performance of the proposed method and three other comparison methods under noise, 15 different levels of multiplicative noise, σ^2 ranging from 0.01 to 0.70, were applied to the synthetic image. Because of limited space, only the results obtained from four different noise levels, whose variances are 0.01, 0.25, 0.45, and 0.70, are presented in this chapter. Figure 15.7 shows the comparison of segmentation results obtained from using the proposed TVOS method and three other comparison methods discussed previously. In addition, the results of the quantitative kappa coefficient analysis (Cohen, 1960) are illustrated in Figure 15.8. The kappa coefficient ranges from 0, being a totally incomplete agreement, to 1, being a totally complete agreement, with respect to the reference data.

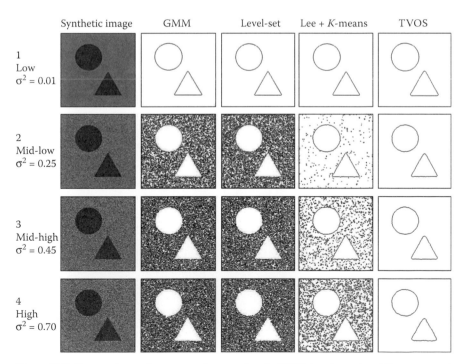

Figure 15.7 Comparison of segmentation results via GMM, level-set, Lee+K-means, and TVOS under different noise levels.

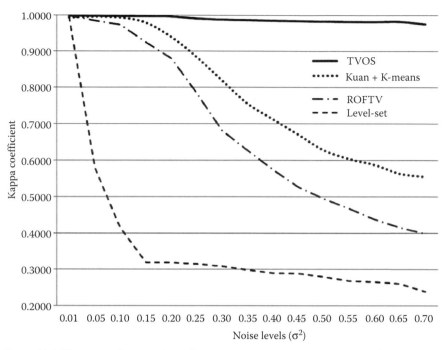

Figure 15.8 Kappa coefficients from four segmentation methods under different noise levels.

As shown in Figure 15.7, the purely global segmentation approach by GMM shows poor segmentation performance. This visual inspection is confirmed numerically by kappa coefficients computed for GMM that were very low in almost all noise levels. Such poor results were obtained because global methods extract tonal and spatial information from the entire image, making them very weak when dealing with highly speckled images. The local based level-set method also showed weakness in correctly segmenting images with high levels of noise. The Lee + K-means method, which extracts pixel information both locally and globally, gives better segmentation results and is less vulnerable to speckle noise. One disadvantage of this method is that as the severity of the noise level increases, its segmentation performance gets considerably worse. The TVOS method, on the other hand, produced robust segmentation results under all tested noise levels, and the computed kappa coefficients were constantly near 1, indicating that the segmentation results were well matched with reference data. Such satisfactory results are obtained by combining the advantages of both global and local methods for which the spatial difference penalty term effectively removed the speckle noise, while intensity and gradient difference term well preserved the boundaries of the dark spots. Hence, the segmentation results achieved using a synthetic test image under different noise levels showed the effectiveness and the robustness of the proposed TVOS method.

15.6.2 Experiment using COSMO-SkyMed imagery

As for the real image testing, Italian X-band (9.6 gigahertz [GHz]) COSMO-SkyMed SAR imagery was utilized. The type of image product used in this chapter is a stripmap whose spatial resolution is 3 m with a swath width of 40 km. The selected image was captured in VV polarization mode and was acquired on September 9, 2009. The size and the location of the image are 20,748 × 21,276 pixels, and the ocean nearby the City of Qingdao, Shandong province, China, respectively. The dark spots in this image have been verified as oil spills by manual analysis.

Four different methods (GMM, level-set, Lee + K-means, and TVOS) were tested using 50 different subsets taken from the original imagery. Among these subsets, only five subsets at different sizes are displayed in Figure 15.9.

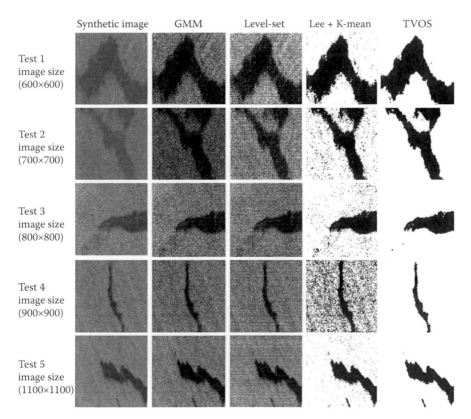

Figure 15.9 Five subset SAR oil-spill images and the segmentation results by four different methods.

The segmentation results obtained from the proposed TVOS method clearly outperformed the other three methods. As seen previously in the synthetic testing section, such satisfactory results were achieved mainly due to the three total variation penalty terms incorporated in the first phase of the proposed method, which helped eliminate the speckle noise significantly while preserving the fine details of the image using a nonlinear combination of pixel values. Another notable achievement of the proposed method is that all of the results of the TVOS method form a homogeneity region within each class, whereas other methods produced substantial amounts of noise within each class. It was learned from this experiment that taking only the global (GMM) or local (level-set) information alone does not produce pleasing segmentation results, especially in a highly speckled SAR images. The Lee + K-means method managed to produce relatively better segmentation results by combining both local- (Lee) and global-based (K-means) schemes. It was discovered, however, that Lee's adaptive filter was insufficiently strong in eliminating speckle noise (at least in the tested images), such that many wrongly classified pixels still appear within each class.

Processing time is another important factor that plays a critical role in determining whether an algorithm can go beyond its prototype and be practically applied in real-world applications. In this aspect, the proposed TVOS method successfully completed the segmentation task in a timely manner, whereas, for instance, the level-set segmentation method unsatisfactorily completed the same task. For example, it took 32 seconds for the proposed TVOS method to process Test-image 1, whereas it took 14 minutes for the level-set method to complete the same segmentation task. The GMM and Lee + K-means methods took 18 seconds and 5 seconds of processing time, respectively, but the segmentation results using these two methods were compromised at the expense of a quick processing time. Only one iteration of the total variation optimization phase of the TVOS was applied to arrive at the "clean" result, while the level-set method took 63 iterations to achieve the final result. The processor of the computer used to test these four methods was equipped with Intel dual core CPUs running at 2.4 GHz with 3 gigabytes of RAM. All the tests were conducted on a PC-based MATLAB platform. It is anticipated that the processing time could be significantly reduced if using other programming languages such as C. Therefore, it is once again confirmed that the proposed TVOS method proved to be fast and robust for processing highly speckled real SAR images.

15.7 CONCLUSION

This chapter has presented a novel two–phase automated segmentation approach, namely TVOS. In its first phase, three total variation penalty terms were implemented iteratively to eliminate the coherent speckle noise by roughly estimating the piecewise smoothness of the original SAR imagery. The total variation minimizers used in this phase include spatial, intensity, and gradient differences to effectively remove the speckles while preserving the boundaries of objects. In its second phase, a purely global classification scheme was implemented because the first-phase processed SAR imagery should contain easily separable classes of dark spots or non-dark spots. The unknown parameters of the GMM were approximated using an expectation-maximization scheme. When the final class likelihoods are estimated, a maximum likelihood classification method was utilized to predict the final class of all pixels. Therefore, the combination of the two aforementioned phases results in a comprehensive segmentation algorithm to efficiently cope with the SAR image segmentation task. The proposed segmentation scheme was designed in a way that it could be utilized in any SAR images regardless of the physical properties of the images. Hence, it can be advantageous and beneficial for a practical and operational use as the proposed algorithm can efficiently detect the dark spots in the ocean using almost any SAR images containing high speckle noises.

The quantitative assessment of the dark spot detection results obtained using synthetic images demonstrated that the TVOS method consistently reached a kappa value near 1 throughout all the noise levels, indicating good matches with the reference data. In contrast, results drawn from other conventional methods showed unpromising segmentation performance because the quantitative kappa measures continually dropped beginning at low to middle noise levels, indicating poor matches with the reference data. The visual inspection of dark-spot detection results obtained using real SAR imagery shows that the TVOS method provides superior segmentation performance in a timely manner compared with other conventional yet well-established methods. It is recommended that further study should be undertaken to incorporate an additional total variation constraint (e.g., texture difference) to enhance segmentation capability to further discriminate dark spots from the background. In addition, the application of dual- or quad-polarimetric SAR images can be particularly beneficial as it provides additional physical parameters to better distinguish between oil spills and look-alikes and possibly even to analyze different oil types.

ACKNOWLEDGMENTS

We wish to acknowledge Dr. Xianfeng Zhang, associate professor at the Institute of Remote Sensing and GIS, Peking University, China, for providing COSMO-SkyMed SAR imagery. We would also like to thank Dr. Alex Wong, assistant professor at the Department of Systems Design Engineering, University of Waterloo, Canada, for his valuable consultation sessions and fruitful discussions.

REFERENCES

Achim, A., P. Tsakalides, and A. Bezerianos. 2003. SAR image denoising via Bayesian wavelet shrinkage based on heavy-tailed modeling. *IEEE Transactions on Geoscience and Remote Sensing* 41(8): 1773–1784.

Alf, M., L. Nieddu, and D. Vicari. 2008. A finite mixture model for image segmentation. *Statistics and Computing* 18: 137–150.

Alpers, W., and H. A. Espedal. 2004. Oils and surfactants. In *Synthetic Aperture Radar Marine User's Manual*, edited by C. R. Jackson and J. R. Apel, 1–23. Washington, DC: U.S. Department of Commerce.

Bannerman, K., R. G. Caceres, M. H. Rodriguez, O. L. Castillo, F. P. de Miranda, and C. Enrico. 2009. Operational applications of RADARSAT-2 for the environmental monitoring of oil slicks in the Southern Gulf of Mexico. *International Geoscience and Remote Sensing Symposium* 3: 381–383.

Bertsekas, D. 1995. *Nonlinear Programming*. Belmont, MA: Athena.

Brekke, C., and A. H. S. Solberg. 2005. Oil spill detection by satellite remote sensing. *Remote Sensing of Environment* 95(1): 1–13.

Brekke, C., and A. H. S. Solberg. 2008. Classifiers and confidence estimation for oil spill detection in ENVISAT ASAR images. *IEEE Geoscience and Remote Sensing Letters* 5(1): 65–69.

Canadian Space Agency (CSA). 2011. RADARSAT constellation. http://www.asc-csa.gc.ca/eng /satellites/radarsat/default.asp (accessed September 15, 2011).

Chan, T. F., and L. A. Vese. 2001. Active contours without edges. *IEEE Transactions on Image Processing* 10(2): 266–277.

Ciappa, A., L. Pietranera, and A. Coletta. 2009. Sea surface transport derived by frequent revisit time series of COSMO SkyMed SAR data. *International Geoscience and Remote Sensing Symposium* 2: 777–780.

Cohen, J. 1960. A coefficient of agreement for nominal scales. *Educational and Psychological Measurement* 20(1): 37–46.

de Miranda, F. P., A. M. Q. Marmol, E. C. Pedroso, C. H. Beisl, P. Welgan, and L. M. Morales. 2004. Analysis of RADARSAT-2 data for offshore monitoring activities in the Cantarell Complex, Gulf of Mexico, using the unsupervised semivariogram textural classifier (USTC). *Canadian Journal of Remote Sensing* 30(3): 424–436.

del Frate, F., A. Petrocchi, J. Lichtenegger, and G. Calabresi. 2000. Neural networks for oil spill detection using ERS-SAR data. *IEEE Transactions on Geoscience and Remote Sensing* 38(5): 2282–2287.

Dempster, A., N. Laird, and D. Rubin. 1977. Maximum likelihood from incomplete data via the EM algorithm. *Journal of the Royal Statistical Society: Series B* 39(1): 1–38.

Etkin, D. S. 2005. Estimating cleanup costs for oil spills. In *2005 International Oil Spill Conference, Miami, May 2005*, 2625–2634.

Frost, V., J. Stiles, K. Shanmugan, and J. Holtzman. 1982. A model for radar images and its application to adaptive digital filtering of multiplicative noise. *IEEE Transactions on Pattern Analysis and Machine Intelligence* 4(2): 157–166.

Gin, K. Y. H., M. D. K. Huda, W. K. Lim, and P. Tkalich. 2001. An oil spill-food chain interaction model for coastal waters. *Marine Pollution Bulletin* 42(7): 590–597.

Girard-Ardhuin, F., G. Mercier, and R. Garello. 2003. Oil slick detection by SAR imagery: potential and limitation. *OCEANS 2003* 1: 164–169.

Grau, M. V., and T. Groves. 1997. The oil spill process: The effect of coast guard monitoring on oil spills. *Environmental and Resource Economics* 10: 315–339.

Karantzalos, K., and D. Argialas. 2008. Automatic detection and tracking of oil spills in SAR imagery with level set segmentation. *International Journal of Remote Sensing* 29(21): 6281–6296.

Karathanassi, V., K. Topouzelis, P. Pavlakis, and D. Rokos. 2006. An object-oriented methodology to detect oil spills. *International Journal of Remote Sensing* 27(23): 5235–5251.

Khan, K., J. Yang, and W. Zhang. 2007. Unsupervised classification of polarimetric SAR images by EM algorithm. *IEICE Transactions on Communications* 90(12): 3632–3642.

Kim, D. J., W. M. Moon, and Y. S. Kim. 2010. Application of TerraSAR-X data for emergent oil-spill monitoring. *IEEE Transactions on Geoscience and Remote Sensing* 48(2): 852–863.

Kuan, D., A. Sawchuk, T. Strand, and P. Chavel. 1987. Adaptive restoration of images with speckle. *IEEE Transactions on Acoustics, Speech and Signal Processing* 35(3): 373–383.

Kubat, M., R. C. Holte, and S. Matwin. 1998. Machine learning for the detection of oil spill in satellite radar images. *Machine Learning* 30(2/3): 195–215.

Lee, J. S. 1980. Digital image enhancement and noise filtering by use of local statistics. *IEEE Transactions on Pattern Analysis and Machine Intelligence* PAMI-2(2): 165–168.

Li, Y., and J. Li. 2010. Oil spill detection from SAR intensity imagery using a marked point process. *Remote Sensing of Environment* 114(7): 1590–1601.

Lira, J., and L. Frulla. 1998. An automated region growing algorithm for segmentation of texture regions in SAR images. *International Journal of Remote Sensing* 19(18): 3595–3606.

Lu, M., Z. He, and Y. Su. 2009. An active contour model for SAR image segmentation. *IET International Radar Conference* 551: 1–5.

Luenberger, D. 1987. *Linear and Non-Linear Programming*, 2nd ed. Reading, MA: Addison-Wesley.

Marghany, M. 2001. RADARSAT automatic algorithms for detecting coastal oil spill pollution. *International Journal of Applied Earth Observation and Geoinformation* 3(2): 191–196.

Marques, R. C. P., E. A. Carvalho, R. C. S. Costa, and F. N. S. Medeiros. 2004. Filtering effects on SAR images segmentation. *Lecture Notes in Computer Science* 3124: 1041–1046.

McCandless, S. W., and C. R. Jackson. 2004. Principles of synthetic aperture radar. In *Synthetic Aperture Radar Marine User's Manual*, edited by C. R. Jackson and J. R. Apel, 1–23. Washington, DC: U.S. Department of Commerce.

McLachlan, G., and T. Krishnan. 1997. *The EM Algorithm and Extensions*. New York: Wiley & Sons.

Monaldo, F. M., D. R. Thompson, and R. C. Beal. 2001. Comparison of SAR-derived wind speed with model predictions and ocean buoy measurements. *IEEE Transactions on Geoscience and Remote Sensing* 39(12): 2587–2600.

Moon, T. K. 1996. The expectation-maximization algorithm. *IEEE Signal Processing Magazine* 13(6): 47–60.

National Oceanic and Atmospheric Administration (NOAA). 2007. *Emergency Response, Responding to Oil Spills.* http://response.restoration.noaa.gov.

Nirchio, F., M. Sorgente, A. Giancaspro, W. Biamino, E. Parisato, R. Ravera, and P. Tribero. 2005. Automatic detection of oil spills from SAR images. *International Journal of Remote Sensing* 26(6): 1157–1174

Otsu, N. 1979. A threshold selection method from gray-level histogram. *IEEE Transactions on Systems, Man and Cybernetics* 9(1): 62–66.

Park, H., G. Roberts, E. Aigner, S. Carter, K. Quealy, and G. V. Xaquin. 2010. The oil spill's effects on wildlife. http://www.nytimes.com/interactive/2010/04/28/us /20100428-spill-map.html (accessed April 10, 2011).

Pelizzari, S., and J. Bioucas-Dias. 2007. Oil spill segmentation of SAR images via graph cuts, In *IGARSS 2007, July 23–28, Barcelona, Spain,* 1318–1321.

Richards, J. A., and X. Jia. 2006. *Remote Sensing Digital Image Analysis,* 4th ed., 67–238. Berlin: Springer-Verlag.

Rudin, L. I., S. Ohser, and E. Fatemi. 1992. Nonlinear total variation based noise removal algorithms. *Physica D: Nonlinear Phenomena* 60(1/4): 259–268.

Shi, L., X. Zhang, G. Seielstad, C. Zhao, and M. He. 2007. Oil spill detection by MODIS images using fuzzy cluster and texture feature extraction. In *OCEANS 2007 Europe, June 18–21, Aberdeen, UK,* 1–5.

Solberg, A. H. S., C. Brekke, and P. O. Husøy. 2007. Oil spill detection in Radarsat and Envisat SAR Images. *IEEE Transactions on Geoscience and Remote Sensing* 45(3): 746–755.

Solberg, A. H. S., G. Storvik, R. Solberg, and E. Volden. 1999. Automatic detection of oil spills in ERS SAR images. *IEEE Transactions on Geoscience and Remote Sensing* 37(4): 1916–1924.

Tomasi, C., and R. Manduchi. 1998. Bilateral filtering for gray and color images. In *Proceedings of the 1998 IEEE International Conference on Computer Vision, Mumbai, January 1998,* 839–846.

Topouzelis, K. N. 2008. Oil spill detection by SAR images: Dark formation detection, feature extraction and classification. *Sensors* 8: 6642–6659.

Trivero, P., W. Biamino, and F. Nirchio. 2007. High resolution COSMO-SkyMed SAR images for oil spills automatic detection. In *IGARSS 2007, July 23–28, Barcelona, Spain,* 2–5.

Wahl, T., T. Anderssen, and A. Skoelv. 1994. *Oil Spill Detection Using Satellite based SAR, Pilot Operation Phase: Final Report,* 153. Försvarets Forskninginstitut: Norwegian Defence Research Establishment,.

Wang, Y., P. Stoica, J. Li, and T. Marzetta T. 2005. Nonparametric spectral analysis with missing data via the EM algorithm. *Digital Signal Processing* 15(2): 191–206.

Yu, Y., and S. Acton. 2002. Speckle reducing anisotropic diffusion. *IEEE Transactions on Image Processing* 11(11): 1260–1270.

Chapter 16

Remote-sensing techniques for natural disaster impact assessment

Piero Boccardo and Fabio Giulio Tonolo

CONTENTS

This chapter provides a thorough review of the remote-sensing methods and techniques currently adopted to monitor a large variety of natural disasters as well as to map their impact, focusing not only on the theoretical aspects but mainly on the operational implications of such topics. The goal is to create reader awareness of what can be achieved by means of remote-sensing data processing, as well as the limits of such methodologies.

A definition of natural disaster is provided to correctly understand which type of disaster will be considered in the chapter. An overview of the relevant statistics related to disasters will highlight the more frequent disasters in

2010, the trend over the past 30 years, and the impact of these disasters in terms of economic and human losses.

Furthermore, the Disaster Risk Management cycle will be described to highlight the activities for which remote sensing is largely adopted and the expected outputs.

The final part of the chapter is dedicated to the description of the main types of calamities (in terms of occurrence frequency and impact on population) and the related information that can be provided by remote sensing.

16.1 INTRODUCTION

The focus of this chapter is on remote-sensing techniques to support natural disaster monitoring and rapid impact assessment activities. Preliminarily, it is crucial to identify which of the various possible definitions of natural disaster is adopted in the following sections.

In the framework of this chapter, the following definition of disasters is considered:

> Situation or event, which overwhelms local capacity, necessitating a request to national or international level for external assistance. An unforeseen and often sudden event that causes great damage, destruction and human suffering. (EM-DAT: The OFDA/CRED International Disaster Database, http://www.emdat.be, Université Catholique de Louvain, Brussels, Belgium)

A similar concept is expressed by the definition of disaster given by the United Nations Office for Disaster Risk Reduction (UNISDR): "A serious disruption of the functioning of a community or a society involving widespread human, material, economic or environmental losses and impacts, which exceeds the ability of the affected community or society to cope using its own resources" (United Nations, 2009).

Disasters can be first classified according to the main causes, that is, natural, technological, or manmade disasters. Only disasters caused by natural causes are analyzed in this chapter. Therefore, events that have human origins are excluded. Natural disasters can be further grouped in disaster types, such as geophysical, meteorological, hydrological, climatological, and biological events.

An overview of the relevant statistics related to disasters allows us to highlight the more frequent disasters in 2010, the trend over the past 35 years, and the impact of these disasters in terms of economic and human losses. To this purpose, information made available by the Centre for Research on the Epidemiology of Disasters (CRED) at the School of Public Health of the Université Catholique de Louvain located in Brussels, Belgium, is analyzed, specifically exploiting the data stored in the EM-DAT database. EM-DAT is a global database on natural and technological disasters that contains essential core data on the occurrence and effects of more than 17,000 disasters in the world from 1900 to present. Figure 16.1 highlights the natural disaster trends over the past 35 years in terms of numbers of disasters as well as killed and affected people.

Regarding data for 2010, a total of 385 natural disasters have been registered, with more than 297,000 casualties. More than 217 million people have been affected worldwide and US$123.9 billion of economic damages have been estimated. In accordance to the trend over the past decade, hydrological disasters (events caused by deviations in the normal water cycle or overflow of bodies of water caused by wind set-up, that is, flood and mass movement events) were by far the most recurrent disasters in 2010. These disasters represented 56.1% of the total disaster occurrence in 2010, and together with meteorological disasters—the second-most frequent disasters—accounted for 79% of total occurrence. The economic impact caused by natural disasters in 2010 was more than 2.5 times higher than in 2009 (US$47.6 billion) and increased by 25.3% compared with the annual average for the period from 2000 to 2009 (US$98.9 billion). From a geographic point of view, about 89% of all people affected by disasters in 2010 lived in Asia; Europe saw the biggest increase in disaster occurrence, and Asia had the largest decrease, counting fewer disasters, victims, and damages compared with the past decade's annual averages (Guha-Sapir, 2011).

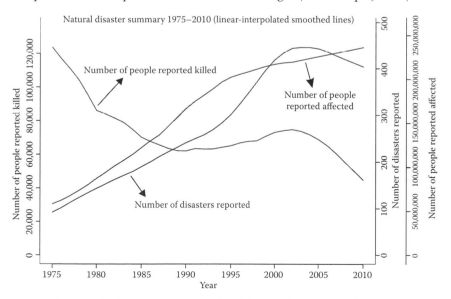

Figure 16.1 Natural disaster summary 1975–2010. (From EM-DAT: The OFDA/CRED International Disaster Database, http://www.emdat.be, Université Catholique de Louvain, Brussels, Belgium. With permission.)

On the basis of these statistics, this chapter focuses on the following:

- The main recurrent natural disaster subgroups, for example, hydrological (floods) and meteorological (storms) events
- Geophysical (earthquakes and tsunamis) events that had a huge impact, in terms of casualties, in 2010 (Haiti earthquake) and 2011 (Japan earthquake and tsunami).
- Climatological events (drought) to give an overview of the potential role of remote sensing in monitoring slow on-set disasters

Each section focuses on the information that can be provided by remote sensing and adopts the following structure:

- *Description*: A brief definition of the disaster and its impact, mainly based on the EM-DAT glossary
- *Rapid impact assessment techniques*: A review of the main remote-sensing techniques currently in use to identify the affected areas and to estimate the affected population

Last, it is important to briefly describe the disaster risk management (DRM) cycle to highlight the phases in which remote sensing is largely adopted. The main goal of DRM and disaster risk reduction (DRR) activities is to limit the impact of a disaster when it occurs, mainly strengthening the community's capabilities to cope with a disaster or to reduce the possibility of a disaster occurring. Different activities can be undertaken to fulfill this goal. A general strategy commonly adopted is based on a disaster cycle including the following phases: risk assessment, mitigation and prevention, preparedness, prediction and warning, response, and recovery.

The focus of the present chapter is on the use of remote sensing in the immediate response phase (rapid impact assessment, especially in terms of affected areas and affected population). Response activities are related to the provision of emergency services and public assistance during or immediately after a disaster to save lives, reduce health impacts, ensure public safety, and meet the basic subsistence needs of the people.

Remote sensing is also exploited in the alert phase, where early warning systems are needed to "to generate and disseminate timely and meaningful warning information to enable individuals, communities and organizations threatened by a hazard to prepare and to act appropriately and in sufficient time to reduce the possibility of harm or loss" (United Nations, 2009). Because such systems are based not only on satellite data but also on ground measurements that are the main input data for detailed forecasting models, they are not discussed in the following sections.

16.1.1 Remote sensing and disasters

Remote-sensing-based analyses are frequently adopted to support both decision makers and responders in the field during disaster management activities, as recently clearly pointed out by the United Nations in the 2011 humanitarian appeal (United Nations, 2011b, "Major Natural Disasters in 2010 and Lesson Learned"):

> At the same time, these crises were crucibles in which new approaches were pioneered with some success. Remote sensing in the hours and days after the Haiti earthquake yielded estimates of numbers of severely affected people that stood the test of time and allowed an unusually rapid flash appeal. Some clusters in Haiti applied advanced methods of mapping the needs so as to orchestrate coverage and identify priorities. (The next step is to translate this into detailed work planning and development of projects for the appeal). . . . Similarly, in Pakistan, the plans in the revised flash appeal were mostly able to encompass the still-expanding scale of needs thanks to information management using remote sensing and other resources necessary for a situation of limited ground access. Moreover, clusters in Pakistan matched this information to their response plans and appeal projects, to a large extent, so as to ensure that their portfolios of projects minimized the gaps and duplications.

To have a clear awareness of the potential of remote sensing, it is necessary to understand its operational implications. The aim of this chapter is to clearly explain the benefits derived from adopting a remote-sensing approach, in particular to support rapid response activities. This will allow end users to know the type of output they can expect, and moreover the related level of accuracy of that output for a specific disaster, in the framework of DM activities. At the same time, the limits of the methodology have to be stressed to avoid any possible misinterpretation of the results.

From a technical point of view, different types of remote-sensing sensors and techniques can be considered, including terrestrial, satellite, and aerial remote sensing (see several other chapters included in this book). The choice is mainly based on the type of disaster and the approximate extent of the affected areas. Generally, the main source of data for response activities is satellite remote sensing, because it allows areas that have accessibility issues with a wide footprint on the ground to be monitored. Furthermore, modern agile satellites can be triggered in a very short time (first images showing the impact of the earthquake that struck Haiti were acquired the morning after the earthquake). As far as the temporal resolution is concerned, the recent availability of constellations drastically decreases the satellite revisiting time, allowing fast dynamic phenomena to be monitored in near real-time.

Very high-resolution (VHR) imagery (with a ground sample distance up to 0.5 meter [m]) is characterized by a very high level of detail that can be exploited not only from a geometric point of view (for map updating and map creation purposes) but especially in terms of semantic content, which is a crucial feature when assessing the impact of a disaster. VHR optical imagery in general has been available for almost a decade and have been proven to be useful in damage assessment in a number of natural disasters (Ehrlich et al., 2009). On the contrary, high-resolution imagery is characterized by a smaller footprint and a limited spectral resolution, characteristics that may limit the effectiveness of automatic classification algorithms as well as the use of this type of data when monitoring countrywide disasters. Radar data are generally exploited when persistent cloud cover makes optical data unusable or when it is necessary to extract accurate information on ground displacements by means of phase data processing.

The licensing policy of the data providers is another crucial feature that must be carefully considered. The aforementioned data belongs to two main families: public-domain data that generally is accessible online (i.e., Moderate Resolution Imaging Spectroradiometer [MODIS], Landsat) or commercial imagery that should be purchased through reseller companies. Sometimes it is possible to freely access satellite data on the basis of scientific agreements between space agencies and principal investigators in charge of specific research projects (e.g., European Space Agency Category-1 users). As far as the disaster response is concerned, several international initiatives have been set up in the recent years to facilitate the access to satellite data in case of major crisis.

16.1.2 International initiatives

The main existing international initiatives aimed at supplying remotely sensed data or related analyses when major disasters occur are described in this section.

The "International Charter Space and Major Disasters," declared formally operational on November 1, 2000, provides a unified system of space data acquisition and delivery to those affected by natural or manmade disasters through authorized users. It is composed of several space agencies and institutions: Each member agency has committed resources to support the provisions of the charter and thus is helping to mitigate the effects of disasters on human life and property (http://www.disasterscharter.org). Authorized users (civil protection, rescue, defense, or security body from the country of a charter member) can request for a charter activation to obtain data and information on a disaster occurrence, which are provided by selected value added resellers supervised by an appointed project manager.

In the framework of the European Programme for the establishment of a European capacity for Earth Observation, GMES (Global Monitoring for

Environment and Security) services are dedicated to the monitoring and forecasting of the Earth's subsystems (http://www.gmes.info). They contribute directly to the monitoring of climate change. SAFER (see GMES, n.d.-a) is a European Union–funded project responsible for the development of the preoperational GMES emergency management service. SAFER reinforces the European capacity to respond to emergency situations, such as fires, floods, earthquakes, volcanic eruptions, landslides, or humanitarian crisis. A priority of the project is to provide rapid mapping (reference maps and assessment maps) during the response phase. SAFER also considers the extension of the use of its products to early warning, before the emergency situation occurs, and to reconstruction, after the emergency situation has occurred (http://www.gmes.info/pages-principales/projects/safer-emergency). G-MOSAIC (GMES, n.d.-b) is the European Union–funded project responsible for the development of the preoperational GMES security service. G-MOSAIC provides intelligence to the European Union and its member states before, during, and after a crisis occurs. The project supports EU intervention activities in the form of preparedness, crisis management, damage assessment, reconstruction, and resilience (http://www.gmes.info/pages-principales/projects/g-mosaic-security).

Google Crisis Response (n.d.) makes critical information more accessible around natural disasters and humanitarian crises, such as hurricanes and earthquakes. This initiative is a project of Google.org, which uses Google's strengths in information and technology to build products and advocate for policies that address global challenges. Activities related to remote sensing include updated satellite imagery and maps of affected areas to illustrate infrastructure damage and to help relief organizations navigate disaster zones (http://www.google.com/crisisresponse/).

Sentinel Asia (n.d.) is a voluntary initiative led by the Asia-Pacific Regional Space Agency Forum (APRSAF) to support disaster management activity in the Asia-Pacific region by applying Web Geographic Information System (WebGIS) technology and space-based technology, such as Earth observation satellites data. In addition to emergency observation of disasters, and based on the requirements of emergency agencies and other key users attending the APRSAF meetings, a top priority was identified for the Sentinel Asia project to emphasize the implementation of satellite-data production systems for wildfire, flooding, and glacier lake outburst flood information (http://sentinel.tksc.jaxa.jp).

16.1.3 Main outputs and dissemination

The main outcomes of remote-sensing-based analysis to support emergency management activities normally take the shape of georeferenced information that can be delivered to users in different formats and modalities.

Regarding the warning phase, the common output is the approximate location of a disaster (e.g., earthquake epicenter, flood-affected country), or in the case of an alert based on forecasting models, the estimated location of the affected areas (e.g., coastal areas hit by a tsunami or a predicted storm path), while also providing a confidence level. The response phase outputs normally are cartographic products, for which the value added information (e.g., flooded areas, collapsed buildings, damaged infrastructures) are integrated with reference cartographic data sets, generally exploiting existing spatial data infrastructures (SDI). Figure 16.2 shows an example of rapid mapping output produced to support the emergency response phase: Road accessibility information (derived by remote-sensing data) is overlaid on a digital map of the affected area. Map readability is a key factor to be taken into account in the map template design phase.

The dissemination process has equal importance to the analysis and map production phases. Final products should reach decision makers at the right time and in the right place. The decision on how to transfer this information must take into consideration environmental factors and network connectivity, considering that those infrastructures may have been seriously affected by the impact of the event.

Normal ways of dissemination include e-mails (including attachments) sent to predefined mailing lists: This is a common option adopted by early warning systems. Another dissemination option is the delivery to specific web portals, focused on emergency management, that aim to gather the relevant and reliable information and analysis related to the latest emergencies. One of the most used portals is ReliefWeb, which was recently renewed. ReliefWeb delivers the most relevant content related to humanitarian emergencies worldwide through really simple syndication (RSS), e-mail, and social media (all useful means for advising potential users of the availability of new and updated information).

With the humanitarian community's improved capacity to handle geographic information, requests for vector data sets to perform analysis better tailored to end user needs are now increasing. Consequently, customized WebGIS applications are being largely implemented (and deployed immediately after specific disaster events) to distribute disaster-related data exploiting Open Geospatial Consortium standards.

16.2 FLOOD AND STORMS

16.2.1 Definition

Floods are hydrological disasters caused when a water body (river, lake) overflows its normal boundaries because of rising water levels. The term general flood is adopted in the case of an accumulation of water on the ground because of long-lasting rainfall and the rise of the groundwater table (also known as water logging). Furthermore, general floods can be the

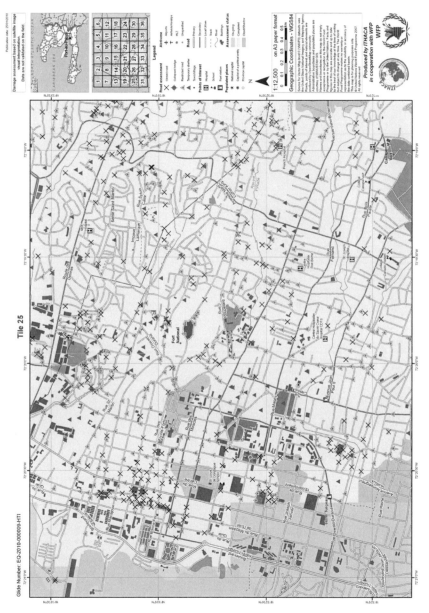

Figure 16.2 Example of Haiti 2010 earthquake road damage assessment map. (Courtesy of ITHACA/WFP, Torino, Italy.)

result of inundation by melting snow and ice, backwater effects, and special causes, such as the breaching of a dam. A particular type of floods are flash flood, or a sudden flooding with short duration. Flash floods are typically associated with thunderstorms and therefore can occur at virtually any place, making it really difficult to forecast or model this type of event.

Floods and flash floods are also a common consequence of severe storms or the landfall of cyclones, which are nonfrontal storm systems characterized by a low pressure center, spiral rain bands, and strong winds. Tropical cyclones usually originate over tropical or subtropical waters. Depending on their location and strength, tropical cyclones are referred to as hurricanes (western Atlantic/eastern Pacific), typhoons (western Pacific), cyclones (southern Pacific/Indian Ocean), tropical storms, and tropical depressions (according to wind speed). As a consequence of storm phenomena, coastal lowlands are threatened by storm surges, which are coastal floods caused by the rise of the water level in the sea as result of strong wind. In 2010, the top-three most significant disasters by number of affected people were flood events in China, Pakistan, and Thailand.

16.2.2 Rapid impact assessment techniques

The rapid impact assessment phase in the case of floods—or floods induced by storms—is mainly focused on the identification of the flood-affected areas. It is therefore necessary to identify the water bodies on postevent images and compare them with the water extent before the event. This goal can be successfully achieved if multitemporal images acquired before and after the event are available or, as an alternative, if reliable and updated reference water-body cartographic data are available (the so-called Normal Water Extent Database [NWED], Wang et al., 2002).

The applied research activities are therefore focused on the identification of proper algorithms or methodologies for the extraction of water bodies by means of satellite data processing, tailored to the technical features of the available images.

Regarding active sensors, synthetic aperture radar (SAR) amplitude images enable the easy identification of water bodies; therefore, they are the main input data for flood analyses (Henry et al., 2003; Aduah et al., 2007; Schumann et al., 2007). The all-weather capability of the radar technology and the possibility to acquire data during the nighttime are crucial advantages of a radar-based approach, especially when assessing the impact of cyclones and hurricanes that are characterized by a persistent and homogeneous cloud coverage (radar attenuation effects should be taken into account). Conversely, the data are affected by geometric distortions (layover, foreshortening, and radar shadows), which are hard to model, especially in mountain regions. The areas presenting as water can be spotted on radar images, exploiting their reflective behavior toward the electromagnetic radiation emitted by the radar sensor, which is roughly that of a specular surface. Water can be easily identified as characterized by low radiometric values. By applying change-detection techniques when multitemporal satellite data are available or exploiting an NWED, it is possible to isolate the flooded areas, distinguishing them from water bodies (see Figure 16.3). Radar- shadowing modeling should be adopted to easily identify areas that can be erroneously classified as water.

SAR images can be exploited for flood-mapping purposes by adopting a methodology based on the use of coherence measurements derived from multipass SAR interferometry. These images also can be used as an indicator of changes in the electromagnetic scattering behavior of the surface (Nico et al., 2000).

A different approach for the definition of flooded areas is based on multispectral optical data processing. Water body identification on a scene acquired by optical sensors can be based on simple but effective histogram threshold techniques; those techniques exploit the behavior of water in the infrared bands, where those surfaces have high absorption rates. Those methods are simple to apply but several disadvantages are evident. Shadows, due to the local morphology or to the presence of clouds, are classified as water bodies, and threshold values cannot be defined uniquely but rather are adapted to the conditions at the moment of the acquisition.

Different water-body classification techniques are mainly based on indexes derived from differential band ratios, to make threshold values independent from image acquisition parameters. In particular, according to literature, the use of the Normalized Differential Water Index (NDWI) is commonly used to identify and classify flooded areas. Unfortunately, a unique definition of this index is not available, probably because of its adaptation to the different characteristics of spectral sensors mounted on a satellite platform normally used for those applications. The most diffuse definition of NDWI, described by (McFeeters (1996); Chatterjee et al. (2005); Jain et al. (2005); Chowdary et al. (2008); and Hui et al. (2008),

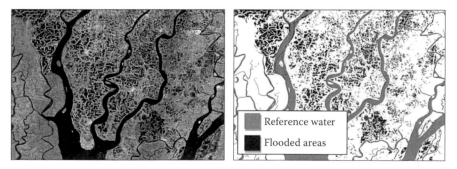

Figure 16.3 (a) Reference water (light gray) and flooded areas (dark gray) identified through (b) radar SAR postevent imagery processing.

is based on reflectivity in the green and near-infrared bands. This index reduces commission errors during classification, mainly because of vegetation and bare soil classes. Hu (2007); Fengming et al. (2008); and Hui et al. (2008) highlighted the low reliability of this index in urban areas, proposing a Modified Normalized Difference Water Index (MNDWI) to minimize errors caused by the presence of shadows. Huggel (2002) proposed an NDWI definition based on reflectivity in the blue and near-infrared bands, especially conceived to the identification of mountain lakes. Fadhil (2006); Sakamoto et al. (2007); Islam et al. (2009); and Mori et al. (2009) used red and short-wavelength infrared (SWIR) to highlight the residual influence of humid soils, while Gao (1996) and DeAlwis et al. (2007) proposed an NDWI definition based on NIR and SWIR.

Finally, a literature review shows a vast heterogeneity of NDWI definition, with, as a common component, the use of differential ratios based on those bands exalting relative reflectivity differences of water spectral signature (visible and infrared bands).

For several reasons, the MODIS sensor is generally used for small-scale flood monitoring (Brakenridge and Anderson, 2003; Aduah, 2007; Voigt et al., 2007). The MODIS mission grants a worldwide coverage, providing daily images and derived products that are public domain. Furthermore, low-geometric-resolution (250, 500, and 1,000 m) MODIS data allow a regional view of the observed phenomena. Therefore, the use of MODIS data permits a multitemporal small-scale analysis of the flood event evolution in the areas of interest, as clearly shown in the example reported in Figure 16.4, related to a flooding in Bangladesh started in June 2007 (water bodies shown in dark gray).

The detection of water bodies and flooded areas is the result of a classifying procedure of MODIS primary reflectance data available in near real-time through the NASA/GSFC MODIS Rapid Response system. For

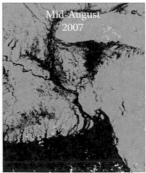

Figure 16.4 2007 Bangladesh flood evolution monitoring based on MODIS multispectral data classification.

classification purposes, the aforementioned indexes useful for water bodies and flooded area detection and for cloud effects reduction are adopted. Once again, by using reliable water bodies data or images acquired before the event, it is therefore possible to isolate only the flooded areas. The main disadvantage of an approach based on optical imagery comes from the cloud coverage, which drastically reduces the areas that can be analyzed and may led to classification errors because of the presence of cloud shadows. To cope with the cloud coverage issue, temporal composite approaches have been developed to minimize the impact of cloud and cloud shadows on the classification accuracy (Ajmar et al., 2010).

The main issues when trying to make the aforementioned procedures unsupervised, generally based on a decision tree approach, is that flood-affected areas are characterized by heterogeneous spectral signatures. Furthermore, given that low- or medium-spatial resolution imagery generally are adopted, the spectral signature of the flood polygons boundaries are spurious. The focus of ongoing research should therefore be on how to automate classification procedures to allow for an operational monitoring of water bodies with global coverage and high-frequency updating.

16.3 EARTHQUAKES

16.3.1 Definition

Earthquakes can be defined as a shaking and displacement of ground caused by a sudden release of stored energy in the Earth's crust that creates seismic waves. They belong to the geophysical natural disasters category and can be of both tectonic or volcanic origin. At the Earth's surface, they are felt as a shaking or displacement of the ground. The released energy can be measured in different frequency ranges using seismometers observations. Different scales for measuring the magnitude of a quake can be adopted, according to a certain frequency range: surface wave magnitude (Ms), body wave magnitude, local magnitude, and moment magnitude. Earthquakes can trigger landslides and, when the epicenter is located offshore, tsunamis (see next section).

In 2009, 22 earthquakes killed 1,888 persons, made 3.2 million victims, and caused US$6.2 billion of damages. In 2010, 25 earthquakes caused 226,735 fatalities, 7.2 million victims, and US$ 46.2 billion of damages (Guha-Sapir, 2011).

16.3.2 Rapid impact assessment techniques

The acquisition of field data supporting the impact assessment in areas hit by severe earthquakes is indeed a hard task, mainly because of the restricted physical accessibility of the affected areas (i.e., unpredictable road conditions, landslides and soil fractures, lack of means of communications with the affected population, panic, growing of diseases, lack of food and water, hazards caused by instable buildings). To cope with the accessibility and time constraints issues, "the use of EO (Earth Observation) data in earthquake contexts, especially for damage assessment purposes, has been widely proposed and a number of results have been presented after every event, mostly based on optical data and manual interpretation" (Polli et al., 2010). The 2010 Haiti earthquake

> will also be known as one of the first events where technology (especially high-resolution imagery) was embraced at such a large scale in a real operational sense. Almost from the very onset of the disaster, high-resolution satellite imagery was available to provide the first glimpse of the devastation caused by this earthquake. (Eguchi et al., 2010)

Both active and passive sensors can be considered for damage identification purposes. Generally, SAR phase imagery is processed to map the surface displacements and precisely identify the location and direction of fault lines as well as vertical displacements, while high resolution optical imagery are needed to assess the building damage grade as well as damage to infrastructures.

As reported by Matsuoka and Yamazaki (2004), SAR systems are capable of recording both the amplitude and phase of backscattering echoes from the objects on the Earth's surface. The SAR signal-processing technique referred to as interferometric SAR (InSAR) is widely used in seismology, volcanology, hydrogeology, and landslide studies. In the past 15 years, InSAR has demonstrated unique capabilities for mapping the topography and the deformation of the Earth's surface. The InSAR approach is based on extracting the phase component of SAR data to compute the pixel-by-pixel difference of the SAR signal relative to a specific area and acquired from nearby geometric conditions. The interferogram—that is, the result of the interferometric processing—contains the measurement of the sensor to target distance and of any possible change distance (Stramondo, 2008). Seismology represents the fieldwork in which InSAR obtains a higher number of results. Since the early 1990s, the capabilities of the InSAR technique, as well as the differential interferomteric SAR (DInSAR) approach, have been exploited to study surface displacement caused by moderate to strong earthquakes (Massonnet et al., 1993; Peltzer and Rosen, 1995; Stramondo et al., 1999; Reilinger et al., 2000; Wegmuller, 2010).

Figure 16.5 shows a differential interferogram derived by means of COSMO-SkyMed SAR data aimed at analyzing the impact of the earthquake that hit central Italy (L'Aquila) in April 2009 (Stramondo et al., 2011).

The SAR amplitude information, which is proportional to the backscattering coefficient of the Earth's surface, proved to be less dependent from the acquisition parameters (satellite geometry, acquisition duration, and

Figure 16.5 Coseismic COSMO-SkyMed differential interferogram. Each phase cycle corresponds with 15 mm satellite-to-ground distance. (Courtesy of Salvatore Stramondo.)

wavelengths of radar; Yonezawa and Takeuchi, 1999) in respect to the phase data. Therefore, the backscattering coefficient derived from SAR intensity images may be used to capture damaged areas by means of multitemporal SAR data processing. Indexes based on the backscattering coefficient and correlation coefficient can be computed for this purposes. Amplitude coherence analysis can be carried out as well to detect low-coherence areas related to high damages on the ground (Matsuoka and Yamazaki, 2004).

Figure 16.6 shows an example of a coherence multitemporal analysis related to the earthquake that hit central Italy (l'Aquila) in April 2009. Blue areas show a high coherence value that can be related to undamaged buildings, whereas low coherence values are depicted in red (higher amplitude values in postevent image) and green areas (higher amplitude values in pre-event image). Specifically, large red areas in Figure 16.6 are due to the presence of temporary shelters set up after the earthquake.

As far as the use optical imagery is concerned, a common approach is based on VHR data, exploiting the availability of satellite missions providing images with a spatial resolution up to 0.5 m. Postdisaster imagery is used to assess the impact of the disaster by detecting, enumerating, and possibly

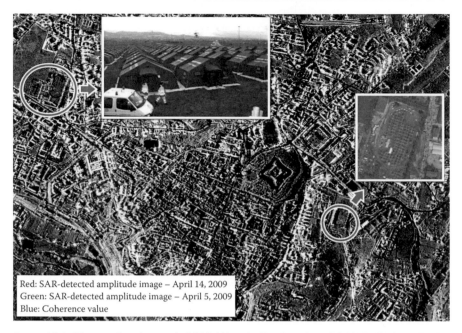

Red: SAR-detected amplitude image – April 14, 2009
Green: SAR-detected amplitude image – April 5, 2009
Blue: Coherence value

Figure 16.6 **(See color insert.)** 2009 L'Aquila Earthquake, COSMO-SkyMed Multi-temporal with Coherence map for postearthquake change detection. The figure highlights new tent camps established to gather displaced persons. (From COSMO-SkyMed Product ©ASI Agenzia Spaziale Italiana, 2009. All rights reserved. Image processed by e-GEOS.)

measuring the size of collapsed buildings and identifying other infrastructures such as interrupted roads. (Ehrlich et al., 2009).

Although several automatic or semiautomatic techniques, which are based on automated (rule-set based) change detection and segmentation algorithms, postclassification comparison, or textural approaches, exist to identify collapsed building after an earthquake, the calibration stage of such methods is a time-consuming procedure, and the accuracy is not completely predictable. Furthermore, the outputs of such approaches generally provide indications of the level of damages at the block or grid level and not at the building level. Therefore, in an operational context, a visual interpretation approach is generally adopted (Chini, 2009; Ehrlich et al., 2009) to produce results as reliable as possible. Generally, the focus is on the following features of interest: collapsed or damaged buildings, road network accessibility, spontaneous camps, and landslides induced by the earthquake (examples provided in Figure 16.2 and Figure 16.7).

The analysis is based on a multitemporal change detection activity between the satellite data acquired before and after the event. When a large amount of data characterized by a high spatial resolution is available, and given the tight time constraint of the emergency response phase,

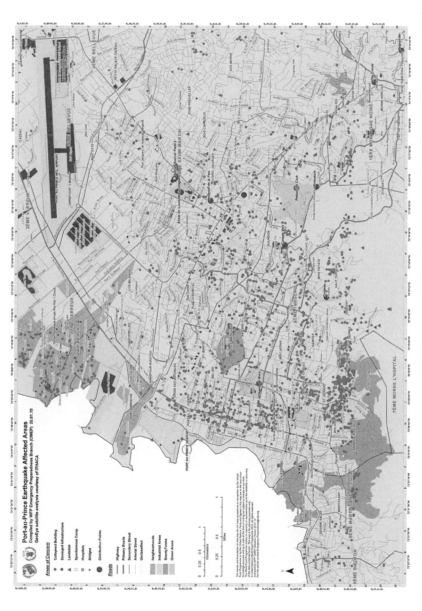

Figure 16.7 Earthquake damage assessment map from Haiti Earthquake 2010. Port-au-Prince affected areas. (Courtesy of ITHACA/WFP, Torino, Italy.)

a coordinated volunteer approach is the most effective to assess the impact of the event.

Concerning the accuracy of the identified damages, recent studies (Saito et al., 2010) highlighted that vertical imagery (and in certain conditions also oblique ones) may be limiting in discriminating the level of damage of some buildings. In the summary of the Second International Workshop on Validation of geoinformation products for crisis management, it is explicitly reported that a validation of a "joint damage assessment (using airborne images) performed with around 6000 geo-tagged photos collected in the field gave an overall accuracy of 60% only" (JRC-ISFEREA and VALgEO, 2010).

Validation activities related to information derived by means of remote sensing are therefore crucial to ensure that end users are aware of the accuracy they can expect from a specific satellite-based analysis, consequently adapting the outputs for a more effective response activity.

16.4 TSUNAMIS

16.4.1 Definition

Tsunami ("harbor wave" in Japanese) is a natural disaster defined as a series of waves caused by a rapid displacement of a body of water (ocean, lake) that can be triggered by earthquakes, volcanic eruptions, or mass movements. The waves are characterized by a very long wavelength, while their amplitude is much smaller offshore. The impact in coastal areas can be destructive as the waves advance inland and can extend over thousands of kilometers.

16.4.2 Rapid impact assessment techniques

In recent years, gigantic earthquakes whose epicenter was located in oceans triggered massive tsunamis, which inundated coastal areas in the surrounding countries. Satellite remote-sensing data were largely used for the first time to assess the tsunami impact after the 2004 Indian Ocean event that mainly affected Indonesia. A common approach was based on visual interpretation of VHR imagery acquired after the event, aimed at identifying the destroyed shorelines, the flooded areas, and the affected rural and urban areas, as reported in several maps produced in the framework of the Disaster Charter initiative. In urban areas, a specific focus was on destroyed buildings. An alternative approach, as mentioned in the flood- and earthquake-related sections, exploits the availability of both pre- and postevent satellite data to carry out change detection analysis, applying unsupervised classification algorithms (Chen et al., 2006). Yamazaki and Matsuoka (2006) showed that the integrated use of satellite-derived indices such as NDVI (Normalized Difference Vegetation Index), NDSI (Normalized Difference Soil Index), and NDWI could be used as an indicator of tsunami affected areas, highlighting the presence of flooded areas as well as loss of vegetation.

More recently, the earthquake that struck Japan on March 2011 triggered a devastating tsunami with 10 m high waves that reached the U.S. west coast. As far as satellite-based damage assessment is concerned, several impact maps were made available the first day after the event. Most of these maps simply added a geographic reference to pre- and post-VHR satellite images, allowing the user to visually interpret the images thanks to the level of detail offered by a 0.5 m spatial resolution image. Some maps, based on the visual analysis of orthorectified and radiometrically enhanced VHR imagery, showed an aggregated damage assessment overview by means of simple but effective color codes highlighting the different inundation levels.

In the days following the tsunami, a huge amount of cartographic products were made available to the humanitarian community, exploiting VHR SAR data to map the flood-affected areas. The map shown in Figure 16.8 is based on the aforementioned SAR-based analysis, where affected coastal areas (standing water and wet areas) have been highlighted in light blue, by overlapping the classification results of the SAR high-resolution data (3 m) acquired by the Italian SAR satellite constellation COSMO-SkyMed to reference cartographic data sets (using optical satellite imagery as backdrop).

SAR images were also processed by means of simple but very effective color composite techniques. Pre- and postdata were assigned to different color channels to create a false color composite highlighting the flood-affected areas.

16.5 DROUGHT

16.5.1 Definition

Conceptual and operational definitions of drought are both available, and they vary depending on the variable used to describe this type of natural disaster. Therefore, it is not possible to provide a unique and accurate definition; generally, however, drought can be classified into four different categories: meteorological, hydrological, agricultural, and socioeconomic drought (Mishra and Singh, 2010). Given the focus of this chapter on the impact assessment phase, attention will be paid to agricultural drought, or a period with declining soil moisture and consequent crop failure. This definition fits the main outcomes of the 2011 Global Assessment Report on Disaster Risk Reduction (United Nations, 2011a), where the following is clearly highlighted:

Figure 16.8 **(See color insert.)** 2011 Japan tsunami. Flood Map derived from the semiautomated analysis of COSMO-SkyMed images. (From COSMO-SkyMed Product ©ASI Agenzia Spaziale Italiana, 2011. All rights reserved. Image processed in very rush mode by e-GEOS Emergency Team.)

- Drought impacts most visibly on agricultural production.
- Meteorological drought is a climatic phenomenon rather than a hazard per se, which only becomes hazardous when it is translated into agricultural or hydrological drought, depending on factors other than just rainfall.

16.5.2 Rapid impact assessment techniques

A common approach to monitor and assess the impact of drought events is based on indices that allow the spatial and temporal dimensions of drought (as well as its intensity) to be detected. The following drought indices, mainly based on precipitation data (also in combination with other climatological variables, such as temperature and evapotranspiration) have been extensively discussed in literature: Palmer Drought Severity Index (PDSI) and Crop Moisture Index (CMI) developed by Palmer, and the Standardized Precipitation Index (SPI) developed by Mckee et al. (Mishra and Singh, 2010). Although largely adopted, this section will be focused on satellite-derived indices related to agricultural drought, allowing for the identification of vegetation anomalies induced by groundwater scarcity.

NDVI is commonly used to monitor vegetation at continental scale. It is based on the reflectance values in the near infrared (NIR) and red bands of the electromagnetic spectrum, which are strictly correlated to vegetation health and density because the chlorophyll absorbs red radiation, whereas the mesophyll leaf structure reflects NIR radiation. NDVI is also effective for accurate description of continental land cover, vegetation vigor assessment and classification, vegetation net primary production estimation, and crop growth condition monitoring (Chopra, 2006). Deshayes et al. (2006) underlined that the relatively low intensity of changes resulting from continuous and progressive mechanism, such as phenological alteration, requires the availability of long time-series, therefore limiting the choice of satellite sensors to be potentially adopted for drought-related analysis (i.e., VEGETATION, NOAA Advanced Very High Resolution Radiometer, MODIS, and Medium Resolution Imaging Spectrometer). A Vegetation Condition Index (VCI) was specifically developed by Kogan (1995) to monitor the effects of drought on vegetation: It is based on actual NDVI values compared with maximal and minimal values observed during a period of interest, allowing for the better identification of rainfall dynamics in geographically nonhomogeneous areas. Furthermore, several methods relate satellite-based drought indices with other meteorological and physical parameters, generally through a weighted linear combination approach in a GIS environment to extract drought risk maps.

Regarding drought impact analysis, a common approach is to identify a historical trend of the aforementioned indices by means of long time-series data processing, possibly taking into account seasonal and interannual dynamics, to assess the actual vegetation condition.

Bellone et al. (2009) adopted a proper statistical method (least squares regression techniques coupled with a robust technique, such as least median of squares regression estimator, which is used for a preliminary outliers detection and removal) to calculate a regression. The defined parameters of such regression, which show the long-term NDVI trends, have been used to calculate and map historical monthly NDVI deviations. Those maps show the pixel-based spatial distribution of vegetation conditions regarding the normal long-term NDVI behavior, modeled by the described trends, with a procedure that allows for the analysis of seasonal variations (every month is compared with the historical trend of the same month). Specifically, a negative departure from the trend describes vegetation condition worse than normal.

Seasonal changes observed in NDVI time-series have proven useful in tracking vegetation phenology, or the study of the annual cycles of green-up, or growth, and senescence of the plants. Specifically, from the yearly NDVI function that best fits the original yearly NDVI time-series, the following phenological parameters or metrics can be extracted for each considered vegetation season: the start of the season, the length of the season, the start to the end of the season, the seasonal amplitude, and the seasonal small integral. Furthermore, Perez et al. (in press) showed that a comparison of values of the considered phenological parameters with the average, minimum, and maximum values computed using NDVI time-series helps to explain the performances of the considered vegetative season and to formulate a judgment on its expected productivity, potentially highlighting droughts related effects. Figure 16.9 shows the prevalent shift of the season in Niger and Chad with respect to the previous 10 years' averages, aggregating the results at administrative boundaries level.

16.6 CONCLUSION

This chapter showed how satellite remote sensing is largely adopted in the rapid impact assessment stage of the DRM cycle. To assess the impact of natural disasters, imagery acquired by both active and passive sensors are processed: The sensor choice is mainly based on data availability, cloud coverage persistence, and type of information required. From a technical point of view, the response to the 2010 Haiti earthquake clearly demonstrated that a modern satellite can be timely triggered to have the first images acquired within a few hours of the event. Furthermore, the availability of constellations drastically decreases the satellite revisiting time, allowing fast dynamic phenomena to be monitored in near real-time. International initiatives support the humanitarian community in accessing remotely sensed data. If images acquired before and after the event are available, different change-detection strategies can be applied to identify the affected areas. A quick-and-dirty methodology, generally

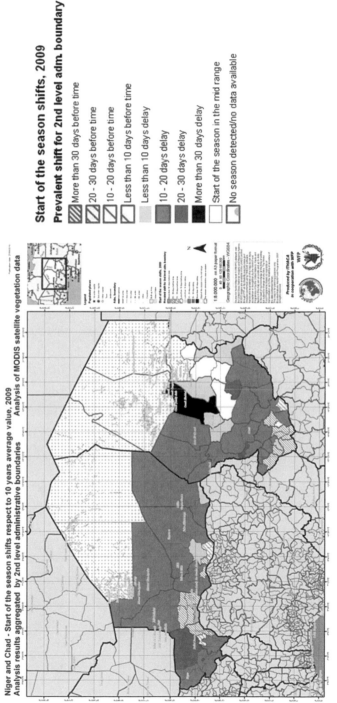

Figure 16.9 Map showing the deviations of the phenological parameter "Start of the Season" in Niger and Chad in the period from June to October 2009. Output aggregated on the second-level administrative boundaries. (Courtesy of ITHACA/WFP, Torino Italy.)

applied for water-related disasters, is based on multitemporal color composites to highlight the difference in the reflectivity or backscattering of the surface, attributable to the impact of the event on the ground (presence of water or loss of vegetation). If only postevent images are available, reference cartographic data are required to correctly identify the affected areas, stressing the importance of operational spatial data infrastructures with a global coverage as well as of volunteer mapping initiatives. VHR resolution imagery, both optical and radar, is the main input data when assessing damages at an urban level: Unfortunately, even if automatic classification techniques (both pixel based and object oriented) have been successfully tested in several case studies, the most common approach in an operational context is visual interpretation. This is confirmed by a quick analysis of the impact maps produced for the latest major emergencies at urban scale and are made available on ReliefWeb, on which only a few products are based on automatic procedures. Applied research therefore should focus on the accuracy and speed of automatic procedures, limiting the time required for the initial calibration step, especially taking into account SAR imagery that could be the only available data when clouds cover the affected areas. Last, it is important to highlight the potential role that volunteers may play in mapping both reference cartographic data as well as damages caused by the disaster, drastically decreasing the response time and ensuring the availability of up-to-date large-scale reference layers for disaster impact assessment at the urban level.

REFERENCES

Aduah, M., B. Maathuis, and Y. Ali Hussin. 2007. Synergistic use of optical and radar remote sensing for mapping and monitoring flooding system in Kafue flats wetland of southern Zambia. ISPRS Commission VII meeting, Instanbul. http://www.isprs2007ist.itu.edu.tr/41.pdf.

Ajmar A., P. Boccardo, F. Disabato, and F. Giulio Tonolo. 2010. Near real time flood monitoring tool. In *Proceedings of the GI4DM Conference, February 2–4, Torino, Italy.*

Bellone T., P. Boccardo, and F. Perez. 2009. Investigation of vegetation dynamics using long-term normalized difference vegetation index time-series. *American Journal of Environmental Sciences* 5(4): 461–467.

Brakenridge, G., and E. Anderson. 2003. Satellite gaging reaches: a strategy for MODIS-based river monitoring. *Ninth International Symposium of the Remote Sensing International Society of Optics of England (SPIE), Crete,* 479–485.

Chatterjee, C., R. Kumar, B. Chakravorty, A. K. Lohani, and S. Kumar. 2005. Integrating remote sensing and GIS techniques with groundwater flow modeling for assessment of waterlogged areas. *Water Resources Management* 19(5): 539–554.

Chen, P., S. C. Liew, and L. K. Kwoh. 2006. Tsunami damage mapping and assessment in Sumatra using remote sensing and GIS techniques. In *Geoscience and Remote Sensing Symposium. IGARSS 2006. IEEE International Conference,* 297–300.

Chini, M. 2009. Earthquake damage mapping techniques using SAR and optical remote sensing satellite data. In *Advances in Geoscience and Remote Sensing*, edited by G. Jedlovec. http://www.intechopen.com/books/advances-in-geoscience-and-remote-sensing/earthquake-damage-mapping-techniques-using-sar-and-optical-remote-sensing-satellite-data.

Chopra, P. 2006. Drought risk assessment using remote sensing and GIS: a case study of Gujarat. Master's thesis, IIRS and ITC. .

Chowdary, V. M., R. V. Chandran, N. Neeti, R. V. Bothale, Y. K. Srivastava, P. Ingle, D. Ramakrishnan, et al. 2008. Assessment of surface and sub-surface water-logged areas in irrigation command areas of Bihar state using remote sensing and GIS. *Agricultural Water Management* 95(7): 754–766.

DeAlwis, D. A., Z. M. Easton, H. E. Dahlke, W. D. Philpot, and T. S. Steenhuis. 2007. Unsupervised classification of saturated areas using a time series of remotely sensed images. *Hydrology and Earth System Sciences Discussion* 4: 1663–1696.

Deshayes, M., D. Guyon, H. Jeanjean, N. Stach, A. Jolly, and O. Hagolle. 2006. The contribution of remote sensing to the assessment of drought effects in forest ecosystems. *Annals of Forest Science* 63(6): 579–595.

Eguchi, R., S. Gill, S. Ghosh, W. Svekla, B. Adams, G. Evans, J. Toro, K. Saito, and R. Spence. 2010. The January 12, 2010 Haiti earthquake: a comprehensive damage assessment using very high resolution areal imagery. In *Proceedings 8th International Workshop on Remote Sensing for Disaster Management*. http://www.enveng.titech.ac.jp/midorikawa/rsdm2010_pdf/19_eguchi_paper.pdf.

Ehrlich, D., H. D. Guo, K. Molch, J. W. Ma, and M. Pesaresi. 2009. Identifying damage caused by the 2008 Wenchuan earthquake from VHR remote sensing data. *International Journal of Digital Earth* 2(4): 309–326.

Fadhil, A. M. 2006. Environmental change monitoring by geoinformation technology for Baghdad and its neighboring areas. In *Proceeding of the International Scientific Conference of Map Asia 2006: The 5th Asian Conference in GIS, GPS, Aerial Photography and Remote Sensing. Bangkok, Thailand.* http://www.geospatialworld.net/index.php?option=com_content&view=article&id=15221%3Aenvironmental-change-monitoring-by-geoinformation-technology-for-baghdad-and-its-neighboring-areas&catid=124%3Aenvironment-conservation-monitoring&Itemid=41.

Fengming, H., X. Bing H. Huabing Y. Qian, and G. Peng. 2008. Modelling spatial-temporal change of Poyang Lake using multitemporal Landsat imagery. *International Journal of Remote Sensing* 29(20): 5767–5784.

Gao, B. C. 1996. NDWI: a normalized difference water index for remote sensing of vegetation liquid water from space. *Remote Sensing of Environment* 58(3): 257–266.

Global Monitoring for Environment and Security (GMES). n.d.-a. G-MOSAIC. http://www.gmes.info/pages-principales/projects/g-mosaic-security.

Global Monitoring for Environment and Security (GMES). n.d.-b. SAFER. http://www.gmes.info/pages-principales/projects/safer-emergency.

Google Crisis Response. n.d. About Google crisis response. http://www.google.com/crisisresponse/about.html.

Guha-Sapir, D., F. Vos, R. Below, and S. Ponserre. 2011. *Annual Disaster Statistical Review 2010: The Numbers and Trends*. Brussels: CRED. http://www.cred.be/sites/default/files/ADSR_2010.pdf.

Henry, J., P. Chastanet, K. Fellah, and Y. Desnos. 2003. ENVISAT multi-polarised ASAR data for flood mapping. *IEEE Transactions on Geoscience and Remote Sensing,* 1136–1138.

Hu, Z., H. Gong, and L. Zhu. 2007. Fast flooding information extraction in emergency response of flood disaster. In *ISPRS Workshop on Updating Geo-spatial Databases with Imagery & The 5th ISPRS Workshop on DMGISs.* http://www.isprs.org/proceedings/XXXVI/4-W54/papers/173-177%20Zhuowei%20Hu.pdf.

Huggel, C., A. Kääb, W. Haeberli, P. Teysseire, and F. Paul. 2002. Remote sensing based assessment of hazards from glacier lake outbursts: a case study in the Swiss Alps. *Canadian Geotechnical Journal* 39(2): 316–330.

Hui, F., B. Xu, H. Huang, and P. Gong. 2008. Modeling spatial-temporal change of Poyang Lake using multi-temporal Landsat imagery. *International Journal of Remote Sensing* 29: 5767–5783.

International Charter Space and Major Disasters. 2000, November 1. About the charter. http://www.disasterscharter.org/web/charter/about.

Islam A. S., S. K. Bala, and A. Haque. 2009. Flood inundation map of Bangladesh using MODIS surface reflectance data. In *Proceedings of the 2nd International Conference on Water and Flood Management, Dhaka, Bangladesh,* Vol. 2. http://teacher.buet.ac.bd/akmsaifulislam/publication/Flood_Map_Paper_112.pdf.

Jain, S., R. D. Singh, M. K. Jain, and A. K. Lohani. 2005. Delineation of flood-prone areas using remote sensing techniques. *Water Resources Management* 19(4): 333–347.

JRC-ISFEREA and VALgEO. 2010. Workshop summary. From *International Workshop on Validation of Geo-Information Products for Crisis Management, Ispra, Italy.* http://globesec.jrc.ec.europa.eu/workshops/valgeo-2010/Summary_of_VALgEO2010.pdf.

Kogan, F. N. 1995. Droughts of the late 1980s in the United States as derived from NOAA polar-orbiting satellite data. *Bulletin of the American Meteorological Society* 76(5): 655–668.

Matsuoka, M., and F. Yamazaki. 2004. Use of satellite SAR intensity imagery for detecting building areas damaged due to earthquakes. *Earthquake Spectra* 20(3): 975–994.

McFeeters, S. K. 1996. The use of Normalized Difference Water Index (NDWI) in the delineation of open water features. *International Journal of Remote Sensing* 17(7): 1425–1432.

Mishra, A. K., and V. P. Singh. 2010. A review of drought concepts. *Journal of Hydrology* 391(1/2): 202–216.

Mori, S., W. Takeuchi, and H. Sawada. 2009. Estimation of land surface water coverage (LSWC) with AMSR-E and MODIS. In *Proceedings of 2nd Joint Student Seminar at AIT: Bangkok, Thailand.* http://stlab.iis.u-tokyo.ac.jp/~wataru/publication/pdf/student2009_paper.pdf.

Nico, G., M. Pappalepore, G. Pasquariello, A. Refice, and S. Samarelli. 2000. Comparison of SAR amplitude versus coherence flood detection methods: A GIS application. *International Journal of Remote Sensing* 21(8): 1619–1631.

Peltzer G., and P. Rosen. 1995. Surface displacements of the 17 May 1993 Eureka Valley, California, earthquake observed by SAR interferometry. *Science* 268: 1333–1336.

Perez F., F. Disabato, M. De Stefano, and R. Vigna. In press. Satellite-based drought analysis for the Niger and Chad 2010 food crisis. *Italian Journal of Remote Sensing*.

Polli, D., F. Dell'Acqua, P. Gamba, and G. Lisini. 2010. Earthquake damage assessment from post-event only radar satellite data. In *Proceedings 8th International Workshop on Remote Sensing for Disaster Management*. http://www.enveng. titech.ac.jp/midorikawa/rsdm2010_pdf/16_fabio_paper.pdf.

Reilinger R. E., S. Ergintav, R. Bürgmann, S. McClusky, O. Lenk, A. Barka, O. Gurkan, et al. 2000. Coseismic and postseismic fault slip for the 17 August 1999, M _ 7.4, Izmit, Turkey earthquake. *Science* 289: 1519–1524.

Saito, K., R. Spence, E. Booth, G. Madabhushi, R. Eguchi, and G. Stuart. 2010. Damage assessment of Port au Prince using Pictometry. In *Proceedings of the 8th International Workshop on Remote Sensing for Disaster Management*. http://www.enveng.titech.ac.jp/midorikawa/rsdm2010_pdf/18_saito_paper.pdf.

Sakamoto, T., N. V. Nguyen, A. Kotera, H. Ohno, N. Ishitsuka, and M. Yokozawa. 2007. Detecting temporal changes in the extent of annual flooding within the Cambodia and the Vietnamese Mekong Delta from MODIS time-series imagery. *Remote Sensing of Environment* 109(3): 295–313.

Schumann, G., R. Hostache, C. Puech, and L. Hoffmann. 2007. High-resolution 3-D flood information from radar imagery for flood hazard management. *IEEE transactions on Geoscience and Remote Sensing* 45(6): 1715–1725.

Sentinel Asia. n.d. About Sentinel Asia. https://sentinel.tksc.jaxa.jp/sentinel2/MB_ HTML/About/About.htm.

Stramondo, S. 2008. 15 years of SAR interferometry. *Bollettino di Geofisica Teorica e Applicata* 49(2): 151–162.

Stramondo, S., M. Chini, C. Bignami, S. Salvi, S. Atzori. 2011. X-, C-, and L-band DInSAR investigation of the April 6, 2009, Abruzzi Earthquake. *IEEE Geoscience and Remote Sensing Letters* 8(1): 49–53.

Stramondo, S., M. Tesauro, P. Briole, E. Sansosti, S. Salvi, R. Lanari, M. Anzidei, et al. 1999. The September 26, 1997 Colfiorito, Italy, earthquakes: Modeled coseismic surface displacement from SAR interferometry and GPS. *Geophysics Research Letters* 26(7): 883–886.

United Nations. 2009. UNISRD terminology on disaster risk reduction. http://unisdr. org/files/7817_UNISDRTerminologyEnglish.pdf.

United Nations. 2011a. Global assessment report on disaster risk reduction, revealing risk, redefining development: summary and main findings. http://www. preventionweb.net/english/hyogo/gar/2011/en/bgdocs/GAR-2011/GAR2011_ ES_English.pdf.

United Nations. 2011b. Humanitarian appeal 2011: consolidated appeal process. http:// ochadms.unog.ch/quickplace/cap/main.nsf/h_Index/CAP_2011_Humanitarian_ Appeal/$FILE/CAP_2011_Humanitarian_Appeal_SCREEN.pdf?openElement.

Voigt, S., T. Kemper, T. Riedlinger, R. Kiefl, K. Scholte, and H. Mehl. 2007. Satellite image analysis for disaster and crisis-management support. *IEEE Transactions on Geoscience and Remote Sensing* 45(6): 1520–1528.

Wang, S., Y. Liu, Z. Yi, and C. Wei. 2002. Study on the method of establishment of normal water extent database for flood monitoring using remote sensing. *Geoscience and Remote Sensing Symposium: Igarss '02, 2002 IEEE International*, Vol. 4, 2048–2050.

Wegmuller, U. 2010. Mapping the Haiti earthquake's co-seismic displacement. *Spier Newsroom.* http://spie.org/documents/Newsroom/Imported/002960/002960_10. pdf.

Yamazaki, F., and M. Matsuoka. 2006. Remote sensing technologies for earthquake and tsunami disaster management. In *Proceedings of the 2nd Asia Conference on Earthquake Engineering, March 10–11, Manila, Philippines.*

Yonezawa, C., and S. Takeuchi. 1999. Detection of urban damage using interferometric SAR decorrelation. *Geoscience and Remote Sensing Symposium: IGARSS '99, IEEE 1999 International,* Vol. 2, 925–927. http://ares.tu.chiba-u. jp/~papers/paper/ACEE/ACEE2006-IA4_Yamazaki.pdf.

Index

Printed and bound by CPI Group (UK) Ltd, Croydon, CR0 4YY

18/10/2024

01776236-0010